Surface Guided
Radiation Therapy

Surface Guided Radiation Therapy

Edited by

Jeremy D. P. Hoisak

Adam B. Paxton

Benjamin Waghorn

Todd Pawlicki

CRC Press
Taylor & Francis Group
Boca Raton London New York

CRC Press is an imprint of the
Taylor & Francis Group, an **informa** business

CRC Press
Taylor & Francis Group
6000 Broken Sound Parkway NW, Suite 300
Boca Raton, FL 33487-2742

First issued in paperback 2021

© 2020 by Taylor & Francis Group, LLC
CRC Press is an imprint of Taylor & Francis Group, an Informa business

No claim to original U.S. Government works

ISBN-13: 978-1-138-59349-7 (hbk)
ISBN-13: 978-1-03-217375-7 (pbk)
DOI: 10.1201/9780429489402

Visit the Taylor & Francis Web site at
http://www.taylorandfrancis.com

and the CRC Press Web site at
http://www.crcpress.com

Collectively, we dedicate this textbook to all of the patients undergoing radiotherapy as well as all of our radiation oncology colleagues over the years who have shaped our understanding of SGRT. Our hope is that this book will play a role in furthering SGRT in the radiotherapy community to benefit the patients that all clinicians serve. Specifically, Dr. Waghorn would like to thank Kristen, Olivia, and Matthew for their ongoing patience, love, and support; Dr. Paxton thanks Julie, Rollie, and Dane – saying I'm looking forward to all our future family adventures doesn't even scratch the surface; Dr. Pawlicki thanks Ally and Nicole for their patience while working on the book during family time on evenings and weekends; and Dr. Hoisak is thankful to his family and Nick for encouragement and support, and extends a special thanks to his cat Taji, for walking across the keyboard when it was time for a break.

Contents

Preface

SURFACE GUIDED RADIATION THERAPY (SGRT) is a rapidly emerging topic in the radiotherapy community. While there are a number of research publications on the topic, we also strove to encompass more technically oriented information related to current SGRT systems on the market. A goal of the textbook was to present a global view of the topics covered through the choice of international expert authors.

This text is divided into seven broad categories. The book begins with a historical overview of non-radiographic localization technologies that are largely precedents of current SGRT systems and practice. This is followed by an overview of the role that SGRT plays in quality radiation treatments and patient safety. The next series of chapters focus on the technical principles of existing SGRT systems. Where possible, we recruited the vendors to write these chapters being careful to provide as much technical detail of their products while avoiding information that is promotional in nature. The technical chapters for each system are paired with a commissioning and quality assurance chapter, written by a clinical physicist expert user. Two clinical areas where SGRT has seen the greatest clinical adoption to date are in breast radiotherapy and cranial radiosurgery. Therefore, each of these disease sites has three dedicated chapters that include a chapter on clinical outcomes. The next section covers SGRT for hypofractionated treatments with and without respiratory motion management, head and neck, extremities, and pediatrics. The last section covers other areas of SGRT that do not neatly fit into the previous sections, such as introducing SGRT into the clinic or forward thinking applications of SGRT such as tattoo-less patient setup. Additionally, discussions of SGRT for proton therapy and bore-type linac systems are provided, as well as custom SGRT solutions with off-the-shelf technologies. We finish the book with a chapter on future possible directions for SGRT.

To encompass the depth and breadth of the material envisioned, we took the approach of brevity – asking all authors for concise chapters. However, eliminating high-quality content to meet a page quota would have been a disservice to our readers, so we were not overly strict about chapter length. Nevertheless, in order to meet the requirement for a book of manageable length, we did shorten some chapters if similar material was covered at different places in the book.

The contributions dealing with topics that are still in development provide our best current knowledge of, and guidance for, SGRT in these areas. Technologies will undoubtedly continue to evolve and recommendations may change in the future but we expect that many components presented here will remain relevant. At a minimum, these chapters will document the current collective state of thought on the topic of SGRT. While some chapters of this book may be superseded in the future, we hope that the majority of the chapters will be useful to the field for many years to come and will help to inspire further clinical development and research.

Acknowledgments

WE WOULD LIKE TO extend our sincerest gratitude to all the authors who gave so much of their time to contribute to this book and who took our edit recommendations and incorporated them where appropriate. We believe this collaboration resulted in the best possible content. Finally, we acknowledge our publisher, Taylor & Francis Group, for their guidance, support, and patience from the inception of this project to its successful completion.

Editors

Jeremy Hoisak, PhD, DABR, is an assistant professor in the Department of Radiation Medicine and Applied Sciences at the University of California, San Diego. Dr. Hoisak's clinical expertise includes radiosurgery, stereotactic body radiation therapy, and respiratory motion management.

Adam Paxton, PhD, DABR, is an assistant professor in the Department of Radiation Oncology at the University of Utah. Dr. Paxton's clinical expertise includes patient safety, motion management, radiosurgery, and proton therapy.

Benjamin Waghorn, PhD, DABR, is the senior director of Clinical Physics at Vision RT, Ltd. Dr. Waghorn's research interests include intensity-modulated radiation therapy, motion management, and surface image guidance systems.

Todd Pawlicki, PhD, DABR, FAAPM, FASTRO, is professor and vice-chair for Medical Physics in the Department of Radiation Medicine and Applied Sciences at the University of California, San Diego. Dr. Pawlicki has published extensively on quality and safety in radiation therapy. He has served on the Board of Directors for the American Society for Radiology Oncology (ASTRO) and the American Association of Physicists in Medicine (AAPM).

Contributors

Hania A. Al-Hallaq, PhD
Department of Radiation and
  Cellular Oncology
The University of Chicago
Chicago, Illinois

Marianne C. Aznar, PhD
Division of Cancer Sciences
Manchester Cancer Research
  Centre
University of Manchester
Manchester, United Kingdom
and
Clinical Trial Service Unit
Nuffield Department of Population
  Health
University of Oxford
Oxford, United Kingdom

Daniel Bailey, PhD
Department of Radiation
  Oncology
Northside Hospital Cancer
  Institute
Atlanta, Georgia

Victoria Bry, BA
Department of Radiation
  Oncology
University of Texas Health Science
  Center at San Antonio
San Antonio, Texas

Rex Cardan, PhD
Department of Radiation
  Oncology
The University of Alabama at
  Birmingham
Birmingham, Alabama

Laura Cerviño, PhD
Department of Medical Physics
Memorial Sloan Kettering Cancer
  Center
New York, New York

Alisha Chlebik, BS, RT(T)
Radiation Oncology Program
Children's Hospital Los Angeles
Los Angeles, California

Jacqueline Dorney, BSc (Hons)
GenesisCare
Waterlooville, United Kingdom

Ryan Foster, PhD
Radiation Oncology Levine Cancer
  Institute
Atrium Health
Charlotte, North Carolina

David P. Gierga, PhD
Department of Radiation
  Oncology
Harvard Medical School
and
Massachusetts General Hospital
Boston, Massachusetts

Alonso N. Gutierrez, PhD, MBA
Department of Radiation
  Oncology
Miami Cancer Institute
Miami, Florida

Lisa Hampton, PhD
Varian Medical Systems
Palo Alto, California

Delena Hanson, MS, RT(R)(T)
Varian Medical Systems
Palo Alto, California

John Heinzerling, MD
Radiation Oncology Levine
  Cancer Institute
Atrium Health
Charlotte, North Carolina

Ellen Herron, BSRT(T)
Department of Radiation
  Oncology
Northside Hospital Cancer
  Institute
Atlanta, Georgi

Jeremy D. P. Hoisak, PhD
Department of Radiation Medicine
  and Applied Sciences
University of California, San Diego
La Jolla, California

Chris Huyghe, MSc
Varian Medical Systems
Palo Alto, California

Orit Kaidar-Person, MD
Radiation Oncology Unit
Oncology Institute
Rambam Medical Center
Haifa, Israel

Grace Gwe-Ya Kim, PhD
Department of Radiation Medicine
  and Applied Sciences
University of California, San Diego
La Jolla, California

Guang Li, PhD
Department of Medical Physics
Memorial Sloan Kettering Cancer
  Center
New York, New York

Lutz Lüdemann, PhD
Department of Radiation Therapy
University of Essen
Essen, German

Ryan Manger, PhD
Department of Radiation Medicine
  and Applied Sciences
University of California, San Diego
La Jolla, California

Icro Meattini, MD
Department of Biomedical,
  Experimental, and Clinical
  Sciences "M. Serio"
University of Florence
and
Radiation Oncology Unit
Oncology Department
Azienda Ospedaliero-Universitaria
  Careggi
Florence, Italy

Arthur J. Olch, PhD
Radiation Oncology Program
Children's Hospital Los Angeles
and
Radiation Oncology Department,
  Keck School of Medicine
University of Southern California
Los Angeles, California

Laura Padilla, PhD
Department of Radiation
  Oncology
Virginia Commonwealth
  University
Richmond, Virginia

Vanessa Panettieri, PhD
Alfred Health Radiation Oncology
The Alfred Hospital
Melbourne, Victoria, Australia

Nikos Papanikolaou, PhD
Department of Radiation
  Oncology
University of Texas Health Science
  Center at San Antonio
San Antonio, Texas

Sandra Paul, BSc(RT)
Alfred Health Radiation Oncology
The Alfred Hospital
Melbourne, Victoria, Australia

**Todd Pawlicki, PhD, FAAPM,
FASTRO**
Department of Radiation Medicine
  and Applied Sciences
University of California, San Diego
La Jolla, California

Adam B. Paxton, PhD
Department of Radiation
  Oncology
University of Utah
Salt Lake City, Utah

Philip Poortmans, PhD, MD
Department of Radiation
  Oncology
Institut Curie
and
Paris Sciences & Lettres
PSL University
Paris, France

Douglas A. Rahn III, MD
Department of Radiation Medicine
   and Applied Sciences
University of California, San Diego
La Jolla, California

Karl Rasmussen, PhD
Department of Radiation
   Oncology
University of Texas Health Science
   Center at San Antonio
San Antonio, Texas

Catherine Russell, BAppSc(RT)
Alfred Health Radiation Oncology
The Alfred Hospital
Melbourne, Victoria, Australia

Bill J. Salter, PhD, FAAPM
Department of Radiation
   Oncology
University of Utah
Salt Lake City, Utah

Raymond Schulz, MSc
Varian Medical Systems
Palo Alto, California

Thomas Speck, BSc
Varian Medical Systems
Palo Alto, California

Dennis N. Stanley, PhD
Department of Radiation
   Oncology
The University of Alabama at
   Birmingham
Birmingham, Alabama

Michael Stead, BSc
Varian Medical Systems
Palo Alto, California

Michael J. Tallhamer, MSc
Department of Radiation
   Oncology
Centura Health
Centennial, Colorado

Xiaoli Tang, PhD
Department of Radiation
   Oncology
Yale New Haven Hospital
New Haven, Connecticut

Benjamin Waghorn, PhD
Vision RT Ltd.
London, United Kingdom

David Wiant, PhD
Department of Radiation
   Oncology
Cone Health Cancer Center
Greensboro, North Carolina

Kenneth Wong, MD
Radiation Oncology Program
Children's Hospital Los Angeles
and
Radiation Oncology Department,
   Keck School of Medicine
University of Southern California
Los Angeles, California

Bo Zhao, PhD
Department of Radiation
    Oncology
University of Texas Southwestern
    Medical Center
Dallas, Texas

Hui Zhao, PhD
Department of Radiation
    Oncology
University of Utah
Salt Lake City, Utah

CHAPTER 1

A History of Surface Guidance Methods in Radiation Therapy

Jeremy D. P. Hoisak and Todd Pawlicki

CONTENTS

1.1 INTRODUCTION

Surface guidance as a concept in radiation therapy can be traced back to the earliest days of external beam treatment. At its essence, the very first form of surface guidance in radiation therapy relies on the eyes of the clinician to set up the patient by visually aligning marks on the patient's skin surface to room lasers that represent the treatment isocenter. Additionally, an optical distance indicator (ODI) is used to measure the source to surface distance (SSD) and the projection of the field light onto the surface of the patient, helping to further confirm the accuracy of the setup. With these visual tools, an internal target is localized by triangulation of points on the patient's skin surface to sufficient accuracy and precision for the treatment techniques in use at the time. Reference pictures of the planned setup are a visual guide to verify that the setup is correct prior to treatment. The patient is then visually monitored via closed-circuit television (CCTV) from outside the treatment vault to ensure that this position is maintained throughout the delivery of radiation. This simple process of patient setup and intra-fraction patient position monitoring using the patient's surface as a surrogate for the internal target is fundamentally the same process still in use today and is the origin of modern surface guided radiation therapy (SGRT).

The introduction of image-guided radiation therapy (IGRT) using orthogonal imaging with kilovoltage (kV) or megavoltage (MV) X-rays,[1] and later cone-beam computed tomography[2] allowed much more accurate alignment of the patient to the planned position through the imaging of bony anatomy and soft tissue. However, IGRT did not supplant the requirement for accurate initial setup of the patient as any improvement to initial setup quality improves the efficiency of the image-guidance process and reduces the need for repeat imaging. Prior to the introduction of 6 degrees-of-freedom (6DOF) couches, any IGRT indication to pitch or roll the patient to the planned position would require the staff to return to the treatment vault and manually adjust the patient. Repeat imaging would then follow these adjustments. As camera technology and computer processing improved, systems emerged with combinations of optical sensors and algorithms to permit 3D imaging of the patient's surface and detection of positional offsets that could be corrected in 6DOF prior to radiographic imaging.

Modern SGRT technology now assumes the role of the clinician's eyes to enable a more robust setup than three points on the patient's skin surface and augments the abilities of the human eye to enable entirely new applications from tattoo-less patient setup to respiratory-correlated treatment techniques. Longitudinal analysis and quantitative manipulation of patient setup and intra-fraction motion data can reveal insights into target localization uncertainty, normal tissue planning margins, and the overall quality of radiation therapy.

This chapter begins with a historical review of optical-based approaches that lead to modern SGRT, from early efforts at verifying patient setup and monitoring patient position with television cameras, to modern real-time in-room guidance and continual localization with advanced 3D imaging and processing technology. The theory and basic principles underpinning current surface imaging technologies are described, including camera calibration and surface registration algorithms. The aim of this chapter is to provide the reader with a thorough appreciation of the history and trajectory of SGRT technologies, including their role in the context of modern IGRT, so that the diverse applications of SGRT described in later chapters of this book can be fully appreciated.

1.2 MANAGING UNCERTAINTY THROUGH LOCALIZATION

In the early days of radiation therapy, treatment fields with large margins relative to the target were used to treat the target area. Opposing the fields ensured coverage along the field axis, and many sites could be effectively treated with parallel-opposed, three-field, or four-field "box" field arrangements. For such techniques, aligning an isocenter marked on the patient's skin to isocentric room lasers and verifying the SSD was sufficient localization to achieve the desired setup accuracy. However, this came at the cost of irradiation of normal tissues.

The drive to minimize normal tissue dose and enable a maximization of tumor dose led to the addition of shielding by blocks, cones and later, the multileaf collimator (MLC) so that the fields could conform to the shape of the target in three dimensions. With 3D conformal radiation therapy, it became both more important and more technically challenging to ensure the target was within the planned high-dose area. Modern linear accelerators equipped with MLCs together with intensity modulated radiation therapy have now made it possible to deliver radiation dose distributions that are highly conformal to the target while sparing adjacent healthy tissues. These finely modulated dose distributions require that the target be localized as accurately as possible to ensure the dose is delivered as intended. Precise and accurate localization of the target can reduce geometric uncertainties and therefore reduce the impact of treatment-limiting side effects. This is a technical problem of considerable difficulty. Many technologies have been developed to help ensure precise and accurate localization of the therapy target, including rigid immobilization devices[3] (e.g., thermoplastic masks and vacuum bags), X-ray image guidance with kV[4,5] or MV[6,7] sources, and nonradiographic localization devices.

1.3 NONRADIOGRAPHIC LOCALIZATION TECHNOLOGIES

Nonradiographic localization technologies have several advantages over radiographic technologies for precise and accurate setup of the patient. In addition to allowing continuous tracking without contributing additional dose to the patient as with radiographic methods, nonradiographic localization can be considered an "always-on" technology, usable by staff while in the treatment room to provide real-time setup guidance for initially positioning and fine-tuning the patient's setup before leaving the room for pretreatment verification imaging and delivery. This in-room ability can potentially reduce radiographic imaging dose and inefficiencies in the clinical workflow by reducing repeat trips into the vault to re-adjust the patient's position or perform repeat imaging to verify an automatic couch shift. Nonradiographic methods can also provide localization of the patient and target when delivery uses noncoplanar couch angles if treating the patient on a linac with a C-arm gantry, which is currently not possible with most gantry-mounted imaging technologies.[8] The elimination of dose from continuous tracking and reduction of dose from verification imaging plus improvements in efficiency and reduced time is of particular benefit in pediatric applications, where patients may not be able to hold still for long, or can reduce anesthesia requirements.

Many nonradiographic localization technologies have been developed and commercialized, using approaches based on optical methods, ultrasound (US), and electromagnetic (EM) principles. Some systems combine elements of these technologies. For example, the SonArray (Varian Medical Systems, Palo Alto, CA), BATCAM (Best Medical, Springfield, VA), and Clarity (Elekta AB, Stockholm, Sweden) all use an infrared (IR) camera to track markers affixed to an US probe to localize subsurface targets and relate the probe's position to the treatment machine's frame of reference, translating US localization of the target into shift instructions for the therapist or automatic couch to align the patient with the planned position.[9]

EM localization technologies employ either active or passive tracking of a beacon or sensor. One such system used active tracking in the form of radiofrequency (RF) emitting beacons that are implanted within the body,[10] later commercialized as Calypso (Varian Medical Systems, Palo Alto, CA). The Calypso system has submillimeter accuracy and has been used primarily for the treatment of localized prostate cancer.[11] Another technology uses wired passive receivers to detect a low-intensity EM signal emitted by a base station and determines their 3D location in space (Flock of Birds, Ascension Technologies, Burlingon, VT). Such a system has been used for a variety of medical applications,[12] including monitoring displacement of the abdominal surface as a surrogate for lung tumor motion tracking.[13] Issues with RF and other EM interference from the linear accelerator (linac) and other equipment in the treatment and simulation rooms as well as the need to place the signal receiver very close to the patient to detect active signals have limited the wider use of EM localization systems in other radiation therapy applications.

Another approach to nonradiographic localization has been to integrate magnetic resonance imaging (MRI) directly into the treatment delivery process. MRI offers volumetric imaging, good soft tissue contrast, and "always-on" imaging with the possibility of providing real-time patient localization and image guidance. Approaches to integration of MRI with RT include having the MR imager on rails[14] and moved into position for imaging, or most recently, integrating the MR imager directly with a Co-60 teletherapy unit[15] or linac.[16–18] It is not yet clear if SGRT has any role to play in MR-guided RT; however, this discussion is beyond the scope of the chapter. The remainder of this chapter discusses optical methods for localization and intra-fraction motion monitoring. These optical methods include video, laser, IR, and 3D surface imaging.

1.4 OPTICAL LOCALIZATION TECHNOLOGIES

1.4.1 Video-Based Approaches

Exploiting the fact that MV treatment vaults have CCTV systems already installed to monitor the patient for movement during treatment imaging and delivery, an early approach to optical localization and tracking used digital subtraction and cancellation of the CCTV feed to verify patient repositioning and detect motion.[19] Another early approach used the CCTV system with a video recorder to obtain a record of patient motion for offline computer analysis, with an accuracy of 1 mm.[20] Since the analysis was performed after treatment, no real-time monitoring or online correction was possible. A later approach to video-based patient monitoring used dedicated cameras in combination with a computer vision algorithm to track the patient. Yan et al. developed a video-based system for tracking regions of interest defined on the patient's skin, with an estimated translational and rotational accuracy of 2 mm and 1.2°, respectively.[21]

1.4.2 Laser-Based Approaches

Motion detection has also been achieved by exploiting the presence of in-room isocentric lasers or by the installation of dedicated laser range finding and scanning systems. Any early version of a patient motion monitoring system used the in-room isocentric lasers as a form of trip-wire, detecting if patients moved during treatment.[22] A more recent implementation used a dedicated laser and camera to actively scan the patient's surface. This system has a 0.1-mm resolution and repositioning accuracy was 0.5 mm when shifts were less than 20 mm. Initially designed to assist with radiation therapy setup, this system was also used to gate treatments and correct for artifacts due to breathing motion during nuclear medicine imaging studies.[23] A similar system was later offered commercially (Galaxy, LAP Laser, Lüneburg, Germany) and has been used to detect patient setup errors after verification imaging.[24,25]

The Sentinel system (C-RAD AB, Uppsala, Sweden) is a commercially available laser-based surface scanning system that uses a laser, camera, and registration software to determine the patient's position in 3D relative to a reference. Reproducibility of Sentinel-based initial patient setups have been shown to be better than 1 mm and 0.4°.[26] The system has also been used to monitor motion during treatment[27] as well as to measure the displacement of the abdominal/chest wall surface as a signal for respiratory-correlated computed tomography.[28]

1.4.3 Marker-Based Approaches

Video cameras can also be used to detect the position of markers placed on the patient's surface. One such system used dedicated dual charge-coupled device (CCD) cameras aimed at the treatment isocenter to detect the location of markers placed on the patient. This CCD system was used for monitoring day-to-day variations in patient setup as well as motion monitoring during treatment, with an accuracy better than 0.5 mm. An additional use of the CCD system was as an independent calibration method for the in-room setup laser and ODI.[29]

The use of markers for tracking has not been restricted to visible wavelengths. Marker-based localization technologies have been developed that passively track reflective markers or active light-emitting diodes (LEDs) with an imaging system operating in the IR spectrum. These systems use a number of markers placed directly on the patient's surface or on a frame or immobilization device that is rigidly attached to the patient's surface. Active and passive IR systems typically employ a small number of markers, inherently limiting their ability to fully describe a patient's position and posture.

One of the first IR localization and tracking systems for radiosurgery employed active or passive IR markers and two ceiling-mounted CCD cameras[30,31] (Polaris, Northern Digital, Waterloo, Canada). This camera system has been used in a variety of applications including image-guided interventions[32] and stereotactic radiosurgery[33] and was eventually commercialized for radiation therapy applications as the Optical Guidance Platform (OGP) (Varian Medical Systems, Palo Alto, CA).

The OGP system employs a rigid array of IR reflective markers that are detected by a ceiling-mounted camera unit consisting of an IR light emitter and camera sensor. OGP could be used for stereotactic and nonstereotactic applications depending on the configuration of the IR marker array. The FrameArray configuration of OGP was used for stereotactic radiosurgery, with markers attached to a rigid head frame. The OGP FramelessArray could be used for stereotactic radiosurgery and fractionated radiotherapy and attached the marker array to a bite block and minimal mask.[34] The requirement for a bite block limited the applicability of the technology in patients with dental issues. Extracranial treatments could be performed by attaching the marker array to an indexed stereotactic couch attachment.

Another IR-based system is the Real-time Position Monitoring (RPM) system (Varian Medical Systems, Inc., Palo Alto, CA), which consists of a ceiling or couch-mounted IR camera and light emitter that is used to detect a reflective marker block placed on the patient.[35] RPM monitors the motion of the abdominal or chest wall surface as a surrogate for breathing. It has been used primarily for guiding respiratory correlated CT,[36] deep inspiration breath hold, and gating during treatment.[37]

1.5 EARLY APPROACHES TO SURFACE IMAGING

Advances in computing power and imaging technology eventually allowed the real-time mapping of many arbitrary points on the patient and simultaneously tracking them over time. If appropriately distributed, a sufficient number of points can describe a surface in 3D. One can therefore consider surface mapping systems to be the logical extension of marker-based approaches, with a corresponding extension in the ability to describe patient position and posture, as well as provide respiratory signals through mapping of the chest wall and abdomen in 3D over time. The next section describes the basic principles for obtaining 3D information from a scene using 2D imaging.

1.5.1 Photogrammetry

From the Greek *photo* meaning "light," *gram* meaning letters or drawing, and *metry* meaning "measure," *photogrammetry* is the process of optically describing a 3D object with information obtained from 2D images. Modern day uses of photogrammetry include the extraction of topographical information from aerial photography[38] and mapping of archeologic excavations.[39] Photogrammetry obtains the distance between two points residing on a plane parallel to the imaging plane by measuring their distance on an image with a known scale. A special case of photogrammetry is stereophotogrammetry, where instead of a single camera, two cameras with a known separation are used to identify common points on the surface. The intersection of their respective line of sight is used to estimate the 3D coordinates of a point on that surface.

Roentgen stereophotogrammetry is a nonoptical approach that uses X-rays, most often for static and dynamic orthopedic studies of prostheses and kinematics.[40,41] Radiation therapy applications of Roentgen stereophotogrammetry have been mostly for stereoscopic localization in spine radiosurgery.[42] Acquisition of kV or MV orthogonal images for localization can also be considered a form a Roentgen stereophotogrammetry.

The earliest application of optical stereophotogrammetry in radiation therapy was for determining a patient's surface contour. One approach used two film cameras and a light pattern projector. A line pattern was projected onto the patient, and the two offset cameras took pictures of the patient while lying on a couch. Points on a patient's surface could be determined to an accuracy of 1–3 mm with the assistance of a computer and the digitizer commonly found in dosimetry departments at the time.[43] Another approach was to project two patterns onto the patient and image the resultant interference pattern with a single film camera to obtain surface contour information.[44] It is important to note that these were not systems for localization and monitoring of patients but for treatment simulation and planning, where obtaining accurate patient contours for accurate dose calculations was difficult before the widespread introduction of CT for simulation and planning.

Marker-based optical stereophotogrammetry was used for patient setup verification in stereotactic radiosurgery via two cameras that imaged a bite block fixed with calibrated landmarks. The accuracy of this system was 0.05 mm with a temporal resolution of 12 ms.[45,46] Another optical method placed the markers directly on the patient's surface and calculated their position using photogrammetry. This system used digital cameras interfaced with a computer and so could be used in near real-time, evaluating initial patient setup and ongoing position monitoring.[47]

1.6 MODERN DIGITAL SURFACE IMAGING APPROACHES

1.6.1 Structured Light Projection

Whereas the original applications of stereophotogrammetry used a projected pattern or interference to extract 3D information from film, later digital imaging versions detected markers placed on the patient's surface. Modern surface imaging systems employ the original method of projecting digitally structured or patterned light onto the patient's surface, thus eliminating the need for external or internal fiducials or reflectors and allowing truly contact-less, noninvasive, nonradiographic imaging, and monitoring of the patient's position. The projected pattern can be achieved by laser interference, by projection of digitally or mechanically structured light projection or by projection of pseudo-random speckle patterns. Laser interference uses two beams projected such that interference between the two produces a regular pattern at a distance.[48] The advantage to this approach is the potential for unlimited depth of field. With light projection, a known structured or pseudo-random optical pattern is projected onto the object surface using a light modulator such as moving mirrors, usually operating in the visual or near-visual spectrum. Some technologies for surface imaging use a pattern of IR dots projected onto the surface.[49] If the imaged surface is planar and orthogonal to the camera, then the reflected pattern will appear to be the same as the projected pattern. If a nonplanar, nonorthogonal object is illuminated, the reflected pattern will appear geometrically distorted, as shown in Figure 1.1a and b. This distortion can be used to determine the 3D coordinates of any point on that illuminated surface. The resolution achievable with a structured light projection approach depends on the width of the optical pattern and the wavelength of light. There are many techniques for structured light projection that depend on the application, such as for imaging moving objects or stationary objects. Light projection systems can also employ

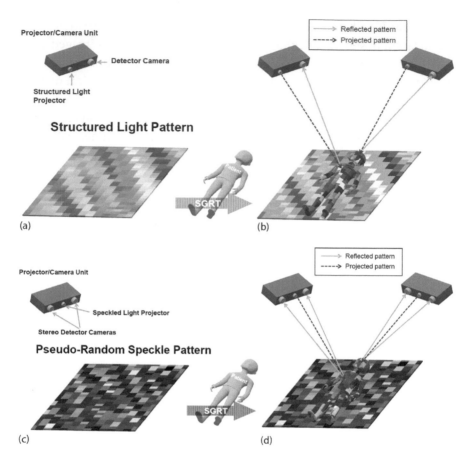

FIGURE 1.1 Approaches to surface imaging: (a) a known pattern of structured light is emitted from the light projector and detected by a monocular detector camera. (b) A nonplanar object, that is, the patient, is placed within that pattern. The surface of the patient is then imaged based on the geometric distortion of the reflected optical pattern detected by the camera. (c) A pseudo-random speckled light pattern is emitted from the light projector and is reflected and detected by stereo detector cameras offset by a known separation within a camera unit. (d) A nonplanar object, that is, the patient, is placed within that speckle pattern of light. The surface of the patient is then imaged based on the recognition of the pattern at a point on the surface and triangulation of that point's location by the camera unit. In (b) and (d), multiple projector/camera units within the treatment room reduce the problem of camera occlusion by objects in the room such as the linac gantry.

grid patterns where the overall pattern consists of pseudo-random patterns forming unique subregions that indicate their 2D position relative to the larger grid pattern. Through pattern matching and triangulation, these subregions can be used to establish point correspondences and increase the robustness of the surface measurement and range determination, as shown in Figure 1.1c and d.

1.6.2 Calibration of Structured Light Projection Surface Imaging Systems

Calibration of a surface imaging camera, including the light projector, is an essential part of determining the overall accuracy of the 3D surface imaging system. Camera and projector calibration procedures establish the relationship between the pixels of the detector camera and a line in 3D space upon which a point on the object's surface exists. Calibration usually involves the acquisition of multiple 2D images of a known calibration phantom at fixed distances and angles from the detector cameras (Figure 1.2). The phantom object commonly used is a checkerboard pattern, as it has relatively easy to detect features to establish correspondences between 2D and 3D spaces. Dot or blob patterns are also used for calibration. In radiation

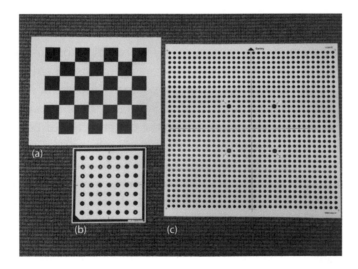

FIGURE 1.2 Examples of vendor-provided phantoms for calibrating surface imaging systems: (a) checkerboard phantom for calibrating Varian IDENTIFY time-of-flight camera, (b) calibration phantom for Varian IDENTIFY surface imaging system, and (c) calibration phantom for Vision RT AlignRT surface imaging system.

therapy applications, the calibration phantom is typically placed on the treatment couch at a known SSD.[50,51] The calibration phantom can be used to initially calibrate the imaging system and to verify that calibration is still valid as part of a periodic quality assurance program or after preventive maintenance.

Calibration of surface imaging systems must also include the light projector. Projector calibration has two aspects, calibration of the light intensity and geometric calibration of the light projection. Intensity calibration typically involves the projection of a dot or blob pattern, and the relationship between the projected intensity and detected pixel intensity is established. It is important that calibration takes place under the same ambient light conditions as will be used for surface imaging or accuracy and resolution performance can be degraded. Geometric calibration of the projector is modeled in a similar manner as the geometric camera calibration, except in inverse with the known values in the image plane and the measured values in 3D space. One approach is to project corner points onto the corner points of the checkerboard phantom and establish point correspondences in a similar way as is done with the detector cameras. Note that this must be done after the geometric calibration of the detector cameras have established the positions of the checkerboard phantom's true corner points.

1.6.3 Time-of-Flight Cameras

An alternative approach to structured light projection for surface imaging is time-of-flight (ToF) technology. ToF camera systems consist of an image sensor, a modulated light source, and a computer processor. As the name suggests, ToF cameras measure the time taken for photons from the projector to travel to the object and return to the detector. The distance traveled by reflected photons is determined by computing their phase shift from the original modulated source. ToF cameras illuminate an entire scene simultaneously and do not require physical or electronic scanning or a second camera as with stereophotogrammetry. A further advantage of ToF cameras over structured light projection is that they do not suffer from loss of energy (in the form of light pattern degradation and reflected intensity) with distance and consequently have much greater depth of field to image objects. ToF cameras have been used in robotics, drones, and industrial applications such as 3D printing. They have also found widespread application in healthcare including guidance of surgical interventions, operating room monitoring, touch-less interaction, and on-patient visualization.[52-54] As costs decrease, ToF cameras are finding consumer applications in smartphones and gaming as they facilitate gesture-based interactions and augmented reality displays. For example, the latest version of the Kinect (Microsoft Corporation, Redmond, WA) depth camera uses ToF technology. ToF camera systems have been applied to radiotherapy applications. One system used ToF sensors to reconstruct the 3D surface of a patient and extract a real-time multidimensional respiratory signal from variations in the surface position due to breathing. The ToF respiratory signal was compared with a commercial system for obtaining a respiratory gating signal, and an average correlation of 0.88 was reported.[55] They also developed the ToF camera to enable patient positioning. The ToF camera sampled 25,000 points on the patient's surface and computed translation and rotations with respect to a reference using an iterative closest point (ICP) algorithm,[56] with an accuracy of 2.8 mm and 0.28° on a phantom, and a mean target registration error (TRE) of 3.8 mm on human subjects. A later version improved phantom results to 1.62 mm and 0.07°, with a temporal resolution of 65 ms.[57,58] Gilles et al. used two ToF cameras to assist with daily patient positioning. The ToF cameras obtained point clouds from the patient's surface and calculated the displacements required to move the patient into position. An evaluation of 150 fractions of head and neck, lung, pelvis, and prostate treatments demonstrated absolute displacement detections for all sites < 1.1 mm.[59]

1.6.4 Surface Registration Algorithms

The primary objective of surface imaging in RT is to compare the current or "live" patient surface to a reference surface and compute the current displacements required to bring the two surfaces into alignment. Several computational approaches to the problem of registering two surfaces have been proposed and implemented; however, the commonly employed method in SGRT systems is based on the ICP algorithm. The ICP algorithm establishes a correspondence between the closest points in the point clouds of their respective surfaces. The root mean square of the estimated translations and rotations required to match the corresponding points is then minimized. The transformation is applied, and new point correspondences are established. The algorithm continues to iterate until a global minimum in the solution space is found.

Variations on the ICP algorithm have been proposed for surface guidance, aimed at improving the speed and robustness of the matching process.[60] For example, a ToF-based surface guidance system used for radiation therapy setup matches prominent surface features prior to point matching and has been shown to be faster than conventional ICP.[58]

Nonrigid versions of ICP have also been developed to improve the degrees of freedom available to perform registration of two surfaces.[61]

1.6.5 Current Commercial SGRT Systems

The first modern SGRT system was the AlignRT system (Vision RT, London, UK), the first prototype being installed in 2001 and receiving regulatory approval in 2006. AlignRT employs ceiling-mounted camera pods consisting of stereo cameras and a light projector. Using the principle of structured light projection, the patient is illuminated, and their surface is mapped using detection of the reflected light pattern and triangulation. Multiple camera pods ensure visibility of the patient through all linac gantry angles and are combined in postprocessing by a dedicated processor to create a 3D map of the patient's surface in real-time. Shift guidance is provided to staff inside and outside the treatment room by dedicated workstations that indicate real-time displacements in 6DOF. The initial design was based on two ceiling-mounted camera pods, while the current system uses three camera pods for the maximum visibility of the patient through all treatment phases, as well as high-definition imaging cameras. Camera pods and software for use in CT simulation are marketed as GateCT and are based on the in-room camera technology. The AlignRT SGRT system is described further in Chapters 5 and 8.

In 2012, the AlignRT system was offered to the radiotherapy community by Varian Medical Systems under the brand name Optical Surface Monitoring System (OSMS). The OSMS was intended partially as a replacement for the now deprecated OGP system used for SRS and other applications. The commercial partnership to sell OSMS has now ended, although OSMS systems are still supported and eligible for ongoing upgrades by Vision RT.

In 2011, another commercial offering for in-room surface guidance became available called Catalyst (C-RAD AB, Uppsala, Sweden). C-RAD previously offered the Sentinel laser scanning-based guidance system in 2006, which is now used to obtain respiratory information during CT and 4D-CT simulation. The Catalyst system uses digital light processing with an LED source and single detector photogrammetry to achieve surface imaging, along with a nonrigid version of the ICP surface registration algorithm. In 2015, the Catalyst HD was introduced, which uses three projector/camera units to improve patient coverage. The Catalyst/Catalyst HD system provides setup guidance to therapists by use of visual displays and projection of color-coded position offset information directly onto the patient. The C-RAD Catalyst/Catalyst HD and Sentinel systems are described further in Chapters 3 and 6.

Another surface guidance system became commercially available in 2017, the IDENTIFY (Varian Medical Systems, Palo Alto, CA), originally a product of HumediQ Global GmbH, a company since purchased by Varian. The IDENTIFY system uses RF identification tags, optical markers, and a ToF camera to guide accessory placement and initial patient setup. The ToF camera system has a wide field of view permitting a full body or "orthopedic" view of the patient while in the initial loading position. The system then transitions to structured light projection-based stereophotogrammetry to achieve surface image guidance during final patient setup, intra-fraction motion monitoring and respiratory management. Setup guidance is provided in-room to therapists by a live video feed augmented with an overlay of the reconstructed surface information, using color coding to indicate position offsets. IDENTIFY is further described in Chapters 4 and 7.

In addition to the commercial offerings, many hospitals and research groups have developed low cost, custom surface guidance solutions with consumer grade depth cameras.[53] The most popular camera for this purpose has proven to be the Kinect, although other camera vendors have been employed in radiation therapy applications.[62] Initially a part of a video game console system to analyze the real-time motion of players and translate these into in-game actions, the Kinect was quickly repurposed to a variety of medical applications through the availability of an application programming interface (API). Subsequent versions improved the performance of the camera, and the latest version, the Kinect Azure Cloud (Microsoft Corporation, Redmond, WA), now exists as a standalone camera and API for third-party applications.[63] Custom applications with consumer-grade depth cameras are described in detail in Chapter 25.

1.7 FUNDAMENTAL CHALLENGES TO SURFACE IMAGING

Optical methods are challenged by the presence of reflective surfaces both on and around the patient that can cause problems from glare. Examples of reflective surfaces in the radiotherapy environment include equipment such as linac covers and flat-panel image detectors. Another challenge is that for the most accurate representation of the patient's surface, the skin must be directly visible to the imaging system, which can be problematic as patients are often covered with clothing, gowns, and sheets.

Immobilization devices and beam modification devices such as custom lead shielding can also interfere with surface imaging. In addition to glare, subsurface scattering from translucent materials such as tissue-equivalent bolus can degrade the quality of the surface image. Variable ambient lighting and differences in skin tone also pose problems for accurate mapping of the patient's surface. Current SGRT systems have methods to mitigate the effects of skin tone and lighting but are currently limited in their ability to adapt to these conditions automatically.

Another challenge to any optical methods in radiation therapy is the potential occlusion of the camera's view of the patient by the treatment equipment, such as the gantry, bore, imaging arms, or other ancillary equipment in the room. The solution to this problem has been to optimize camera placement and to employ multiple cameras so that the patient is always visible, however, with this flexibility comes an increased cost to install the system, and additional calibration maintenance complexity.

1.8 CLINICAL VERSATILITY OF SGRT SYSTEMS

The widespread adoption of surface imaging compared with other localization systems demonstrates the versatility of the technology and its applications. Whereas previous localization technologies tended to be highly focused on a particular clinical requirement, such as IR-based head tracking for SRS, US-based prostate localization, or respiratory motion management, an SGRT system has sufficient spatial and temporal accuracy for use in SRS localization as well as continuous position monitoring, conventional patient setup, and the delivery of respiratory-gated techniques.

The clinical versatility of SGRT systems also allows their application toward guidance in most anatomic treatment sites. This consistent and widespread use can in turn produce widespread gains in clinical efficiency and other quality and safety benefits. As well be discussed in later chapters of this book, SGRT improves clinical efficiency by eliminating the need for trial and error in the IGRT verification loop as therapists have guidance in the room and reduces imaging dose from repeat acquisitions. SGRT can also reduce the need for immobilization, improving patient comfort. SGRT systems also play a role in improving treatment quality, patient safety, and the reduction of errors. Although many vendors are currently exploring patient identification solutions, these currently employ ancillary systems such as palm or facial recognition outside the treatment room. High-resolution surface imaging opens up the possibility of automatic patient identification using the guidance cameras themselves. Another promising use of SGRT to improve patient safety is for collision avoidance. Currently implemented in treatment planning systems as a means of predicting collisions during treatment delivery, and at the machine console to identify hardware collisions between gantry and couch, there is currently a need for an in-room system that can detect the potential for collisions between equipment and the patient in nontreatment geometries. The role of SGRT in improving quality and safety is discussed in Chapter 2.

As was discussed earlier in the chapter, modern SGRT technology transforms the process of visual initial setup of the patient and on-going positioning monitoring via CCTV into a quantitative, digital process. Quantitative analysis of continuous SGRT data opens up the possibility of new applications beyond patient setup and intra-fraction motion monitoring. Continuous logging of patient position data creates opportunities for the analysis of setup and intra-fraction motion uncertainty, which may in turn allow customization of normal tissue margins, and the resultant benefits to treatment outcomes this can create. High-resolution imaging of the patient's surface could be a highly sensitive tool for detecting changes in the patient's morphology and could be used for automatic weight monitoring and/or tumor response. Future applications of SGRT are discussed at the end of the book in Chapter 26.

Finally, another reason SGRT has found widespread adoption is that the technology is being developed and promoted by large, dedicated companies that offer industrial scale reliability and service support. This makes investment in SGRT technologies a more attractive prospect for administrators if there is the assurance of continued upgrades and product improvement.

1.9 THE ROLE OF SGRT IN THE ERA OF IGRT

As discussed, SGRT has fundamental limits to what it can achieve in the context of radiation therapy, foremost of these being that surface imaging cannot image subsurface targets and relies on the surface as a surrogate for internal target position. However, radiographic IGRT also has fundamental limits to its application in radiation therapy, in that it cannot be used in real-time during in-room patient setup and is currently used in an iterative process to align the patient and cannot be used to continuously monitor for intra-fraction motion without adding additional radiation dose to the patient. Given their respective limitations, SGRT and IGRT should be viewed not as rivals, but partners, each serving a complementary role.

Despite their complementary nature, a common argument when allocating healthcare resources is that SGRT adds little to the most important metric in radiation therapy, that of treatment outcomes. It should be pointed out that this criticism was once leveled at IGRT.[64] As studies and evidence emerged, it is now clear that IGRT can improve outcomes and reduce toxicities through improved quality of radiation therapy delivery.[65] Later chapters in this book will explore in detail how SGRT can have many wide-ranging impacts on the quality of radiation therapy delivery for most sites, including faster and more efficient patient setup, reduction in the potential for errors, reduced need for imaging, and less invasive immobilization for improved patient comfort. SGRT can also offer psychosocial benefits from reduced skin marking and may permit new techniques that were previously difficult or not feasible with available localization technology. In the near future, SGRT may play a role in adaptive therapy strategies. Current and future workflows for adaptive therapy have the potential to be time consuming. For example, online treatment adaptation, in which a patient's plan is modified by the clinician while the patient is on the treatment table, can take considerable time during which it is essential to monitor the patient's position and ensure they do not deviate from the last radiographic image position. Offline adaptive strategies could also benefit from SGRT. Quantitative analysis of inter- and intra-fraction localization data could lead to the design of custom margins, a method by which SGRT could directly impact the therapeutic ratio and thus treatment outcomes. SGRT's ability to improve the quality of radiation therapy delivery may yet yield a similar impact as IGRT on treatment outcomes and toxicities.

KEY POINTS

- SGRT originated with the human eye aligning patients to the treatment geometry by reference to surface marks on the skin.

- Nonradiographic localization and position monitoring systems using optical, EM, and US methods allow continuous real-time guidance in initial patient setup and on-going patient monitoring without incurring extra radiation dose as with radiographic systems.

- Modern optical surface imaging systems employ stereophotogrammetry and structured light projection coupled with point-cloud registration algorithms to register a live patient surface map to a reference surface.

- SGRT is complementary to IGRT and has an important role in the accurate and precise delivery of radiation therapy. SGRT may become an important component of emerging adaptive radiation therapy strategies.

REFERENCES

1. Antonuk LE. Electronic portal imaging devices: a review and historical perspective of contemporary technologies and research. *Phys Med Biol.* 2002;47:R31.
2. Jaffray D, Siewerdsen JH. Cone-beam computed tomography with a flat-panel imager: initial performance characterization. *Med Phys.* 2000;27(6):1311–1323.
3. Verhey LJ. Immobilizing and positioning patients for radiotherapy. *Semin Radiat Oncol.* 1995;5(2):100–114.
4. Jin JY, Yin FF, Tenn SE, Medin PM, Solberg TD. Use of the BrainLAB ExacTrac X-ray 6D system in image-guided radiotherapy. *Med Dosim.* 2008;33(2):124–134.
5. Shirato H, Shimizu S, Kunieda T, et al. Physical aspects of a real-time tumor-tracking system for gated radiotherapy. *Int J Radiat Oncol Biol Phys.* 2000;48(4):1187–1195.
6. Wiersma RD, Mao W, Xing L. Combined kV and MV imaging for real-time tracking of implanted fiducial markers. *Med Phys.* 2008;35(4):1191–1198. doi:10.1118/1.2842072.
7. Cho B, Poulsen PR, Sloutsky A, Sawant A, Keall PJ. First demonstration of combined kV/MV image-guided real-time dynamic multileaf-collimator target tracking. *Int J Radiat Oncol Biol Phys.* 2009;74(3):859–867. doi:10.1016/j.ijrobp.2009.02.012.
8. D'Ambrosio DJ, Bayouth J, Chetty IJ, et al. Continuous localization technologies for radiotherapy delivery: report of the American Society for Radiation Oncology Emerging Technology Committee. *Pract Radiat Oncol.* 2012;2(2):145–150.
9. Lachaine M, Falco T. Intrafractional prostate motion management with the Clarity Autoscan system. *Med Phys Int J.* 2013;1(9).
10. Balter JM, Wright JN, Newell LJ, et al. Accuracy of a wireless localization system for radiotherapy. *Int J Radiat Oncol Biol Phys.* 2005;61(3):933–937. doi:10.1016/j.ijrobp.2004.11.009.
11. Willoughby TR, Kupelian PA, Pouliot J, et al. Target localization and real-time tracking using the Calypso 4D localization system in patients with localized prostate cancer. *Int J Radiat Oncol Biol Phys.* 2006;65(2):528–534. doi:10.1016/j.ijrobp.2006.01.050.
12. Franz AM, Haidegger T, Birkfellner W, Cleary K, Peters TM, Maier-Hein L. Electromagnetic tracking in medicine—A review of technology, validation, and applications. *IEEE Trans Med Imaging.* 2014;33(8):1702–1725. doi:10.1109/TMI.2014.2321777.

13. Hoisak JDP, Sixel KE, Tirona R, Cheung PCF, Pignol J-P. Prediction of lung tumour position based on spirometry and on abdominal displacement: accuracy and reproducibility. *Radiother Oncol.* 2006;78(3):339–346. doi:10.1016/j.radonc.2006.01.008.

14. Jaffray D, Carlone M, Milosevic M, et al. A facility for magnetic resonance–guided radiation therapy. *Semin Radiat Oncol.* 24(3):193–195.

15. Mutic S, Dempsey JF. The ViewRay system: magnetic resonance-guided and controlled radiotherapy. *Semin Radiat Oncol.* 2014;24(3):196–199.

16. Lagendijk JJW, Raaymakers BW, Raaijmakers AJE, et al. MRI/linac integration. *Radiother Oncol.* 2008;86(1):25–29. doi:10.1016/j.radonc.2007.10.034.

17. Raaymakers BW, Lagendijk JJW, Overweg J, et al. Integrating a 1.5 T MRI scanner with a 6 MV accelerator: proof of concept. *Phys Med Biol.* 2009;54(12):N229.

18. Winkel D, Bol GH, Kroon PS, et al. Adaptive radiotherapy: The Elekta Unity MR-linac concept. *Clin Transl Radiat Oncol.* 2019.

19. Connor WG, Boone MLM, Veomett R, et al. Patient repositioning and motion detection using a video cancellation system. *Int J Radiat Oncol Biol Phys.* 1975;1(1–2):147–153.

20. Norwood HM, Stubbs B. Patient movements during radiotherapy. *Br J Radiol.* 1984;57(674):155–158. doi:10.1259/0007-1285-57-674-155.

21. Yan Y, Song Y, Boyer AL. An investigation of a video-based patient repositioning technique. *Int J Radiat Oncol Biol Phys.* 2002;54(2):606–614.

22. Simpkins GS, Ascoli FA, Sherman DM, Kadish SP. A simple system for detecting patient movement during radiation therapy. *Radiology.* 1981;138(3):735–736. doi:10.1148/radiology.138.3.7465859.

23. Brahme A, Nyman P, Skatt B. 4D laser camera for accurate patient positioning, collision avoidance, image fusion and adaptive approaches during diagnostic and therapeutic procedures. *Med Phys.* 2008;35(5):1670–1681. doi:10.1118/1.2889720.

24. Moser T, Habl G, Uhl M, et al. Clinical evaluation of a laser surface scanning system in 120 patients for improving daily setup accuracy in fractionated radiation therapy. *Int J Radiat Oncol Biol Phys.* 2013;85(3):846–853.

25. Moser T, Fleischhacker S, Schubert K, Sroka-Perez G, Karger CP. Technical performance of a commercial laser surface scanning system for patient setup correction in radiotherapy. *Phys Med.* 2011;27(4):224–232.

26. Pallotta S, Marrazzo L, Ceroti M, Silli P, Bucciolini M. A phantom evaluation of Sentinel™, a commercial laser/camera surface imaging system for patient setup verification in radiotherapy. *Med Phys.* 2012;39(2):706–712. doi:10.1118/1.3675973.

27. Pallotta S, Simontacchi G, Marrazzo L, et al. Accuracy of a 3D laser/camera surface imaging system for setup verification of the pelvic and thoracic regions in radiotherapy treatments. *Med Phys.* 2013;40(1):011710. doi:10.1118/1.4769428.

28. Jönsson M, Ceberg S, Nordström F, Thornberg C, Bäck SAJ. Technical evaluation of a laser-based optical surface scanning system for prospective and retrospective breathing adapted computed tomography. *Acta Oncol.* 2015;54(2):261–265. doi:10.3109/0284186X.2014.948059.

29. Gerig L, El-Hakim S, Szanto J, Girard A. The development and clinical application of a patient position monitoring system. *Int J Radiat Oncol Biol Phys.* 1993;27:163.

30. Bova FJ, Buatti JM, Friedman WA, Mendenhall WM, Yang CC, Liu C. The University of Florida frameless high-precision stereotactic radiotherapy system. *Int J Radiat Oncol Biol Phys.* 1997;38(4):875–882. doi:10.1016/S0360-3016(97)00055-2.

31. Meeks SL, Tomé WA, Willoughby TR, et al. Optically guided patient positioning techniques. *Semin Radiat Oncol.* 2005;15(3):192–201. doi:10.1016/j.semradonc.2005.01.004.

32. Cleary K, Peters TM. Image-guided interventions: technology review and clinical applications. *Annu Rev Biomed Eng.* 2010;12(1):119–142. doi:10.1146/annurev-bioeng-070909-105249.

33. Wagner TH, Meeks SL, Bova FJ, et al. Optical racking technology in stereotactic radiation therapy. *Med Dosim.* 2007;32(2):111–120. doi:10.1016/j.meddos.2007.01.008.

34. Wang JZ, Rice R, Pawlicki T, et al. Evaluation of patient setup uncertainty of optical guided frameless system for intracranial stereotactic radiosurgery. *J Appl Clin Med Phys.* 2010;11(2):92–100. doi:10.1120/jacmp.v11i2.3181.

35. Rietzel E, Rosenthal SJ, Gierga DP, Willett CG, Chen GT. Moving targets: detection and tracking of internal organ motion for treatment planning and patient set-up. *Radiother Oncol.* 2004;73:S68–S72. doi:10.1016/S0167-8140(04)80018-5.

36. Rietzel E, Chen GT. Improving retrospective sorting of 4D computed tomography data. *Med Phys.* 2006;33(2):377–379. doi:10.1118/1.2150780.

37. Pedersen AN, Korreman S, Nyström H, Specht L. Breathing adapted radiotherapy of breast cancer: reduction of cardiac and pulmonary doses using voluntary inspiration breath-hold. *Radiother Oncol.* 2004;72(1):53–60. doi:10.1016/j.radonc.2004.03.012.

38. Westoby MJ, Brasington J, Glasser NF, Hambrey MJ, Reynolds JM. Structure-from-Motion' photogrammetry: a low-cost, effective tool for geoscience applications. *Geomorphology.* 2012;179:300–314. doi:10.1016/j.geomorph.2012.08.021.

39. Fussell A. Terrestrial photogrammetry in archaeology. *World Archaeol.* 1982;14(2):157–172. doi:10.1080/00438243.1982.9979857.

40. Kärrholm J. Roentgen stereophotogrammetry: review of orthopedic applications. *Acta Orthop.* 1989;60(4):491–503. doi:10.3109/17453678909149328.

41. Selvik G. Roentgen stereophotogrammetry: a method for the study of the kinematics of the skeletal system. *Acta Orthop Scand.* 1989;60(S232):1–51. doi:10.3109/17453678909154184.

42. Medin PM, Solberg TD, De Salles AAF, et al. Investigations of a minimally invasive method for treatment of spinal malignancies with LINAC stereotactic radiation therapy: accuracy and animal studies. *Int J Radiat Oncol Biol Phys.* 2002;52(4):1111–1122. doi:10.1016/S0360-3016(01)02762-6.

43. Velkley DE, Oliver GD. Stereo-photogrammetry for the determination of patient surface geometry. *Med Phys.* 1979;6(2):100–104. doi:10.1118/1.594538.

44. Chu J, Bloch P. Application of Moire patterns for obtaining surface contour information on patients receiving radiotherapy. In: *Three-Dimensional Machine Perception* (Vol. 283, pp. 2–6). International Society for Optics and Photonics; 1981. doi:10.1117/12.931982.

45. Menke M, Hirschfeld F, Mack T, Pastyr O, Sturm V, Schlegel W. Photogrammetric accuracy measurements of head holder systems used for fractionated radiotherapy. *Int J Radiat Oncol Biol Phys.* 1994;29(5):1147–1155. doi:10.1016/0360-3016(94)90412-X.

46. Schlegel W, Pastyr O, Bortfeld T, Gademann G, Menke M, Maier-Borst W. Stereotactically guided fractionated radiotherapy: technical aspects. *Radiother Oncol.* 1993;29(2):197–204. doi:10.1016/0167-8140(93)90247-6.

47. Rogus RD, Stern RL, Kubo HD. Accuracy of a photogrammetry-based patient positioning and monitoring system for radiation therapy. *Med Phys.* 1999;26(5):721–728. doi:10.1118/1.598578.

48. Geng J. Structured-light 3D surface imaging: a tutorial. *Adv Opt Photonics.* 2011;3(2):128–160. doi:10.1364/aop.3.000128.

49. Das A, Galdi C, Han H, Ramachandra R, Dugelay JL, Dantcheva A. Recent advances in biometric technology for mobile devices. In: *2018 IEEE 9th International Conference on Biometrics Theory, Applications and Systems, BTAS 2018.* IEEE; 2019. doi:10.1109/BTAS.2018.8698587.

50. Zheng F, Kong B. Calibration of linear structured light system by planar checkerboard. In: *International Conference on Information Acquisition, 2004. Proceedings* (pp. 344–346). IEEE. doi:10.1109/ICIA.2004.1373385.

51. Zhang L, Garden AS, Lo J, et al. Multiple regions-of-interest analysis of setup uncertainties for head-and-neck cancer radiotherapy. *Int J Radiat Oncol Biol Phys.* 2006;64(5):1559–1569.

52. Mirota DJ, Ishii M, Hager GD. Vision-based navigation in image-guided interventions. *Annu Rev Biomed Eng.* 2011;13(1):297–319. doi:10.1146/annurev-bioeng-071910-124757.

53. Bauer S, Seitel A, Hofmann H, et al. Real-time range imaging in health care: a survey. In: *Time-of-Flight and Depth Imaging. Sensors, Algorithms, and Applications* (pp. 228–254). Springer, Berlin, Germany.

54. Maier-Hein L, Mountney P, Bartoli A, et al. Optical techniques for 3D surface reconstruction in computer-assisted laparoscopic surgery. *Med Image Anal.* 2013;17:974–996.

55. Schaller C, Penne J, Hornegger J. Time-of-flight sensor for respiratory motion gating. *Med Phys.* 2008;35(7):3090–3093. doi:10.1118/1.2938521.

56. Besl PJ, McKay ND. Method for registration of 3-D shapes. In: *Sensor Fusion IV: Control Paradigms and Data Structures* (Vol. 1611, pp. 586–606). International Society for Optics and Photonics; 1992. doi:10.1117/12.57955.

57. Schaller C, Adelt A, Penne J, Hornegger J. Time-of-Flight sensor for patient positioning. In: *Medical Imaging 2009: Visualization, Image-Guided Procedures, and Modeling* (Vol. 7261, p. 726110). SPIE; 2009. doi:10.1117/12.812498.

58. Placht S, Stancanello J, Schaller C, Balda M, Angelopoulou E. Fast time-of-flight camera based surface registration for radiotherapy patient positioning. *Med Phys.* 2012;39(1):4–17. doi:10.1118/1.3664006.

59. Gilles M, Fayad H, Miglierini P, et al. Patient positioning in radiotherapy based on surface imaging using time of flight cameras. *Med Phys.* 2016;43(8):4833–4841. doi:10.1118/1.4959536.

60. Rusinkiewicz S, Levoy M. Efficient variants of the ICP algorithm. *Proc Int Conf 3-D Digit Imaging Model 3DIM* (Vol. 1, pp. 145–152). 2001. doi:10.1109/IM.2001.924423.

61. Erdi YE, Rosenzweig K, Erdi AK, et al. Radiotherapy treatment planning for patients with non-small cell lung cancer using positron emission tomography (PET). *Radiother Oncol.* 2002;62(1):51–60.

62. Jenkins C, Xing L, Yu A. Using a handheld stereo depth camera to overcome limited field-of-view in simulation imaging for radiation therapy treatment planning. *Med Phys.* 2017;44(5):1857–1864. doi:10.1002/mp.12207.

63. Jana A. *Kinect for Windows SDK Programming Guide.* Packt Publishing Ltd, 2012.

64. Bujold A, Craig T, Jaffray D, Dawson LA. Image-guided radiotherapy: has it influenced patient outcomes? *Semin Radiat Oncol.* 2012;22(1):50–61. doi:10.1016/j.semradonc.2011.09.001.

65. Zelefsky MJ, Fuks Z, Hunt M, et al. High-dose intensity modulated radiation therapy for prostate cancer: early toxicity and biochemical outcome in 772 patients. *Int J Radiat Oncol Biol Phys.* 2002;53(5):1111–1116.

CHAPTER 2

Safety and Quality Improvements with SGRT

Hania A. Al-Hallaq and Bill J. Salter

CONTENTS

2.1 INTRODUCTION

Radiation has played an important role in cancer treatments since its discovery. This role is expected to grow given recent mounting evidence that implicates radiation in amplifying the innate immune response[1] in addition to halting tumor growth. In both applications, accurate targeting is necessary to protect healthy tissues. This has fueled technical advancements in delivery that have increased the complexity of radiation therapy (RT). As in many other disciplines, increasing complexity has been accompanied by an increase in the rate of treatment errors and in the risk associated with such errors.[2,3] To reduce these errors, the field of RT has prioritized safety initiatives particularly within the last decade.[4,5]

While various strategies exist to reduce errors, identifying, quantifying, and characterizing them are often the first necessary steps. This is especially important when multiple steps of the treatment process are customized, as in RT. Surface imaging is an optical modality that provides three-dimensional data in real-time to quantify the deviation from a reference surface.[6,7] As a result, it can provide noninvasive monitoring of the entire treatment positioning and delivery process. In this chapter, the role of surface image guidance as an integral tool for improving safety and quality of external beam radiation therapy (EBRT) will be described.

2.2 ERRORS IN RT

2.2.1 Brief History of Errors in RT

As in other healthcare fields, RT has suffered from treatment errors. Some are of minor clinical impact such as neglecting to place bolus for a treatment fraction (e.g., dosimetric order of magnitude <1% for a single treatment fraction).[8] Other errors are moderate such as neglecting to use a physical wedge as planned (e.g., dosimetric order of magnitude ~30%–50% for a single treatment fraction).[9] Unfortunately, there have also been some catastrophic errors such as those highlighted in 2010 by the *New York Times*.[5,10]

One such error involved a computer malfunction in which the multileaf collimator (MLC) instructions were deleted from intensity modulated radiation therapy (IMRT) treatment fields in the record and verify (R&V) system. This resulted in large doses delivered to healthy tissue over several fractions that ultimately proved fatal. In 2010, the rate of treatment errors in RT that rise to a level requiring some form of formal reporting to a government agency (i.e., medical events) was quantified to affect 1 in 600 patients or 0.2%.[8,11,12] In comparison, the serious error rate is often much lower in industries that are considered "ultrasafe," such as aviation where the error rate affects 1 in 10 million people.[8] Moreover, it has been conjectured that errors in RT may be underreported as there is no federal body monitoring EBRT as there is in aviation such as the National Transportation Safety Board.[8,10]

Highlighting these errors in the popular press has spurred an increase in the field of RT to instigate new safety and quality initiatives including encouraging reporting of all errors into a national Radiation Oncology-Incident Learning System (RO-ILS) database[13] and emphasizing the importance of creating a departmental culture that revolves around safety.[13,14] And as in aviation, learning from "near-miss" scenarios has been emphasized as a tool to improve safety and quality.[13] Prior to these initiatives, the field of RT had recognized that errors associated with treatment could be reduced by adopting strategies such as automation, checklists, and standardization.[5,15] Not only must errors be identified, but they must also be quantified in terms of severity. Recently, the RT community has adopted risk analysis methods pioneered in fields outside of healthcare.[5,14] One such method is failure mode and effects analysis, wherein failure modes are enumerated for a particular workflow and the risks of these failures are assigned a risk score. Other strategies are also under investigation.[16]

2.2.2 Mitigating Errors in RT

Institutional[2,17,18] and national databases[13] have been used to study and learn from the most common and critical errors in RT. Broadly, they can be categorized as occurring during simulation, treatment planning, or treatment delivery. There is some evidence that time-intensive procedures such as stereotactic body radiation therapy (SBRT) are particularly vulnerable.[3,14] Even in routine treatments, examples have been documented in which errors occur due to the care team rushing to deliver a treatment due to patient discomfort.[11] In general, some of the most serious errors identified can be categorized as treating the incorrect patient, incorrect anatomic site, or delivering the wrong dose.[11–13] RO-ILS has recently reported on errors submitted to their national database.[13] Of 2344 events reported, 44% of the 396 errors of the highest priority could be grouped into three categories: "problematic plan," "wrong shift instructions," and "wrong shift performed," with error rates of 4.2%, 1.7%, and 1.5%, respectively. Thus, errors related to incorrect isocenter localization comprised 3.2% of the total.

To best identify errors in a complex workflow that involves multiple teams, all personnel should be engaged to harness the power of independent review and redundancy. The Swiss cheese model postulates that multiple domains such as engineering, people, procedures, and administration (i.e., layers of Swiss cheese) are stacked together in such a way as to minimize the likelihood of holes (i.e., failures) being aligned.[19,20] Thus, multiple strategies within each layer are typically implemented concurrently in RT, and the layers that are typically used will be reviewed briefly here. Although these strategies are commonly employed in RT, this list is by no means exhaustive. Also, some error strategies address latent failures (i.e., errors that occur due to omission or inaction), while others addresses active failures (i.e., errors that occur because the wrong action/choice was selected).[19] Finally, there is some evidence that identifying errors by verifying the inputs to the treatment workflow is more effective than by verifying the outputs.[13] In other words, prospective identification may be more effective than retrospective.[5]

2.2.2.1 Automation

This error reduction strategy with the engineering layer is motivated by the knowledge that humans will make errors as a result of lapses in attention or memory due to stress and/or fatigue or due to an inability to respond appropriately when deviations from standard operating procedures occur.[19,21] Automating steps of the process can reduce the rate of occurrence of some of these causes of error but may also introduce additional errors because automation can reduce the level of engagement by the human user with the process.[22] To counteract this potential disengagement, mental models have been used to prepare personnel for responding to catastrophic events in aviation.[23] A mental model refers to simulating a catastrophic event in order to prepare personnel to recognize and respond appropriately to one if it were to occur. Examples of automation in RT include R&V systems,[11,24] knowledge-based planning,[25] automatic segmentation of organs-at-risk (OARs),[26] and automated treatment plan evaluation.[27] However, systems of automation such as R&V have also been reported as contributing factors for some errors due to operator disengagement and/or other human–machine interface-related errors.[11] In fact, the incident reported by the *New York Times* in which the MLC sequence was deleted from the R&V software has occurred at other centers (personal communication, the University of Chicago, Chicago, IL). Fortunately, in those cases, the human operators noticed that the beam was not being delivered in a modulated fashion and interrupted delivery before the patient was harmed. In general, it is accepted that automation leads to a net reduction of errors compared to manual processes but does not necessarily eliminate them altogether.[24]

2.2.2.2 Checklists

The aviation industry has successfully used checklists to improve safety.[15,28] Subsequently, this strategy has been adopted in healthcare. For manual procedures such as surgeries, checklists have been shown to be a valuable alternative that can impact outcomes and patient survival.[29] Based on these positive results, RT has likewise adopted the use of checklists, including time-outs. Checklists are intended to slow down the human operator(s) while performing a highly technical process to ensure that no omissions have been made and adequate communication has occurred.[15] For example, these strategies are particularly useful to verify patient identity, the laterality of the treatment site, the presence of a cardiovascular implantable electronic device prior to imaging/treatment, and the intended dose to be delivered. Because checklists rely on verification by the human user, errors may still arise due to inattention, disengagement, or miscommunication.[5,15] However, when layered with other safety strategies, checklists have proved to be useful and are currently recommended by professional societies such as American Society of Radiologic Technologists (ASRT)[30] and American Association of Physicists in Medicine (AAPM).[5,15]

2.2.2.3 Forcing Functions

Forcing functions, which actively disable further action, are more likely to eliminate errors than strategies that rely on manual compliance such as checklists.[5] This is because they actively prevent the undesired workflow from continuing. One of the pitfalls of forcing functions is that they cannot be applied to manual processes. Forcing functions (i.e., interlocks) are integrated into linear accelerator (linac) operation. They have been used to ensure that physical blocks or wedges have been inserted into the treatment field in the correct orientation. More recently, physical wedges have fallen out of favor altogether in many clinics in order to eliminate the potential risks that accompany their use,[17,31] but the example still applies. Another example of forcing functions is the use of tolerance tables to ensure that treatment couch coordinates are within a specified distance from the desired position. Tolerance tables are used as a surrogate to quantify isocenter localization accuracy by acquiring the initial couch position following verification of the isocenter accuracy by alternate means such as X-ray imaging.[32] Without perfect patient immobilization and indexing, tightening tolerances can decrease specificity without a corresponding increase in sensitivity[32,33] and furthermore can decrease users' sensitivity to warning messages if they are constantly overriding tolerance table violations. Thus, tolerance tables are typically used in conjunction with other safe practices to ensure accurate isocenter localization without compromising efficiency.

2.2.2.4 Standardization

Humans are capable of reflexively performing a task with which they are familiar (i.e., "on autopilot"). This is in part because they create "mental models" that can instinctively focus their attention when something strays from an expected workflow. The standardization of processes harnesses this innate and instinctual ability. Standardization can be applied at all stages of the RT process and can improve quality[5,34] in addition to reducing errors.[2] National recommendations have been made for standardization of the following workflow elements: the written directive,[35] the IMRT planning and treatment process,[36] and nomenclature for organ and target delineation.[37] The use of checklists is one method by which to standardize processes but written policies and procedures as well as a culture that values standardization are also important.[12,38,39]

2.2.3 Role of Surface Imaging in Error Prevention

Surface imaging provides a real-time, 3D surface model of the patient throughout treatment, an RT delivery process known as surface-guided radiation therapy (SGRT). In a sense, SGRT can be considered as an independent, second observer of the treatment process. This is analogous to the use of computer-aided diagnosis systems that have been successfully designed to serve as a second opinion for radiologists in the identification of suspicious lesions on screening mammograms.[40] When performing a repetitive task in which the error and detection rates are low, automated systems could direct the attention of the human operator appropriately. In order for such tools to be utilized to their full potential, the human operator should be trained to create mental models in which they will seek out additional help or halt a treatment if the independent system indicates that something is amiss (i.e., error management[22]). In summary, surface imaging can serve as a quantitative second observer of all external beam RT treatments provided that the human operator is trained to properly interpret the results.

2.3 SAFETY IMPROVEMENTS WITH SGRT

SGRT has the potential to reduce errors in a variety of ways.[41] While surface imaging cannot provide insight into the accuracy of the delineated treatment volume or the dose distribution, it could potentially address some of the sources of error related to patient identification, immobilization accuracy, treatment isocenter location, or patient motion during treatment. Because SGRT is a fairy new modality, there are relatively few publications in the literature on this topic. These will be summarized when available. Otherwise, case studies will be presented in an effort to support the hypothesis that SGRT will become an increasingly important tool in the RT safety tool belt.

2.3.1 Patient Identification

One of the most critical safety components across healthcare fields is correctly identifying the patient to be treated.[29] In addition to using time-outs,[30] it may be advantageous to incorporate forcing functions or quantitative methods into the process of verifying patient identity. Wiant et al. retrospectively investigated whether the uniqueness of each patient's surface could be used for quantitative identification in a limited cohort of patients.[42] The surface generated from the computed tomography (CT) simulation scan for 16 breast cancer patients was compared to the surfaces acquired during daily treatment using a distance-to-agreement (DTA) function. A similarity score extracted from a histogram of DTA data (Figure 2.1) could successfully distinguish between patients because a larger percentage (i.e., >80%) of points met either the 3-mm or 5-mm DTA criterion for the correct patient while this percentage was much lower (<50%) for the incorrect patient. SGRT was found to be highly specific (99%) and sensitive (99%) if a 55% threshold were used to assess the 3-mm DTA histogram data in these 16 patients. Although these results are

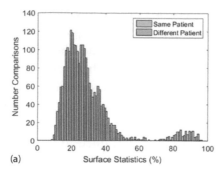
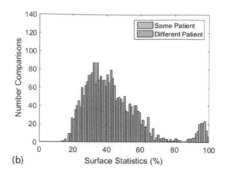

FIGURE 2.1 Histograms of intra-patient and inter-patient DTA comparisons for 3-mm (a) and 5-mm (b) thresholds demonstrate a clear separation between the percentages of points meeting the DTA criteria for the same patient versus a different patient. (Adapted from Wiant, D. et al., *J. Appl. Clin. Med. Phys.*, 17, 271–278, 2016.)

promising, further studies must be conducted to determine whether this high success rate would translate to identifying patients using anatomy such as the torso or pelvis, which do not possess as many topographically salient features. With the recent integration of SGRT system software with the treatment console (e.g., Varian OSMS), the patient's CT surface could serve as a quantitative check of patient identity that could be an added layer to current forms of verification such as the date of birth or medical record number.

2.3.2 Measurement of Immobilization Accuracy

SGRT has provided clinics with the ability to characterize their institution-specific immobilization accuracy, particularly for frameless stereotactic radiosurgery (SRS) and RT for head and neck (H&N) and breast cancer. To implement frameless SRS, Li et al. investigated patient motion within an open-face mask tracked with SGRT and found that after deliberate forced movement, patients' surfaces returned to within 1–2 mm of the original position.[43] Zhao et al. compared shoulder immobilization with a moldable cushion to shoulder stirrups using SGRT and found that a custom mold was more accurate in reproducing the CT-simulated position.[44] In both instances, SGRT helped to quantify intermittent motion in real-time. Currently, SGRT is the only clinically available modality that provides this type of feedback without administering radiation dose.

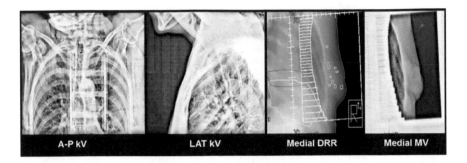

| A-P kV | LAT kV | Medial DRR | Medial MV |

FIGURE 2.2 Orthogonal kV X-ray and MV portal images obtained during verification imaging with the patient positioned on a breast board at a 10° angle indicate no obvious discrepancy between the intended and implemented positions.

Errors in immobilization device location and orientation can also be detected by surface guidance. This is a particularly concerning failure mode because it has been shown to result in errors that tend to propagate further into the workflow.[45] Figure 2.2 shows images from a case in which a breast cancer patient treated with deep inspiration breath hold (DIBH) was positioned by surface guidance for first treatment day imaging. Because the surface guidance software reported that the patient was pitched by ~5° relative to the expected surface orientation, the treatment team investigated the accuracy of the breast board angle setting. The CT simulation documentation indicated that the breast board angle should be 10°, which is what the treatment team had used to position the patient for pretreatment imaging. Visual inspection of the tangential images obtained at this breast board angle setting did not indicate any obvious discrepancy (Figure 2.2), but in order to allay the concerns of the treatment team, a CT scan was repeated of the patient using the 10° breast board setting. Comparison of the two CT scans demonstrated (Figure 2.3) that the breast board angle used for the original simulation was actually 5°, thus revealing that the breast board angle had been recorded incorrectly at the time of simulation. Once the corrected breast board angle of 5° was used, the surface images showed much better agreement to the CT surface (i.e., pitch < 1°). This near-miss demonstrated that SGRT was much more sensitive to the breast board angle than kV/MV images. This failure mode would have benefited from prospective detection at the point of origin to prevent its propagation to the initial imaging verification session. With the recent development of surface imaging

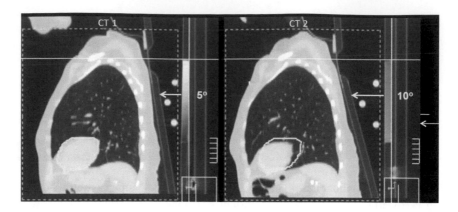

FIGURE 2.3   The CT scan used for treatment planning was acquired at a 5° breast board angle (CT1) while the CT scan of the pretreatment filming position was acquired at the documented 10° breast board angle (CT2).

systems that can track immobilization devices and patients in the CT room, this is now possible. Because detection of immobilization device accuracy could benefit all EBRT patients, select clinics have adopted it for initial positioning of all EBRT patients.[6]

2.3.3 Isocenter Localization Accuracy

The most commonly reported treatment errors are related to incorrect isocenter localization.[13,14] This may be due to the fact that the isocenter does not always have a readily visible anatomic surrogate. Or perhaps, as reliance on image-guided radiation therapy (IGRT) has increased, the treatment team may have fallen out of the habit of visually confirming the isocenter location relative to patient anatomy. The IGRT process has also become quite sophisticated with availability of high-resolution 3D data from cone-beam computed tomography (CBCT), necessitating a careful review of image registration prior to finalization of IGRT shifts. These complex, multistep processes can lead to necessary pauses in workflow that can result in errors due to lapses in memory or inattention. Treatment of multiple targets also increases the probability of localization errors particularly if the targets are within close proximity of each other.

A secondary IGRT modality such as SGRT could be used to verify the isocenter location to within a certain threshold (e.g., <1 cm). For example, Ford and Dieterich reported that in a case study of errors associated with SBRT, one such error was caused by a miscalibration of the imaging isocenter by several centimeters.[14] This could have been detected by SGRT since its isocenter is initially calibrated independently using the light field and lasers. In another case reported to the RO-ILS database, the IGRT reference images used for the entirety of an EBRT treatment were incorrectly exported to the treatment console such that they did not reflect the treatment isocenter.[13] This error could also have been prevented by SGRT as the treatment isocenter is attached to the reference DICOM surface that is exported from the plan to the surface imaging software. Errors that affect the entire treatment course, such as this one, are extremely concerning and thus are most in need of additional precautions. For SGRT to be useful in preventing such errors, the treatment team must learn to rely on this added information by incorporating it into the mental models that they use for troubleshooting.

2.4 QUALITY IMPROVEMENTS WITH SGRT

According to AAPM TG-100,[5] "not meeting the desired level of quality is a failure." While safety is one aspect of overall quality, some of the failures in RT are due to low quality and may not necessarily constitute an error that would be reported to RO-ILS. Thus, improving quality should be an associated and ongoing goal of RT. SGRT can help to accomplish this goal by improving initial patient positioning accuracy, correcting for intra-fraction motion, detecting anatomic changes throughout the course of treatment, and increasing efficiency.

2.4.1 Improving Initial Patient Positioning Accuracy

SGRT has been shown to provide higher patient positioning accuracy compared to localization with lasers. Stanley et al. demonstrated that compared to positioning of skin marks/tattoos to lasers, SGRT reduced the residual shifts detected by CBCT in over 6000 treatment fractions across multiple anatomic sites include the pelvis, abdomen, extremities, chest, and breast.[46] But even when IGRT is used to position patients, SGRT can provide complementary information due to its typically larger field of view compared to X-ray modalities. Figure 2.4 shows multiple surfaces of a single breast cancer patient that are overlaid in different colors, each of which represents a different treatment fraction. The figure demonstrates that although the patient is immobilized on a breast board and positioned with daily X-ray guidance, there is wide day-to-day variability in the positioning of the ipsilateral arm that correspondingly affects the reproducibility of the breast tissue.[47]

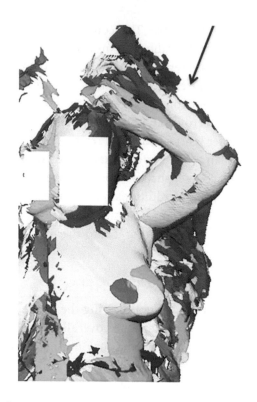

FIGURE 2.4 Surface images acquired of the same patient on different treatment days, each represented in a different color, demonstrate wide variability in arm positioning from day to day. (Adapted from Bert, C. et al., *Int. J. Radiat. Oncol.*, 64, 1265–1274, 2006.)

When SGRT has been used to observe breast cancer treatments, it has prompted several institutions to alter at least one aspect of their immobilization strategy.[6] According to Padilla et al., SGRT prompted two workflow changes for patients receiving whole-breast RT: (1) MV orthogonal filming was replaced with OBI kV imaging to reduce interoperator subjectivity when assessing the images and (2) the external contour was outlined on all digitally reconstructed radiographs (DRRs) to focus the attention of the treatment team to the breast surface in addition to bony landmarks.[48] Batin et al. found large inter-fraction variability in the chin and arm positions in their postmastectomy patients that were observed by SGRT.[49] As a result, they added a chin strap and hand grips to the breast board for immobilization which resulted in better reproducibility of the clavicle.[49] Even when no changes to the immobilization approach were made, SGRT led to improved accuracy. Shah et al. reported that the therapy team focused longer on chin and arm positioning accuracy after surface guidance was implemented.[50] They conjectured that this was due to a combination of the real-time feedback provided by surface guidance in addition to the fact that the system served as a second observer of the entire positioning and treatment process. In summary, the daily use of SGRT has been shown to be capable of improving the quality of initial patient positioning for EBRT.

2.4.2 Detecting Anatomic Changes

A subset of patients experience anatomic changes such as weight loss, swelling, or changes to reconstructed tissue or implants during the course of RT. Some of these changes may be so gradual that they are not readily detectable by physical examination. If these changes significantly alter the anatomy, they may degrade the dosimetric quality of the intended treatment plan. SGRT may be able to detect such anatomic changes and thus alert the treatment team as to when a new plan may be required.

Patients receiving RT for H&N cancer, in particular, tend to suffer from weight loss throughout treatment.[51] Such changes can be detected by surface guidance. Several studies of the accuracy of SGRT for initial positioning in an open-face mask found a significant spike in the discrepancy of the patient surface at time of treatment from the time of CT simulation and attributed this to weight loss.[44,52,53] While these registration discrepancies could be as large as 1 cm when a patient loses a significant amount of weight,[53] more studies are needed to determine whether SGRT is sensitive enough to detect clinically relevant levels of weight loss.

Multiple cases have been reported of breast cancer patients monitored by SGRT for DIBH treatment that demonstrate the power of surface guidance to quantitatively detect anatomic changes. In one case, a discrepancy between bony alignment on X-ray imaging and the breast surface on SGRT was found for a patient with a reconstructed left breast. To understand whether this discrepancy could have been caused by a change in breath-hold amplitude, a new CT scan was acquired and registered to the original scan. As shown in the subtraction of the registered CT scans in Figure 2.5, the patient was able to reproduce her breath-hold consistently, but the reconstructed breast tissue was found to have swollen by approximately 1 cm. While projection of the intended plan onto the new CT scan indicated that tangential irradiation

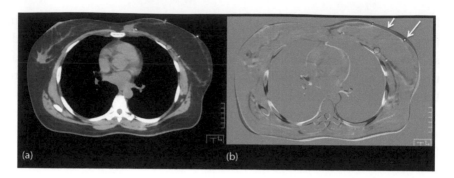

FIGURE 2.5   CT scans acquired at DIBH of a patient with a reconstructed left breast demonstrate that the treatment planning scan (a) and registered repeat CT scan (b) match when considering the heart and lung anatomy but demonstrate swelling of the reconstructed breast surface, as shown by the white arrows.

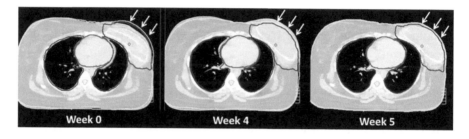

FIGURE 2.6   CT scans of a patient with a left-sided tissue expander at various time points during which considerable change in the surface, as shown by the white arrows, was detected by surface imaging due to filling of the expander with 300 cc between weeks 0 and 4 and subsequent resorption between weeks 4 and 5.

remained robust to these anatomic changes, this is not always the case. Figure 2.6 shows another case in which a patient with a reconstructed breast and tissue expander was treated with DIBH. At the first session of verification images, a discrepancy was detected between alignment of bony anatomy as verified with X-ray imaging and the breast surface monitored with SGRT. Upon further examination of the patient's hospital record, it was discovered that the expander had been filled with an additional 300 cc of saline after the CT simulation session. This large change in anatomy necessitated a new treatment plan to adequately cover the target while protecting OARs. Without SGRT, these changes may not have been detected.

2.4.3 Monitoring and Correcting for Intra-fraction Motion

Some of the first applications of SGRT were to enable accurate frameless SRS and DIBH treatments by detecting and correcting for intra-fraction motion. Because SGRT provides real-time data at multiple frames per second, it was thought that combining it with an open-face mask could provide comparable accuracy to the use of a stereotactic frame affixed to the patient's skull.[54] For such applications, SGRT is referenced to another IGRT modality; first the internal target is verified with CBCT and then a reference surface is acquired of the patient at this position and a beam-hold is triggered if the patient moves out of tolerance. This is possible because the surface of the chin/nose/cheek bones is a good surrogate for targets within the brain due to the rigid anatomy of the skull. In the case of DIBH treatments, SGRT provides an alternative to spirometry-based methods by enabling intra-fraction monitoring of the breast surface during voluntary breath-holds. Voluntary DIBH guided by surface imaging has been demonstrated to provide comparable intra-DIBH variability and intra-fraction stability to spirometry.[55] These two clinical applications were adopted because of the obvious ability of the patient's surface to serve as a surrogate for the target. However, this does not mean that other clinical workflows would not also benefit from intra-fraction monitoring using SGRT. Wiant et al. observed breast cancer patients treated during free breathing with SGRT throughout treatment and learned that the root-mean squared deviation of the chest surface correlated significantly with treatment time indicating that patients drift from the initial position the longer that they remain on the treatment table.[56] These findings imply that any patient receiving EBRT may benefit from intra-fraction monitoring to disable the beam if they move out of tolerance. Lengthy treatments such as SBRT, in which there are often multiple targets to be treated accompanied by the necessarily complex/lengthy IGRT workflows, can benefit greatly from intra-fraction monitoring. This sort of information could only be practically gathered with a real-time, noninvasive imaging modality such as SGRT.

2.4.4 Increasing Efficiency

As technical improvements are added to enable complex radiation treatments, such as IGRT and motion management, efficiency has suffered. Longer treatment times reduce throughput and compromise the quality of treatment as patient discomfort increases.[56] SGRT can arguably be considered as a 3D replacement for lasers that are currently used to position patients, with the added advantage of potentially improving positioning efficiency. Batin et al. investigated if SGRT could be used instead of daily X-ray imaging for positioning postmastectomy breast cancer patients for proton treatment.[49] Not only did they find that positioning of the chest wall was more accurate with SGRT than with orthogonal radiographs, but it also reduced the time required to position patients for treatment from an average of 11 min with radiographs to an average of 6 min (Figure 2.7). Given the dose reduction afforded by replacing X-ray imaging with SGRT coupled with the increased accuracy and efficiency, Batin et al. have replaced daily X-ray imaging with SGRT in their clinic. SGRT can also improve efficiency for positioning intact breast cancer patients. Table 2.1 shows institutional data for shifts from tattoos/skin marks following X-ray imaging for 100 patients positioned for their initial isocenter localization session (verification with images). For the 50 patients in which positioning was augmented by SGRT, the shifts necessary for

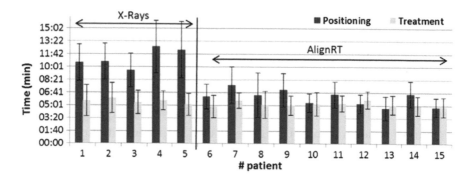

FIGURE 2.7   Use of SGRT (e.g., AlignRT) to position postmastectomy breast cancer patients for proton treatments reduces the time required to position patients compared to the use of X-rays but does not affect the treatment time. (Adapted from Batin, E et al., *Pract. Radiat. Oncol.*, 6, e235–e241, 2016.)

TABLE 2.1   Percentage of Patients during Initial Verification Imaging Session with Post-IGRT Shifts < 1 cm and Total Time < 30 min

	Patients with Post-IGRT Shifts <1 cm (%)	Patients with Total Time <30 min (%)
Patients positioned *without* SGRT (*n* = 50)	64%	44%
Patients positioned *with* SGRT (*n* = 50)	92%	72%

accurate localization of the isocenter as well as the time required to complete the filming session were reduced compared to the 50 patients who were positioned without SGRT.

More recently, surface guidance systems are offering solutions that incorporate a virtual representation[57] of the patient surface at the loading position (i.e., couch fully retracted and lowered). Presentation of a virtual reality representation of the patient as a silhouette, visible on an in-room monitor, enables the patient to assist in their own positioning. This has, in turn, led to faster and higher quality initial orientation of the patient, for subsequent alignment at the isocenter, and to a feeling of increased control by patients over their situation by virtue of being able to participate in their own alignment.

While efficiency improvements can be quantified, other associated benefits are less quantifiable, yet equally important. For example, increased efficiency implies an improvement in adherence to the treatment schedule that can subsequently reduce the stress felt by the treatment team, potentially reducing the probability of errors. Thus, improved efficiency, along with its less tangible by-products, should be an ongoing quality improvement goal in clinics. It appears clear that SGRT could potentially expedite these efficiency improvements for all EBRT patients.

2.5 FUTURE APPLICATIONS

Because SGRT is a new imaging modality, some of its applications are under development or theoretical at this stage. With the 3D real-time tracking afforded by SGRT, there is the potential to provide collision prediction or even to perform machine geometry checks. Furthermore, surface imaging could be used to track physiologic and/or biometric parameters in patients. These potential applications will be briefly reviewed and are discussed further in Chapter 26.

2.5.1 Collision Detection

As the complexity of EBRT increases, particularly for the treatment of multiple metastases, couch rotations are often employed to achieve improved dose conformity and sparing of OARs. The likelihood of potential collisions with the patient or couch increases with the increased use of couch rotations as for 4-Pi[58] or SRS[59] treatments. Furthermore, many traditional treatments that do not require couch rotations also suffer from potential collisions that are only discovered when the treatment is attempted. These situations are exacerbated for treatment isocenters that are not centered within the patient habitus or when extremities are positioned far from the torso. One example is for breast or lung cases that are treating a lateral target and in which patients have both arms above the head.[60] While these treatments are often identified and revised before a true collision occurs, this decreases clinical efficiency and increases patient anxiety.

Both graphical and analytical methods have been used to predict and reduce potential collisions in RT.[59] Graphical methods provide a way to visualize the treatment beam relative to the linac[61] while analytical methods calculate planes of intersection between the linac couch and gantry.[62] However, neither of these methods accounts for the presence of the patient, which can be an additional and common source of potential collisions. Moreover, a real-time collision prediction method would provide the safest solution because it would account for the daily variation in the patient's position relative to the linac couch and gantry, even accounting for intra-fraction variations.[63]

SGRT provides a 3D model of the patient including immobilization devices. Such information could be used to predict permutations of the gantry and couch positions that would not be achievable in practice for a given treatment isocenter as shown in Figure 2.8. Moreover, such a model could be used to select a beam isocenter that would maximize the clearance of the gantry a priori. While these methods are still in development,[64] they could provide real-time feedback during treatment, which would prevent collisions by interfacing directly with the linac. This is in addition to the previously mentioned efficiency

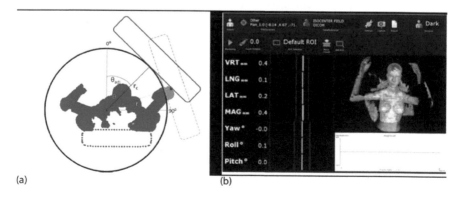

(a) (b)

FIGURE 2.8 Potential collisions were modeled (a) using a mannequin's 3D surface (b) to determine the angles that would collide with the elbow, as shown in green colorwash. (Adapted from Padilla, L. et al., *Med. Phys.*, 42, 6448–6456, 2015.)

improvements afforded by SGRT that result from reducing the rate of plan modification when clearance issues are encountered. While such efficiency improvements, which lead to complete avoidance of an error, may be challenging to quantify they are, nonetheless, of significant clinical value.

2.5.2 Biometric Measurements

Surface cameras can be used to perform contactless detection of physiologic parameters such as respiratory and heart rates because they measure both depth and color changes.[6] The respiratory rate can be calculated by tracking the change in depth of pixels with surface imaging systems to an accuracy that is comparable to currently used motion management methods in RT.[65] The heart rate can be tracked by detecting changes in the patient's skin color caused by changes in blood flow.[66]

Surface scanners have also been studied for the task of facial recognition.[67] This task is more challenging because facial recognition can be affected by the ambient light conditions, the patient's pose, or accessories (i.e., glasses, hair style) resulting in false positives that are not tolerable in RT.[67] Infrared facial recognition solutions, however, are insensitive to ambient light conditions and overcome many of these limitations, providing an attractive contact-free option for RT applications. As an additional alternative, patient identification using palm vein scanners that also rely on surface imaging technology provides higher specificity.[57] Some of these emerging applications such as heart rate tracking and facial recognition have not yet been applied to RT. As long as they are capable of meeting the sensitivity requirements of RT applications, it is quite possible that they will be adopted in the future.

2.5.3 Machine Performance Checks

Isocentricity of the couch rotation (i.e., runout) is an important geometric characteristic that must be quantified especially in light of the increased use of couch kicks in modern RT. While more manual processes are currently used, it is not unreasonable to expect that aspects of surface imaging could be used to automate these tests, making it not only more efficient but also able to be performed more regularly, perhaps even during regular patient treatments. This could lead to more timely detection of a problem, versus being identified at the time of the next machine geometry QA. Moreover, other aspects of couch geometry may lend themselves to measurement and QA by surface imaging techniques, and such improvements are additional examples of how SGRT will surely continue to evolve as we learn to more fully exploit this relatively new and very capable modality.

2.6 CONCLUSIONS

SGRT has the potential to improve safety and quality in RT by quantifying many aspects of the treatment workflow and serving as an independent, secondary observer. Improvements afforded by SGRT are often intertwined. For example, improvements in immobilization quality not only result in increased positioning accuracy but also in overall treatment efficiency. In turn, these improvements in efficiency can decrease the likelihood of errors. Although SGRT has initially been adopted for intra-fraction motion management, including frameless SRS and DIBH treatment of breast cancer, the potential safety, and quality improvements described in this chapter suggest that every external beam patient could potentially benefit from this noninvasive, real-time technology. Finally, the widespread clinical adoption of SGRT is still in progress and, as such, the opportunities for realizing the full benefit of the modality are still under investigation. It is clear, however, that SGRT has enabled and will continue to enable significant improvements in the safety, quality, and efficiency of modern RT.

KEY POINTS

- Surface imaging provides intra-fraction motion monitoring, with both high temporal and spatial resolution but without any added dose from ionizing radiation, which has considerable safety and quality implications.

- Surface imaging has enabled effective and efficient delivery of new treatment approaches (e.g., DIBH) that have significantly improved treatment quality (e.g., for left-sided breast cancer treatments).

- Efficiency improvements can translate directly into safety and quality improvements.

- The role of SGRT is still evolving as the RT community continues to learn how to fully exploit the unique information provided by surface imaging.

REFERENCES

1. Luke JJ, Lemons JM, Karrison TG, et al. Safety and clinical activity of pembrolizumab and multisite stereotactic body radiotherapy in patients with advanced solid tumors. *J Clin Oncol Off J Am Soc Clin Oncol.* 2018;36(16):1611–1618. doi:10.1200/JCO.2017.76.2229.

2. Huang G, Medlam G, Lee J, et al. Error in the delivery of radiation therapy: results of a quality assurance review. *Int J Radiat Oncol Biol Phys.* 2005;61(5):1590–1595. doi:10.1016/j.ijrobp.2004.10.017.

3. Gensheimer MF, Zeng J, Carlson J, et al. Influence of planning time and treatment complexity on radiation therapy errors. *Pract Radiat Oncol.* 2016;6(3):187–193. doi:10.1016/j.prro.2015.10.017.

4. Zietman A, Palta J, Steinberg M, eds. *Safety Is No Accident: A Framework for Quality Radiation Oncology and Care.* Arlington, VA: ASTRO; 2012. https://www.astro.org/uploadedFiles/Main_Site/Clinical_Practice/Patient_Safety/Blue_Book/SafetyisnoAccident.pdf.

5. Huq MS, Fraass BA, Dunscombe PB, et al. The report of Task Group 100 of the AAPM: application of risk analysis methods to radiation therapy quality management. *Med Phys.* 2016;43(7):4209–4262. doi:10.1118/1.4947547.

6. Al-Hallaq, H., Gutierrez, A., and Padilla, L. Surface Image-guided radiotherapy: clinical applications and motion management. In: *Image Guidance in Radiation Therapy: Techniques, Accuracy, and Limitations: American Association of Physicists in Medicine 2018 Summer School Proceedings,* Vanderbilt University, Nashville, Tennessee, July 26–28, 2018, P. Alaei and G. Ding, eds., Madison, WI: Medical Physics Publishing; 2018.

7. Gutierrez, A., Al-Hallaq, H., and Stanley, D. Surface Image-guided radiotherapy: overview and quality assurance. In: *Image Guidance in Radiation Therapy: Techniques, Accuracy, and Limitations: American Association of Physicists in Medicine 2018 Summer School Proceedings,* Vanderbilt University, Nashville, Tennessee, July 26–28, 2018, P. Alaei and G. Ding, eds., Madison, WI: Medical Physics Publishing; 2018.

8. Ford E, Terezakis S. How safe is safe? Risk in radiotherapy. *Int J Radiat Oncol.* 2010;78(2):321–322. doi:10.1016/j.ijrobp.2010.04.047.

9. Klein E, Drzymala RE, Purdy JA, Michalski J. Errors in radiation oncology: a study in pathways and dosimetric impact. *J Appl Clin Med Phys.* 2005;6(3):81–94. doi:10.1120/jacmp.v6i3.2105.

10. Bogdanich W. Radiation offers new cures, and ways to do harm. *The New York Times.* https://www.nytimes.com/2010/01/24/health/24radiation.html. Published January 23, 2010. Accessed October 25, 2018.

11. Patton G, Gaffney D, Moeller J. Facilitation of radiotherapeutic error by computerized record and verify systems. *Int J Radiat Oncol Biol Phys.* 2003;56(1):50–57.

12. Terezakis SA, Pronovost P, Harris K, Deweese T, Ford E. Safety strategies in an academic radiation oncology department and recommendations for action. *Jt Comm J Qual Patient Saf.* 2011;37(7):291–299.

13. Ezzell G, Chera B, Dicker A, et al. Common error pathways seen in the RO-ILS data that demonstrate opportunities for improving treatment safety. *Pract Radiat Oncol.* 2018;8(2):123–132. doi:10.1016/j.prro.2017.10.007.

14. Ford E, Dieterich S. Safety considerations in stereotactic body radiation therapy. *Semin Radiat Oncol.* 2017;27(3):190–196. doi:10.1016/j.semradonc.2017.02.003.

15. Fong de Los Santos L, Evans S, Ford EC, et al. Medical Physics Practice Guideline 4.a: development, implementation, use and maintenance of safety checklists. *J Appl Clin Med Phys.* 2015;16(3):5431.

16. Pawlicki T, Samost A, Brown DW, Manger RP, Kim G-Y, Leveson NG. Application of systems and control theory-based hazard analysis to radiation oncology. *Med Phys.* 2016;43(3):1514–1530. doi:10.1118/1.4942384.

17. Bissonnette J-P, Medlam G. Trend analysis of radiation therapy incidents over seven years. *Radiother Oncol.* 2010;96(1):139–144. doi:10.1016/j.radonc.2010.05.002.

18. Ford E, Gaudette R, Myers L, et al. Evaluation of safety in a radiation oncology setting using failure mode and effects analysis. *Int J Radiat Oncol Biol Phys.* 2009;74(3):852–858. doi:10.1016/j.ijrobp.2008.10.038.

19. Collins SJ, Newhouse R, Porter J, Talsma A. Effectiveness of the surgical safety checklist in correcting errors: a literature review applying Reason's Swiss cheese model. *AORN J.* 2014;100(1):65–79.e5. doi:10.1016/j.aorn.2013.07.024.

20. Reason J. Human error: models and management. *BMJ.* 2000;320(7237): 768–770.

21. Institute of Medicine (US) Committee on Quality of Health Care in America. In: Kohn LT, Corrigan JM, Donaldson MS, eds. *To Err Is Human: Building a Safer Health System.* Washington, DC: National Academies Press (US); 2000. http://www.ncbi.nlm.nih.gov/books/NBK225182/. Accessed November 10, 2018.

22. McBride SE, Rogers WA, Fisk AD. Understanding human management of automation errors. *Theor Issues Ergon Sci.* 2014;15(6):545–577. doi:10.1080/1463922X.2013.817625.

23. Kontogiannis T, Malakis S. A proactive approach to human error detection and identification in aviation and air traffic control. *Saf Sci.* 2009;47(5): 693–706. doi:10.1016/j.ssci.2008.09.007.

24. Fraass BA, Lash KL, Matrone GM, et al. The impact of treatment complexity and computer-control delivery technology on treatment delivery errors. *Int J Radiat Oncol Biol Phys.* 1998;42(3):651–659.

25. Berry SL, Ma R, Boczkowski A, Jackson A, Zhang P, Hunt M. Evaluating inter-campus plan consistency using a knowledge based planning model. *Radiother Oncol J Eur Soc Ther Radiol Oncol.* 2016;120(2):349–355. doi:10.1016/j.radonc.2016.06.010.

26. Sharp G, Fritscher KD, Pekar V, et al. Vision 20/20: perspectives on automated image segmentation for radiotherapy. *Med Phys.* 2014;41(5):050902. doi:10.1118/1.4871620.

27. Covington E, Chen X, Younge KC, et al. Improving treatment plan evaluation with automation. *J Appl Clin Med Phys.* 2016;17(6):16–31. doi:10.1120/jacmp.v17i6.6322.
28. Gerstle CR. Parallels in safety between aviation and healthcare. *J Pediatr Surg.* 2018;53(5):875–878. doi:10.1016/j.jpedsurg.2018.02.002.
29. Haynes AB, Weiser TG, Berry WR, et al. A surgical safety checklist to reduce morbidity and mortality in a global population. *N Engl J Med.* 2009;360(5):491–499. doi:10.1056/NEJMsa0810119.
30. American Society of Radiologic Technologists. The practice standards for medical imaging and radiation therapy. https://www.ncbi.nlm.nih.gov/pubmed/2198797.https://www.asrt.org/docs/default-source/practice-standards-published/ps_rt.pdf?sfvrsn=18e076d0_16.
31. Njeh CF, Suh TS, Orton CG. Radiotherapy using hard wedges is no longer appropriate and should be discontinued. *Med Phys.* 2016;43(3):1031–1034. doi:10.1118/1.4939262.
32. Hadley SW, Balter JM, Lam KL. Analysis of couch position tolerance limits to detect mistakes in patient setup. *J Appl Clin Med Phys.* 2009;10(4):2864.
33. Saenz DL, Astorga NR, Kirby N, et al. A method to predict patient-specific table coordinates for quality assurance in external beam radiation therapy. *J Appl Clin Med Phys.* 2018;19(5):625–631. doi:10.1002/acm2.12428.
34. Ohri N, Shen X, Dicker AP, Doyle LA, Harrison AS, Showalter TN. Radiotherapy protocol deviations and clinical outcomes: a meta-analysis of cooperative group clinical trials. *J Natl Cancer Inst.* 2013;105(6):387–393. doi:10.1093/jnci/djt001.
35. Evans SB, Fraass BA, Berner P, et al. Standardizing dose prescriptions: an ASTRO white paper. *Pract Radiat Oncol.* 2016;6(6):e369–e381. doi:10.1016/j.prro.2016.08.007.
36. Moran JM, Dempsey M, Eisbruch A, et al. Safety considerations for IMRT: executive summary. *Pract Radiat Oncol.* 2011;1(3):190–195. doi:10.1016/j.prro.2011.04.008.
37. Mayo CS, Moran JM, Bosch W, et al. American Association of Physicists in Medicine Task Group 263: standardizing nomenclatures in radiation oncology. *Int J Radiat Oncol Biol Phys.* 2018;100(4):1057–1066. doi:10.1016/j.ijrobp.2017.12.013.
38. Albuquerque K, Miller A, Roeske J. Implementation of electronic checklists in an oncology medical record: initial clinical experience. *J Oncol Pract.* 2011;7(4):222–226. doi:10.1200/JOP.2011.000237.
39. Santos EF de los, Evans S, Ford EC, et al. Medical Physics Practice Guideline 4.a: development, implementation, use and maintenance of safety checklists. *J Appl Clin Med Phys.* 2015;16(3):37–59. doi:10.1120/jacmp.v16i3.5431.
40. Giger ML, Chan H-P, Boone J. Anniversary paper: history and status of CAD and quantitative image analysis: the role of Medical Physics and AAPM. *Med Phys.* 2008;35(12):5799–5820. doi:10.1118/1.3013555.

41. Hoisak JDP, Pawlicki T. The role of optical surface imaging systems in radiation therapy. *Semin Radiat Oncol.* 2018;28(3):185–193. doi:10.1016/j.semradonc.2018.02.003.

42. Wiant D, Verchick Q, Gates P, et al. A novel method for radiotherapy patient identification using surface imaging. *J Appl Clin Med Phys.* 2016;17(2):271–278. doi:10.1120/jacmp.v17i2.6066.

43. Li G, Lovelock DM, Mechalakos J, et al. Migration from full-head mask to "open-face" mask for immobilization of patients with head and neck cancer. *J Appl Clin Med Phys.* 2013;14(5):243–254.

44. Zhao B, Maquilan G, Jiang S, Schwartz D. Minimal mask immobilization with optical surface guidance for head and neck radiotherapy. *J Appl Clin Med Phys.* 2018;19(1):17–24. doi:10.1002/acm2.12211.

45. Novak A, Nyflot MJ, Ermoian RP, et al. Targeting safety improvements through identification of incident origination and detection in a near-miss incident learning system. *Med Phys.* 2016;43(5):2053. doi:10.1118/1.4944739.

46. Stanley DN, McConnell KA, Kirby N, Gutiérrez AN, Papanikolaou N, Rasmussen K. Comparison of initial patient setup accuracy between surface imaging and three point localization: a retrospective analysis. *J Appl Clin Med Phys.* 2017;18(6):58–61. doi:10.1002/acm2.12183.

47. Bert C, Metheany KG, Doppke KP, Taghian AG, Powell SN, Chen GTY. Clinical experience with a 3D surface patient setup system for alignment of partial-breast irradiation patients. *Int J Radiat Oncol.* 2006;64(4):1265–1274. doi:10.1016/j.ijrobp.2005.11.008.

48. Padilla L, Kang H, Washington M, Hasan Y, Chmura SJ, Al-Hallaq H. Assessment of interfractional variation of the breast surface following conventional patient positioning for whole-breast radiotherapy. *J Appl Clin Med Phys.* 2014;15(5):177–189. doi:10.1120/jacmp.v15i5.4921.

49. Batin E, Depauw N, MacDonald S, Lu H. Can surface imaging improve the patient setup for proton postmastectomy chest wall irradiation? *Pract Radiat Oncol.* 2016;6(6):e235–e241. doi:10.1016/j.prro.2016.02.001.

50. Shah AP, Dvorak T, Curry MS, Buchholz DJ, Meeks SL. Clinical evaluation of interfractional variations for whole breast radiotherapy using 3-dimensional surface imaging. *Pract Radiat Oncol.* 2013;3(1):16–25. doi:10.1016/j.prro.2012.03.002.

51. Langius JAE, Twisk J, Kampman M, et al. Prediction model to predict critical weight loss in patients with head and neck cancer during (chemo) radiotherapy. *Oral Oncol.* 2016;52:91–96. doi:10.1016/j.oraloncology.2015.10.021.

52. Wiant D, Squire S, Liu H, Maurer J, Lane Hayes T, Sintay B. A prospective evaluation of open face masks for head and neck radiation therapy. *Pract Radiat Oncol.* 2016;6(6):e259–e267. doi:10.1016/j.prro.2016.02.003.

53. Stieler F, Wenz F, Shi M, Lohr F. A novel surface imaging system for patient positioning and surveillance during radiotherapy. *Strahlenther Onkol.* 2013;189(11):938–944. doi:10.1007/s00066-013-0441-z.

54. Li G, Lovelock DM, Mechalakos J, et al. Migration from full-head mask to "open-face" mask for immobilization of patients with head and neck cancer. *J Appl Clin Med Phys.* 2013;14(5):243–254.

55. Bartlett F, Colgan R, Carr K, et al. The UK HeartSpare Study: randomised evaluation of voluntary deep-inspiratory breath-hold in women undergoing breast radiotherapy. *Radiother Oncol.* 2013;108(2):242–247. doi:10.1016/j.radonc.2013.04.021.

56. Wiant D, Wentworth S, Maurer JM, Vanderstraeten CL, Terrell JA, Sintay BJ. Surface imaging-based analysis of intrafraction motion for breast radiotherapy patients. *J Appl Clin Med Phys.* 2014;15(6):147–159. doi:10.1120/jacmp.v15i6.4957.

57. Zhao H. Verification of Patient Treatment Accessories and Posture. Presented at the: AAPM Annual Meeting; 2018; Nashville, TN. https://aapm.onlinelibrary.wiley.com/doi/abs/10.1002/mp.12938. Accessed December 16, 2018.

58. Yu VY, Tran A, Nguyen D, et al. The development and verification of a highly accurate collision prediction model for automated noncoplanar plan delivery. *Med Phys.* 2015;42(11):6457–6467. doi:10.1118/1.4932631.

59. Hua C, Chang J, Yenice K, Chan M, Amols H. A practical approach to prevent gantry-couch collision for linac-based radiosurgery. *Med Phys.* 2004;31(7):2128–2134. doi:10.1118/1.1764391.

60. Davis AM, Pearson EA, Pan X, Pelizzari CA, Al-Hallaq H. Collision-avoiding imaging trajectories for linac mounted cone-beam CT. *J X-Ray Sci Technol.* October 2018. doi:10.3233/XST-180401.

61. Humm JL, Pizzuto D, Fleischman E, Mohan R. Collision detection and avoidance during treatment planning. *Int J Radiat Oncol Biol Phys.* 1995;33(5):1101–1108. doi:10.1016/0360-3016(95)00155-7.

62. Becker SJ. Collision indicator charts for gantry-couch position combinations for Varian linacs. *J Appl Clin Med Phys.* 2011;12(3):3405.

63. Cardan R, Popple RA, Fiveash J. A priori patient-specific collision avoidance in radiotherapy using consumer grade depth cameras. *Med Phys.* 2017;44(7):3430–3436. doi:10.1002/mp.12313.

64. Padilla L, Pearson EA, Pelizzari CA. Collision prediction software for radiotherapy treatments. *Med Phys.* 2015;42(11):6448–6456. doi:10.1118/1.4932628.

65. Silverstein E, Snyder M. Comparative analysis of respiratory motion tracking using Microsoft Kinect v2 sensor. *J Appl Clin Med Phys.* 2018;19(3):193–204. doi:10.1002/acm2.12318.

66. Gambi E, Agostinelli A, Belli A, et al. Heart rate detection using Microsoft Kinect: validation and comparison to wearable devices. *Sensors.* 2017;17(8). doi:10.3390/s17081776.

67. Silverstein E, Snyder M. Implementation of facial recognition with Microsoft Kinect v2 sensor for patient verification. *Med Phys.* 2017;44(6):2391–2399. doi:10.1002/mp.12241.

CHAPTER 3

Technical Overview and Features of the C-RAD Catalyst™ and Sentinel™ Systems

Karl Rasmussen, Victoria Bry, and Nikos Papanikolaou

CONTENTS

Editor's Note: SGRT vendors have a variety of technologies in their product portfolio, and some of those technologies are integrated into their SGRT systems to different degrees. Since this textbook is focused on SGRT, the editors have asked the authors to limit their chapter content specifically to surface image guidance for patient setup and inter- and intra-fraction patient monitoring.

3.1 INTRODUCTION

C-RAD is based in Uppsala, Sweden, and was founded in 2004, launching the Sentinel surface guidance system along with the c4D software platform in 2006. In 2011, the Catalyst (single camera) surface guidance system was released, followed up with the Catalyst HD system, which offers three cameras. C-RAD released the Catalyst PT version designed for proton/particle therapy in 2015. The technology uses either laser scanning (Sentinel) or single/multiple camera digital light processing (DLP) surface scanning at the near-visible light range (Catalyst/Catalyst HD/Catalyst PT) for the purpose of providing surface-guided radiation therapy (SGRT) for patients during computed tomography (CT) simulation, treatment setup, and/or treatment delivery.

C-RAD's Sentinel laser surface scanning unit is ceiling-mounted in either treatment and simulation/diagnostic rooms, while the Catalyst/Catalyst HD surface camera units are ceiling-mounted in treatment rooms. C-RAD systems are compatible with a wide range of treatment devices, including all major linear accelerator (linac) vendors, all major imaging vendors, as well as custom designs for proton and heavy ion facilities.

The C-RAD technology uses a nonrigid version of the iterated closest point (ICP) registration algorithm that analyzes data structures and performs a deformable image registration to calculate deviations in patient setup and during treatment delivery monitoring. The deformable registration has the benefit of facilitating alignment of nonrigid treatment areas, such as the breast or extremities. A separate stereotactic radiosurgery (SRS) module is available that incorporates a higher resolution mesh combined with a rigid version of the ICP registration algorithm.

The cPosition software addresses inter-fractional setup accuracy and monitors patient surface changes, while the cMotion software monitors intra-fraction motion and quantifies inadvertent movements.

The cRespiration software utilizes an external gating box that interfaces directly with the linac and video/feedback goggles that allow for synchronized breathing techniques and respiratory motion management. C-RAD provides routine and daily quality assurance (QA) to account for couch sag and to maintain a stable coordinate system.

The C-RAD systems are intended to provide a noninvasive and nonionizing method to improve the overall accuracy of radiation delivery through real-time surface monitoring during patient setup and treatment delivery. A retrospective study by Walter et al. observed the theoretical setup error that would have occurred if patients were positioned by Catalyst and found surface registration to be less reliable in the longitudinal direction for pelvic localization.[1] However, a study conducted by Stanley et al. found that the overall three-dimensional (3D) shift corrections for patients initially aligned with the C-RAD Catalyst system were significantly smaller than those who were aligned only using tattoos, over a wide range of treatment sites.[2] This demonstrates the strength of SGRT for the management of inter-fraction setup uncertainty. This chapter provides a technical overview of the C-RAD SGRT technologies, including a thorough description of the hardware and software algorithms used in the systems. Additionally, a brief discussion of the general workflows used with the systems and recommended QA procedures is given.

3.2 SYSTEM DESIGN

3.2.1 Hardware

C-RAD has two key products for surface guidance. The Sentinel (Figure 3.1) is a laser surface scanning system that is designed to be installed within a CT or radiotherapy treatment room and is used to assist with patient setup, respiratory-correlated imaging, and delivery techniques such as 4D-CT and DIBH. The Catalyst products (Figure 3.2) consist of one or three optical surface imaging cameras that are ceiling-mounted in the radiotherapy treatment room and are used to assist with patient setup, positioning, and motion monitoring during treatment delivery. Table 3.1 lists some of the key technical specifications for the Sentinel and Catalyst/Catalyst HD systems.

The Sentinel and Catalyst/Catalyst HD systems both maintain uniform readings over time, that is, long-term stability, within 0.3 mm and a measurement reproducibility or reliability of 0.2 mm. The Catalyst has a positioning and motion detection accuracy within 1 mm for a fixed

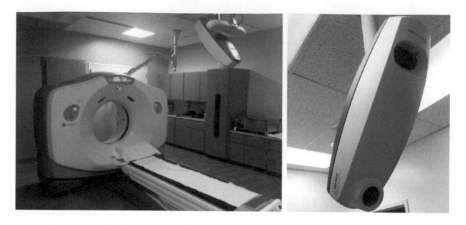

FIGURE 3.1   Ceiling-mounted laser scanning Sentinel system installed in a CT simulation room (left) and a close-up view of Sentinel system (right).

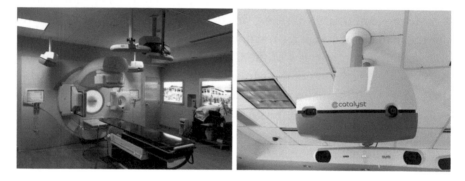

FIGURE 3.2   Ceiling-mounted three-camera Catalyst HD system setup in a linear accelerator treatment room (left) and a close-up view of one of the Catalyst HD cameras (right).

couch position. The Catalyst HD has greater accuracy as it incorporates two extra camera units compared to the Catalyst, while the Sentinel is based on a single scanning laser. The Catalyst HD's positioning and motion detection accuracy are within 0.5 mm for rigid body and SRS modes.

In addition to the camera units, the systems employ couch tracking and gating devices that are critical components for managing patient motion. A laser couch tracking device (Figure 3.3) is used within the CT room to avoid system drift of the surface imaging system and to minimize effects caused by couch sag for isocenter adjustments used for gating treatment planning.

TABLE 3.1   Technical Specifications of the C-RAD Sentinel and Catalyst/Catalyst HD
Surface Guidance Systems

	Technical Specifications		
	Sentinel	**Catalyst**	**Catalyst HD/PT**
Configuration	One laser/detector camera unit	One light projector/detector camera unit	Three light projectors/detector camera units
Description	Laser-based surface scanning system for respiratory-correlated image acquisition and reconstruction	Digital light projection for 3D surface reconstruction with real-time, nonrigid iterated closest point registration for computing 6 degree-of-freedom isocentric shifts. Wavelength of light projection: ~528 nm (green) and ~624 nm (red)	
Primary use	4D-CT (in CT simulation room)	Patient setup and intra-fraction motion monitoring	Patient setup and intra-fraction motion monitoring including SRS and proton/particle treatments
Long-term stability	Within 0.3 mm	Within 0.3 mm	
Measurement reproducibility	0.2 mm	0.2 mm	
Positioning accuracy	Within 1 mm for rigid body	Within 1 mm for rigid body	Within 0.5 mm for rigid body For SRS: within 0.5 mm and 1° (mean value + standard deviation) in all directions for rigid body and open mask
Motion detection accuracy	Within 1 mm	Within 1 mm for rigid body when coach is in fixed position during treatment	Within 0.5 mm for rigid body when coach is in fixed position during treatment For SRS: within 0.5 mm and 1° (mean value + standard deviation) in all directions for rigid body and open mask for both coplanar and noncoplanar fields

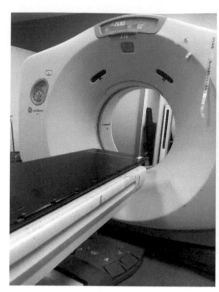

FIGURE 3.3 The laser couch tracker is placed behind the CT couch (left), scans for the stepped pattern on the attached plate (right), and uses the laser reflection to track couch distance.

The couch tracking device consists of a laser unit that moves vertically and a white reflector mounted to the end of the moving couch. Couch distance is measured by the laser in real-time from the reflector. Prior to each CT acquisition, the couch position is scanned vertically by the laser to establish the virtual couch height, and the laser is positioned in a vertically stable region to track the couch throughout the full range of travel and compensate for couch sag. The couch sag correction process will be described later in this chapter.

In order to gate or hold a treatment beam or to perform either prospective or retrospective 4D-CT scanning, an external gating interface (EXGI) box is required to provide a connection with the CT scanner. Additionally, goggles with audio and visual coaching capabilities are provided and can be used to help improve breath synchronization with the cRespiration software by coaching the patient when to breath in and out. Freislederer et al investigated the overall system latency of a Catalyst optical scanner and Elekta linac and found the time delay for "beam on" to be 851 ± 100 ms.[3]

The c4D Catalyst software is used to perform respiratory gating in the treatment room. This software is used to select the patient and control the cRespiration and cMotion applications. Gating interfaces have currently been developed for Elekta (Elekta AB, Stockholm, Sweden) and Varian (Varian Medical Systems, Palo Alto, CA) linacs. For Varian linacs, the EXGI hardware/software interface is used to perform respiratory gating of the treatment beam and is provided by C-RAD. The Elekta Response interface is a treatment beam gating interface for Elekta linacs and is provided by Elekta. The components include a response control module and a relay module. Freislederer et al. also observed that when using the Catalyst system for gated and ungated treatments with an Elekta linac, the total delivered dose decreased with the size of the gating window.[3] This study was consistent with the current literature confirming dose delivery accuracy remains below 1% for clinically relevant gating levels of about 30%.

3.2.1.1 Sentinel™

The Sentinel surface scanning unit is mounted to the ceiling (typically 3 m above the floor and 2–3 m from the laser source) of simulation or treatment rooms (Figure 3.4). Sentinel scanning unit placement will affect the ability to generate surface images, and specific use cases will have custom installations. This installation can be optimized based on the available field of view (FOV) of the equipment used as well as the specific treatment sites to be treated. The Sentinel system has been found to be a good supplement for the CBCT system for accurate daily treatment setup.[4]

The Sentinel scanning unit consists of a red line laser and a rapidly rotating mirror, or galvanometer, powered by an optical fiber connected to the computer. As the galvanometer projects laser lines on to the surface of the target, part of the light is reflected back to the Sentinel system. An optical bandpass filter mounted on to a monochrome charge-coupled device (CCD) camera detects the reflected light. Known angles of the laser light and reflections are used to produce a 3D surface image of the object through a triangulation technique.

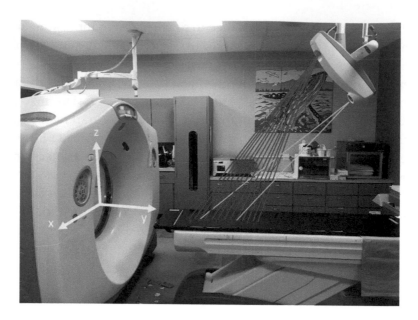

FIGURE 3.4 Sentinel system setup demonstrating projected lines from laser (red) and reflected lines to camera (blue) installed with a CT scanner.

3.2.1.2 Catalyst/Catalyst HD

The Catalyst system is mounted to the ceiling in a configuration aimed to avoid interference and to provide an unobstructed view from the rotating gantry (Figure 3.5). The original Catalyst system consists of a single camera unit, while the Catalyst HD/PT uses three camera units for optimal patient coverage.

The Catalyst camera units consist of a DLP projector, composed of high-speed switching DLP chips made up of micro-mirrors, and a CCD camera calibrated to the same coordinate system. The Catalyst projector performs 3D surface measurements by performing DLP, which is the projection of light in a sequence of patterns, at multiple wavelengths, on to the surface of the patient. Similar to the Sentinel, the monochrome CCD camera makes continuous measurements of the reflected light patterns. A reconstruction algorithm compares projected and captured patterns to identify the coordinates of each pixel on the captured image (Figure 3.6).

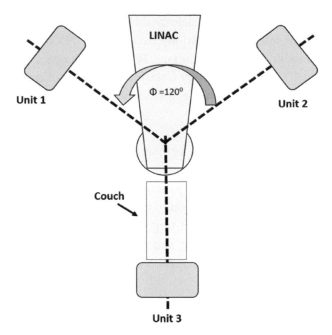

FIGURE 3.5   Catalyst HD camera unit architecture. Camera units are spaced at 120° intervals for 360° coverage of the patient.

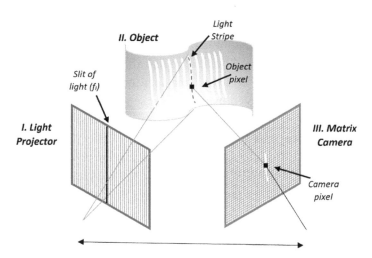

FIGURE 3.6   A triangulation technique is used to identify coordinates of each pixel.

The Catalyst system uses a deformable image registration algorithm to continuously calculate deviations between the 3D live image of the patient's surface and a reference image. Reference images can be acquired by Sentinel at CT simulation or by Catalyst at the time of treatment setup. This algorithm is described in the next section.

3.2.2 Software/Algorithm

The Catalyst software uses a deformable ICP algorithm that initializes data structures and performs an image registration between the reference (source) surface and the current patient surface (target). The classic ICP algorithm[5,6] uses local optimization to (1) select points on the source surface and pairs these points with a target surface, (2) minimize the mean square error between the source and target points by computation of a rigid transformation, and (3) apply this transformation to the source surface.

The use of the modified deformable ICP algorithm by Catalyst allows for regions of interest (ROI) that are distant from the isocenter to have a reduced impact on isocentric shifts relative to ROI near the isocenter. The key components of the nonrigid ICP algorithm are the source surface, the deformable transformation algorithm's node graph, and the target surface. Initially, a deformable node graph is made from the reference surface, which creates specific points to be used in the modified ICP algorithm. Initially, these points are rigidly aligned to the overall target image from the real-time imaging data. The final registration is then applied using the modified ICP algorithm, and the time averaged (user configurable) real-time image is then displayed over the user selected reference as shown in Figure 3.7. This deformable computation is invisible to the user, but the resultant rigid 6 degrees-of-freedom (6DOF) transformation is provided to the user. An additional module for SRS treatments incorporates a similar algorithm as listed above with a higher degree of rigidity and a higher resolution node graph to accommodate the limited FOV for open-face masks.

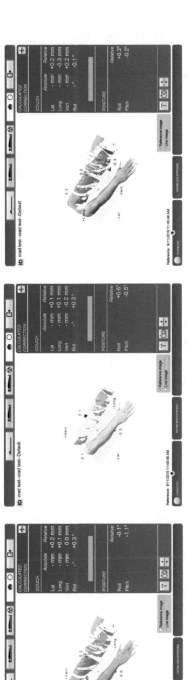

FIGURE 3.7 Live image (blue, left). Reference image (green, middle). Live with image registration (blue/green, right).

3.3 GENERAL CLINICAL WORKFLOWS

While detailed clinical workflows are discussed later in this book, brief descriptions of general clinical workflows with C-RAD systems are described here.

3.3.1 CT Simulation with Sentinel™

Prior to treatment planning, the patient is set up on the CT couch in the same position that will be used for treatment. A scan volume that encompasses, at a minimum, the treatment region that will be acquired by the CT scan for an individual patient is selected using the Sentinel system. This area will be used to evaluate patient motion characteristics.

The Sentinel captures a reference image encompassing the region visible to the cameras. This image can be used to later assist in patient alignment at treatment setup. In order to track respiratory motion, the CT technician or simulation therapist is able to create virtual tracking points within the software to visualize the patient's chest movement at all times during the CT scan. These tracking points are used to evaluate the patient's breathing cycle using the cRespiration application. With the aid of audio/video feedback coaching goggles, the patient can be guided to breathe in and out to accommodate for a retrospective or prospective study.

3.3.2 Retrospectively Sorted 4D-CT Simulation

A 4D-CT acquired with Sentinel allows for a contactless measurement, capturing the free breathing respiratory trace signal of a patient with no additional device being placed on the patient. The Sentinel system supports amplitude-based triggering (SNR = 14.6 dB) and is more beneficial for radiotherapy because it uses an absolute respiratory signal versus a relative signal.[7] The primary point, sent to CT and used to gate the beam, and optional secondary point, used for visual monitoring, signals correspond to the selected virtual tracking points and provide feedback to the system on the characteristics of the patient's breathing (see Figure 3.8). Only a primary point is required so care must be taken to place the virtual tracking point in a location that remains visible throughout the entire scan length.

While maximum and minimum breathing levels can be set for retrospectively sorted 4D-CT scanning, patients can also breathe freely without the need for coaching as the software will retrospectively calculate minimum and maximum breathing phases, as shown in Figure 3.8. This information can be used by third-party software to bin the acquired raw data.

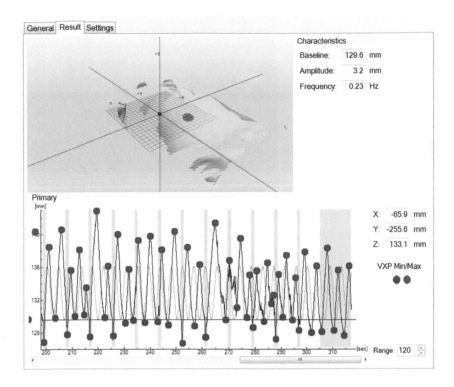

FIGURE 3.8 Patient results screen from C-RAD Sentinel software. The lower graph demonstrates the detected minimum (red dots) and maximum (blue dots) breathing phase for each respiratory cycle.

3.3.3 Breath-Hold CT Simulation

A prospective breath-hold study using Sentinel allows coaching of the patient through a reproducible exhalation breath hold or deep inspiration breath hold (DIBH) technique. A study by Schönecker et al. studied the C-RAD system for guiding DIBH and found that the heart completely moved out of the treatment field for six of nine patients and DIBH enabled a reduction of the maximum dose to the heart and left arterial descending artery by 59% and 75%, respectively.[8]

Using primary and secondary virtual tracking points, the patient's breath-hold position is monitored during the CT scan. The recommended point placement for the primary signal is 0.5–1.5 cm below the xiphoid process (midline, below the breast line). The secondary point can be used to address specific patient concerns, including arching of the back to achieve the breath-hold (by placement on the shoulder) or determining whether

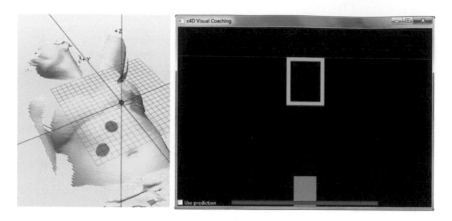

FIGURE 3.9 (Left) Sentinel reference image shows primary (red) and secondary (blue) tracking point used to evaluate the patient's breathing characteristics. The recommended point placement for the primary signal is typically on the midline, 0.5–1.5 cm below the xiphoid process. (Right) Visualization of the primary tracking point to the patient. The max expiration baseline (blue line), current breathing level (orange bar), and gating window (green box) are shown.

the patient is a "belly-breather" who may possibly not derive therapeutic benefits of certain techniques such as DIBH. An example of point placement is shown in Figure 3.9.

The audio/video feedback goggles provide the patient with a visual representation of their current breathing level, as well as a graphical representation of the "ideal" breathing window for the patient. When coaching is activated, a computer-recorded voice or the CT technician will coach the patient to breathe in and out according to the selected inspiration or expiration time. The CT can be acquired when the patient's breathing level is within the desired breathing window.

3.3.4 In-Room Guidance with Catalyst/Catalyst HD

3.3.4.1 Initial Patient Setup

On a treatment day, the goal of initial patient setup is to position the patient in the same position as was used for CT simulation. The cPosition application is used to verify precise patient positioning prior to irradiation. One of the three reference surfaces can be used by cPosition to perform initial patient positioning (Figure 3.10): (1) a reference image captured by the Sentinel system during CT simulation, (2) the external body structure imported by DICOM RT from the treatment planning system, or (3) a reference image captured by the Catalyst/Catalyst HD system in the treatment room.

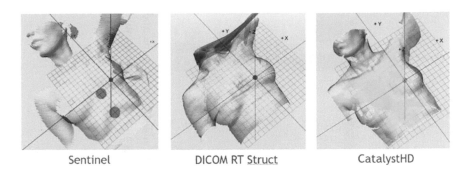

| Sentinel | DICOM RT Struct | CatalystHD |

FIGURE 3.10 Reference surfaces that can be used in cPosition to set up a patient. On the left is a reference surface obtained by Sentinel during simulation. The middle image is the external body contour imported via DICOM RT from the treatment planning system. On the right is a surface acquired by the Catalyst/Catalyst HD camera(s) in the treatment room.

Settings in the software can optimize scan volume size and camera parameters for individual patients. The scan volume is the region that the algorithm will track during patient setup and treatment delivery. If necessary, manual edits can be made to crop ROIs of the reference volume that are unwanted or unnecessary for patient setup, such as immobilization devices or oxygen tubes or intravenous (IV) lines. In order to reduce over- and underexposure of the surface images from each camera unit, the acquisition time and gain values for each camera are optimized. Preset time/gain values can be used; however, individual settings should be adjusted on the first day of treatment to accurately visualize the patient across all cameras. Figure 3.11 demonstrates the importance of camera optimization with over- and under-exposure. If care is not taken to appropriately visualize the patient, the accuracy of patient setup and tracking will be limited. Generally, settings should be optimized for the region of the patient that will be receiving treatment.

As described previously, the cPosition registration algorithm calculates surface-based positioning offsets in 6DOF. The absolute and relative deviations of the live image compared to the reference image can be shown on the cPosition interface or can be projected directly onto the patient's surface to provide real-time guidance to therapists in the treatment room. These calculated corrections are used to create a color map for the therapists in the treatment room, providing them with a tool to assist in making adjustments to the patient's posture as shown in Figure 3.12.

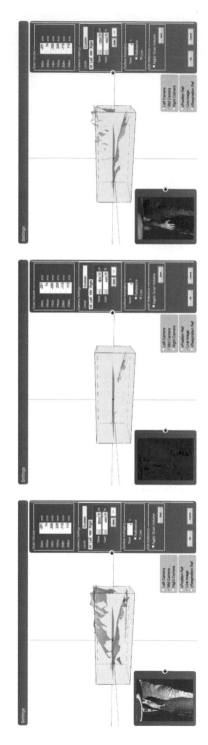

FIGURE 3.11 Examples of overexposed (red) and underexposed (blue) areas of the surface image where adjustments to time and gain values allow for desired exposure. Shown is over-exposure (left image), under-exposure (middle image), and optimized exposure (right image).

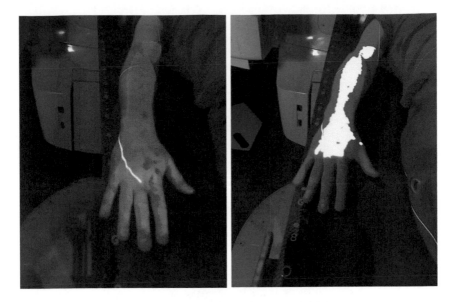

FIGURE 3.12 Examples of in-room, on-patient projection of incorrect patient positioning. The patient's arm is placed too high, shown in red (left), and placed too low, shown in yellow (right).

3.3.4.2 Patient Motion Monitoring

Using the previously mentioned algorithm, the cMotion (Catalyst/Catalyst HD only) application monitors patient motion during treatment delivery, with a sample result shown in Figure 3.13. The green bars in the lower region of the figure represent the total vector shift of the patient during treatment using the cMotion deformable registration algorithm. This vector can have a user-defined maximum value that will trigger a beam hold if the tolerance is exceeded.

Additionally, the cRespiration application can be used during respiratory-managed treatments for tracking the location of the primary and secondary virtual tracking points relative to the desired breathing positions, an example of which is demonstrated in Figure 3.14. Beam hold can be set for a maximum user-defined isocenter shift vector displacement (cMotion), a user-defined gating window (cRespiration), or both methods.

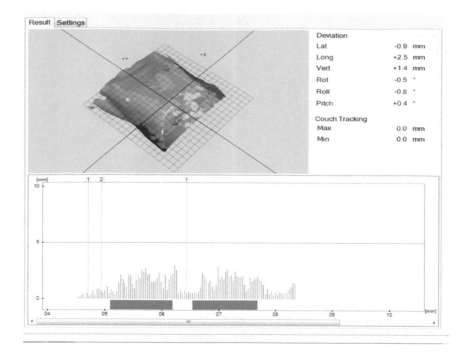

FIGURE 3.13   Example of cMotion (Catalyst/Catalyst HD only) results for a patient treatment. (Top) A three-dimensional overlay can be used to retrospectively review the treatment. (Bottom) Beam on time is noted with the dark gray bars along the *x*-axis (time) with the number of the beam being delivered listed at the top of the graph. Green vertical lines demonstrate total vector shift as a function of time.

3.4 QUALITY ASSURANCE

The report of the American Association of Physicists in Medicine (AAPM) Task Group (TG) 147 provides recommendations on QA methods for nonradiographic localization and positioning systems, including SGRT systems.[9] At the time of writing, the report of AAPM Task Group 302, surface image-guided radiotherapy, was in progress. Supporting the recommendation made by TG-147, C-RAD provides several methods for routine system QA. The commissioning process and routine QA of Catalyst and Sentinel is described further in Chapter 6.

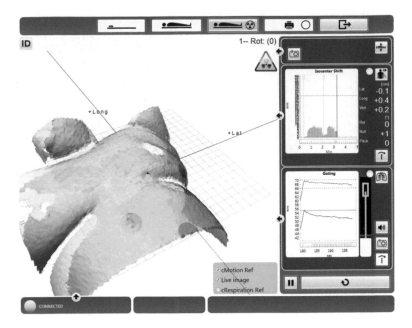

FIGURE 3.14 View of the cMotion/cRespiration application during treatment delivery shows the patient surface overlays (left), overall vector displacement (top right), and real-time virtual tracking points signals (bottom right). Shown are the real-time trace signal lines (red primary and blue secondary) as well as the shift from the minimum breathing baseline (horizontal blue line) to a DIBH window (dashed green lines).

3.4.1 Routine QA of the Sentinel™ System

Daily QA of the Sentinel system in the CT room is recommended before operating the system to avoid any drift in signal acquisition during couch movement. Initial calibration of the system includes aligning the daily check device (Figure 3.15).

The first QA check (check mode) verifies the couch profile without a patient on the treatment couch. The treatment couch begins in the "start position" where the patient is set up and is then moved forward, away from the Sentinel to the stop position, replicating couch motion during a CT scan. A calibration mode is available that repeats this same

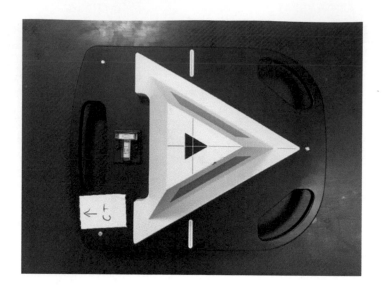

FIGURE 3.15   Daily check device for Sentinel.

process; however, an additional measurement is made with 70 kg of weight added on to the CT couch to simulate the additional couch sag from a patient.

3.4.2 Routine QA of the Catalyst™ System

C-RAD recommends routine QA of the Catalyst/Catalyst HD camera unit(s) to evaluate any potential system drift that may occur to the coordinate system. While a full-system calibration is performed by the service engineer, isocenter calibration can be performed by the on-site clinical physicist. Similar to the Sentinel system, daily QA of the system includes aligning the Catalyst daily check device with three spheres (Figure 3.16) to the in-room lasers. At initial installation, a mathematical model of the daily check device is created, and each daily check refers to this as the reference image of the daily check device.

Finally, for SRS treatments, an additional QA mode is available, which uses the QUASAR Penta-Guide phantom (Modus Medical Devices, London, Canada) to correlate in room kilovoltage (kV) or megavoltage (MV) imaging to the Catalyst HD isocenter. Deviations are recommended to be $\leq \pm 0.5$ mm for the X, Y, and Z directions.

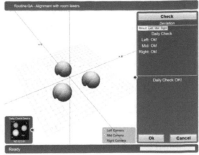

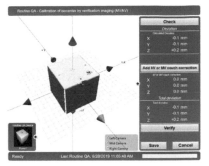

FIGURE 3.16 Catalyst daily check device (top left) and routine QA daily check display (top right) after being aligned according to in-room laser system. Penta-Guide phantom (bottom left) and routine QA display (bottom right) after in room kV/MV imaging verification. These tests verify the spatial accuracy of the C-RAD Catalyst/Catalyst HD system for standard and stereotactic-specific applications, respectively.

KEY POINTS

- C-RAD's Sentinel and Catalyst/Catalyst HD systems use laser and near-visible light projectors, respectively, in combination with calibrated cameras to provide real-time information about a patient's surface position in relation to a reference position for initial setup and treatment delivery. This allows for evaluation of patient setup, management of intra-fraction motion, and coordination of respiratory motion management during treatment.

- Camera settings are optimized to each specific patient, including the monitored ROI and skin tone.

- Sentinel can be used during CT simulation to provide breathing information to retrospectively reconstruct a 4D-CT dataset or prospectively gate the patient's breathing level for breath-hold techniques such as DIBH.

- While the surface image acquisition methods differ from other manufacturers, the most unique aspect of the C-RAD system is the deformable algorithm. This allows for a full analysis of the patient surface for both setup and tracking. This difference becomes most apparent with regions of high variability such as the breast and extremities.

REFERENCES

1. Walter F, Freislederer P, Belka C, Heinz C, Söhn M, Roeder F. Evaluation of daily patient positioning for radiotherapy with a commercial 3D surface-imaging system (Catalyst™). *Radiat Oncol.* 2016;11(1):154.
2. Stanley DN, McConnell KA, Kirby N, Gutiérrez AN, Papanikolaou N, Rasmussen K. Comparison of initial patient setup accuracy between surface imaging and three point localization: A retrospective analysis. *J Appl Clin Med Phys.* 2017;18(6):58–61.
3. Freislederer P, Reiner M, Hoischen W, et al. Characteristics of gated treatment using an optical surface imaging and gating system on an Elekta linac. *Radiat Oncol.* 2015;10(1):68.
4. Wikström K, Nilsson K, Isacsson U, Ahnesjö A. A comparison of patient position displacements from body surface laser scanning and cone beam CT bone registrations for radiotherapy of pelvic targets. *Acta Oncol.* 2014;53(2):268–277.
5. Besl PJ, McKay ND. Method for registration of 3-D shapes. *Sensor Fusion IV: Control Paradigms and Data Structures.* 30 April 1992. doi:10.1117/12.57955.
6. Zhang, Z. Iterative point matching for registration of free-form curves and surfaces. *Int J Comput Vis.* 1994;13(2):119–152.
7. Heinz C, Reiner M, Belka C, Walter F, Söhn M. Technical evaluation of different respiratory monitoring systems used for 4D CT acquisition under free breathing. *J Appl Clin Med Phys.* 2015;16(2):334–349.
8. Schönecker S, Walter F, Freislederer P, et al. Treatment planning and evaluation of gated radiotherapy in left-sided breast cancer patients using the Catalyst™/Sentinel™ system for deep inspiration breath-hold (DIBH). *Radiat Oncol.* 2016;11(1):143.
9. Willoughby T, Lehmann J, Bencomo JA, et al. Quality assurance for non-radiographic radiotherapy localization and positioning systems: Report of Task Group 147. *Med Phys.* 2012;39(4):1728–1747.

CHAPTER 4

Technical Overview and Features of the Varian IDENTIFY™ System

Raymond Schulz, Chris Huyghe, Lisa Hampton,
Delena Hanson, Michael Stead, and Thomas Speck

CONTENTS

Editor's Note: SGRT vendors have a variety of technologies in their product portfolio and some of those technologies are integrated into their SGRT systems to different degrees. Since this textbook is focused on SGRT, the editors have asked the authors to limit their chapter content specifically to surface image guidance for patient setup and inter- and intra-fraction patient monitoring.

4.1 INTRODUCTION

IDENTIFY (Varian Medical Systems, Palo Alto, CA) is a surface image guidance solution for radiation therapy patient setup verification, inter-fraction positioning, and intra-fraction motion management. IDENTIFY utilizes time-of-flight (ToF) camera technology for accessory navigation and initial head-to-toe orthopedic patient setup verification on the couch top, and stereo vision cameras for surface image guidance during final patient positioning at isocenter and intra-fraction motion monitoring. IDENTIFY also includes a biometric patient identity tool based on the palm scanning and accessory verification tools employing radio-frequency identification (RFID). As these technologies do not employ surface imaging technology, they are beyond the scope of this chapter. Further details can be found elsewhere.[1,2] A central data management system manages the records associated with the patient simulation and treatment sessions. After each session, automated reports are published and transferred to the user's oncology information system (OIS).

4.2 SYSTEM DESIGN

4.2.1 Hardware Design

4.2.1.1 ToF Setup Camera

Accessory placement and initial patient loading are performed with the IDENTIFY setup camera. This camera includes an RGB sensor for capturing the video stream and to detect optical markers placed on setup accessories. A second sensor within this camera unit is used for ToF measurements.

4.2.1.2 Surface Imaging Cameras

The surface measurements for final patient setup are performed by using a ceiling mounted stereo camera system and a random pattern projector that project a static random pattern onto the patient's surface (Figure 4.1). Each projector unit contains two stereoscopic cameras (calibrated both intrinsically, by Varian, and to one another, by the clinical user), which captures the reflected illuminated scene and calculates 3D point clouds. The software triggers the simultaneous acquisition of both camera images. The image data are transferred to the PC, which houses two graphic cards for GPU calculation and processing. In both camera images, corresponding points are searched, based on the method of triangulation and a 3D point cloud is calculated.

The 3D data from multiple cameras is combined in one 3D point cloud. The measured 3D point cloud is processed, and a surface mesh is generated. Currently measured surface meshes are compared to reference surface

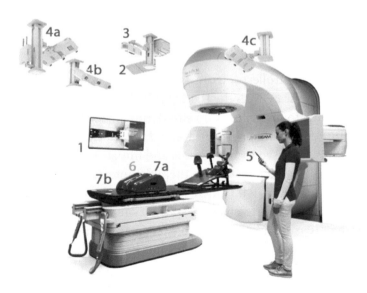

FIGURE 4.1   The Varian IDENTIFY system component parts.

meshes, and the deviations are visualized in virtual reality by overlaying colors representing offsets onto a live video feed, which is displayed on monitors in the treatment room and the control room. A rigid registration algorithm calculates the translational and rotational errors of the registered point surfaces. It is possible to reduce the registration region by defining a specific region of interest (ROI). The algorithm compares two surface meshes in the ROI and calculates a transformation with the 6 degrees-of-freedom (6DOF) axis data to transform one mesh to the other. The output transformation values of the registration algorithm are shown on the IDENTIFY user interface.

IDENTIFY employs a proprietary blue LED light technology to illuminate the patient. The system has been optimized so that function is not dependent on specific ambient lighting settings as changing lighting conditions are often found in the radiation therapy clinic.

The surface image camera's temporal resolution is up to 10 Hz, depending on the size of the tracking ROI (the yellow-highlighted part on the surface). It has been optimized for real-time inter- and intra-fraction motion management. Each camera captures 3D data at about 6–10 frames per second. The ROI comes into play when the 6D shifts are computed. Without setting an ROI, the 6D shift is computed over the whole surface at about 3 fps. For a face ROI, the frame rate is about 7–10 fps. The smaller the ROI, the faster the computation of the 6DOF values.

4.2.1.3 Handheld Controllers

Small handheld controllers (Figure 4.2) drive much of the system interaction and communicate with the IDENTIFY client via wireless connectivity. At computed tomography (CT) simulation, the handheld controller is used to capture the orthopedic surface and accessories for use throughout treatment. It is also used to take photos and initiate the respiratory coaching tool. In the treatment vault, the therapist interacts with the handheld controller to enable SGRT camera functionality, to select surfaces for patient monitoring, and to enable respiratory management. All messages and commands sent and received are encrypted both in the network and in the application connection layer. Multiple handheld controllers are provided for each IDENTIFY client system: two (2) in CT simulation and three (3) in the treatment vault. A full list of all IDENTIFY components is given in Table 4.1.

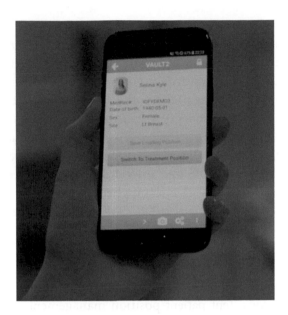

FIGURE 4.2   The IDENTIFY handheld controller.

TABLE 4.1   IDENTIFY System Components—Detailed Listing

No.	Description
1	*In room monitor* Displays the patient ID, patient photo, setup photos, setup accessories, patient surface image for setup on couch top, supplemental respiration coaching signal
2, 6	*RFID reader and antennas, RFID tagged device* Detects patient (optional) and RFID tagged set up devices on the couch top
3	*Orthopedic setup time-of-flight camera system* Tracks the position of set up devices and the initial head to toe set up position of the patient on the couch top
4 a, b, c	*3D surface imaging camera system* Tracks the position of the patient at isocenter, utilizing a specified region of interest for interfraction positioning, intrafraction motion and respiration monitoring
5	*Handheld controller* Portable, wireless handheld controller for acquiring patient and setup photos, and for interacting with IDENTIFY
7a	*Position marker* This is used for on-couch top device position verification
7b	*Table reference position marker* Used to determine the position of the accessories relative to the reference marker

4.2.1.4 Data Management

IDENTIFY has been developed to run on a central server/multiple client architecture. The benefit of this architecture is the automatic access and transfer of information from one client (simulation room client or treatment room client system) to another without any additional patient data entry. The only prerequisite is that the machines share a single OIS database/server. Using HL7, IDENTIFY will automatically update its scheduling data. The central server manages importing patient data and appointments from the OIS and provides management and storage of general settings for IDENTIFY room client systems. The central server also acts as a continuously active DICOM server and a report generator server.

4.3 OVERVIEW OF CLINICAL WORKFLOW

IDENTIFY was developed with the primary clinical goal of efficient workflow and consistent patient position management. The IDENTIFY workflow typically begins in the CT simulation room with the fast capture of the head-to-toe patient "orthopedic" surface. The orthopedic setup surface image has a large field of view (FOV) capable of visualizing the entire patient from "head to toe," and hence the name. Within the treatment room, and with the treatment couch in the loading position, the accessories are verified, and the head-to-toe orthopedic patient surface is visualized on the large in-room display screen. This is accomplished using augmented reality, overlaying computed setup information onto a live video feed to assist with orthopedic patient setup. The full IDENTIFY workflow is shown in Figure 4.3. A "single-click" operation switches the system from orthopedic setup to the surface image-guided functionality for inter-fraction positioning and intra-fraction motion management. These workflows apply to, but are not limited to, techniques such as stereotactic body radiation therapy (SBRT), stereotactic radiosurgery (SRS), and deep inspiration breath hold (DIBH). At the end of each treatment session, an automated report (PDF format) is created that, optionally, can be transferred to the OIS.

4.3.1 CT Simulation

4.3.1.1 Patient Selection and Data Management

IDENTIFY is designed as a secondary data management system, complementing the user's OIS while simultaneously managing patient biometric data and IDENTIFY-specific data. In this context, the patient is required

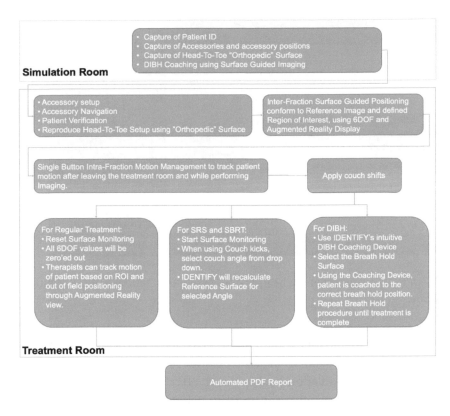

FIGURE 4.3 A high-level overview of IDENTIFY's clinical workflow.

to be previously registered in the OIS system in order to prevent duplicate data entry and ensure the integrity of the entered data.

Once the patient is scheduled in the OIS, the patient dataset is automatically registered in IDENTIFY. Automated registration is achieved using the HL7 SUI (schedule) outbound protocol. In the absence of HL7 functionality, it is also possible to register a patient manually in IDENTIFY. The patient scheduled for CT simulation is selected in IDENTIFY through use of a handheld controller that automatically displays all patients scheduled for CT simulation. Once selected, patient-specific details are loaded to the IDENTIFY simulation client.

4.3.1.2 Patient Setup

Once the patient is setup for CT simulation, planning images are acquired. At the completion of image acquisition, the therapist retracts the CT couch and lowers it to the patient loading position. Additional setup photos may be taken, as required. The therapist then acquires a head-to-toe

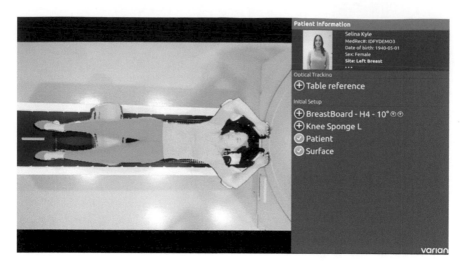

FIGURE 4.4   Capture of the head-to-toe "orthopedic" surface in CT simulation mode.

3D reference surface of the patient on the couch (i.e., the orthopedic view) (Figure 4.4) in the simulation setup position. This allows for reproduction of the patient's simulation position in the treatment room. "Out-of-field" positioning of the patient may assist in patient setup by minimizing tilts and shifts and changes in posture (e.g., arm positions or the amount the head is turned to the side).

The head-to-toe surface is captured by a ToF camera. It sends out a time-modulated IR light signal and measures the round-trip time the light takes to reflect from the surface and return to the camera sensor. A point cloud comparison algorithm computes the z-distance between the reference image and the live surface image and creates a heat map, which is used during patient setup on the linac to color the surface. The surface is then rendered using augmented reality techniques onto the live video image to show deviations of the surface to the user. Note that the reference image should be captured during CT simulation or on the first day of treatment. At any point during treatment, a new reference surface can be captured.

4.3.1.3 Patient Setup Device Detection and Placement

Once the patient is unloaded from the CT couch, IDENTIFY uses an optical positional marker on the immobilization devices and a reference position marker on the couch top to record the position of the devices in the correct setup configuration. For accessories that can be inclined (e.g., breast board), an additional marker is placed to detect inclination. Each accessory with optical markers is calibrated by a separate accessory calibration routine.

The optical markers each consist of a single circle that is detected by the setup camera. IDENTIFY uses monocular (i.e., nonstereoscopic) video tracking to detect the index position of accessories on the treatment couch. It segments circular markers on the treatment couch and positioning accessories and computes the 3D position of the couch and the accessories using 2D–3D point matching. Note that this mode does not employ a point cloud-based calculation as with the patient surface registration. Instead, the method relies on a 3D model (the four center points of the circles of the table marker) and the corresponding four points in 2D that are visible to the camera. Estimating this position is done by the perspective n-point algorithm[3] and includes a calibration step involving nonlinear optimization. The computed 3D position is then mapped onto the corresponding index position. Following capture, the therapist confirms the device setup on the couch top using the handheld controller. A list of the detected devices is displayed on the in-room monitor. The recorded setup will then be available for the subsequent treatment delivery.

4.3.1.4 Respiratory Management

If the patient is to be treated with breath-hold respiratory motion management, the surface image guidance is utilized to monitor chest wall surface excursion during DIBH CT simulation.

When using breath-hold respiratory management, IDENTIFY is trained on the breathing pattern of the patient. The areas, with the most motion, are identified within the software in *red* and the areas with no motion are identified in *gray*.

The desired amplitude for the DIBH, the breath-hold window (Figure 4.5), is defined as a percentage of the total amplitude, plus a user-defined tolerance. The breathing amplitude is measured from a user-defined area on the patient's surface. A visual coaching device (VCD) is available for the patient to view the ideal breath-hold setting. While the CT couch and patient are in motion during CT acquisition, the patient surface location must be updated to account for the couch motion during DIBH CT acquisition. IDENTIFY uses the full field of view to compute the CT couch speed by tracking the *gray* static areas and correct for that in the algorithm. IDENTIFY uses a local workstation with dual graphics cards to update the acquired breathing pattern in near real-time.

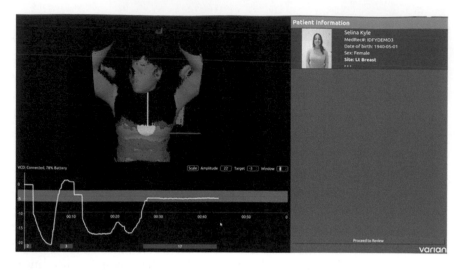

FIGURE 4.5 Display of respiratory coaching in IDENTIFY during CT simulation.

4.3.2 Treatment Planning

Once the treatment plan is complete, the (1) planning CT images, (2) the DICOM RT structure set, and (3) the DICOM RT Plan file are exported from the treatment planning system (TPS) to IDENTIFY. Imported DICOM surface structure(s), the plan isocenter, and treatment fields are automatically associated with the patient in IDENTIFY. The desired patient DICOM surface from the TPS system is established as the surface image-guidance reference image. A specified ROI for monitoring patient position and motion during treatment delivery is then defined on the reference surface using the IDENTIFY planning tool.

4.3.2.1 Planning Tool

The IDENTIFY planning tool is a Microsoft Windows application and can be used for new patient preparation (Figure 4.6). The planning tool is pre-programmed with the generic positions of the surface imaging cameras and provides the user feedback on the quality of the drawn ROI. In order to determine strong versus poor ROI coverage, the algorithm takes into consideration the number of points in the ROI and analyzes the resulting 3D shape to ensure there are enough 3D features to deliver a valid result.

The planning tool allows for multiisocenter treatments by creating separate disease-specific sites. Each site has its own surfaces, ROIs, accessory

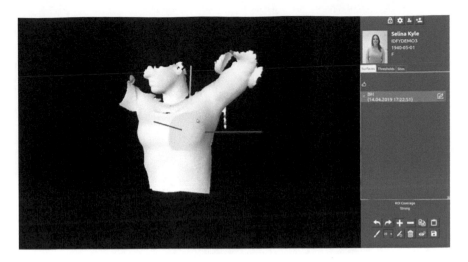

FIGURE 4.6 The IDENTIFY planning tool user interface for delineating surface ROIs.

lists, and notes. This information can also be obtained with the handheld device at the time of treatment. A 6DOF tolerance table is configured in the data manager to allow auto-population of default motion limit thresholds based on those disease-specific treatment sites. Patient-specific position and rotational offset thresholds may be changed at any time for each unique patient and disease condition in the planning tool or using the handheld controller.

4.3.3 Treatment Delivery

4.3.3.1 Patient Selection and Identification Verification

Once the patient is added to the treatment machine schedule in the OIS, the patient is automatically tagged as "scheduled for treatment delivery" in IDENTIFY. When the therapist loads a patient's plan at the treatment delivery system, the selected patient is automatically opened in IDENTIFY, utilizing the HL7 SUI (protocol).

4.3.3.2 Accessory Verification and Patient Set Up to the Couch Top

The setup camera automatically detects the correct device and its position and inclination. Once all devices are set up, and the patient's identity has been successfully verified, the head-to-toe setup surface is automatically displayed over a live video feed using augmented reality (Figure 4.7).

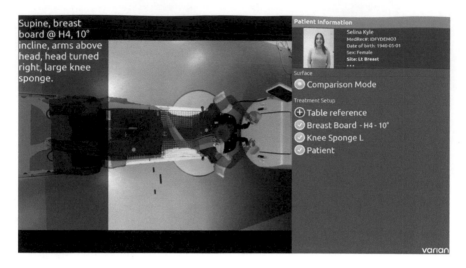

FIGURE 4.7 Head-to-toe orthopedic surface created during CT simulation, displayed in the treatment room with accessory verification for initial patient setup.

The patient is then positioned on the couch, with the initial patient position verified against the head-to-toe image acquired at simulation. As the patient lays on the couch-top at the loading position, the initial patient setup ToF surface image guidance system attempts to match the head-to-toe "orthopedic" reference image (Figure 4.8) to the current patient position. Patient positioning errors are most easily corrected at initial patient loading, eliminating more challenging repositioning later in the workflow. Positioning accuracy thresholds for this initial setup alignment are generally set to 10–20 mm.

IDENTIFY's augmented reality technology (Figure 4.8) uses blue and red color schemes and gradients to display the translation and rotation offsets of the current patient position from the reference image, overlaid with a live video feed of the patient on the in-room monitor and on the monitor at the console. A *blue* color indicates the pixel is below the reference image and *red* indicates the pixel is above the reference image in the z-axis (vertical).

4.3.3.3 Patient Set Up to Isocenter

Three-dimensional surface image tracking is employed for setup of the patient at isocenter. The surface imaging system consists of three stereo vision cameras with an overall system submillimeter accuracy.[4] Each camera has a 16-mm lens. The live surface of each camera is registered

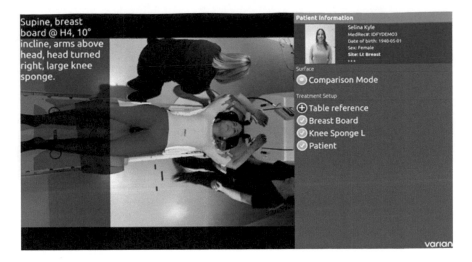

FIGURE 4.8 IDENTIFY's augmented reality aids the therapists with repositioning the patient in the loading position by overlaying setup information onto a live video feed.

to the reference surface using an iterative closest point (ICP) algorithm.[5] Overlapping points from the three cameras are weighted, and a point-to-plane error metric[6] is used to compute the distance between the reference and live surfaces. A nonlinear rigid optimization algorithm finds the optimal match.

The FOV of each camera is approximately $50 \times 50 \times 50$ cm. Using the DICOM structure set obtained from the TPS as a reference, a live, 3D surface of the body is matched to the DICOM reference surface. The matching can also be performed against a newly captured reference image.

Translational and rotational offsets are displayed with complimentary color overlays, with the reference surface displayed in *purple* and the actual live surface displayed in *green*, facilitating patient setup (Figure 4.9). IDENTIFY's projected LED light is *blue* and is focused around the isocenter using a lens caps generates a narrow, collimated beam of light.

Patient surface image-based positioning relative to the CT reference surface is displayed in a 6DOF format, with display of the current position offset from the reference for the three translational axes (vertical, longitudinal, and lateral) and three rotational axes (pitch, yaw, roll). The offset relative to the reference surface is based on the defined ROI, which was defined in the planning tool, but also can be created or ad hoc updated within IDENTIFY at the linac console. The displayed FOV encompasses

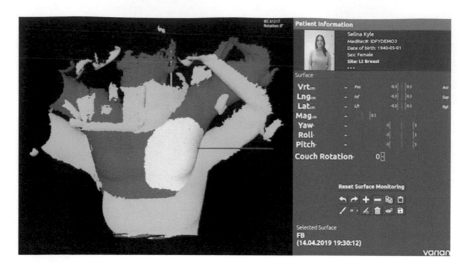

FIGURE 4.9 Patient surface image-based positioning using virtual reality displaying offsets in all 6 degrees of freedom. The user-defined ROI is in yellow. The purple surface is the reference surface and the green surface is the actual surface.

the combined FOV of the three high-resolution cameras. Once the alignment is within 2 cm of the reference image, IDENTIFY will automatically switch back from the two-surface purple/green view to the *red-blue* virtual reality view. Patient motion outside of the ROI is displayed as *red* (higher than the CT reference image) and *blue* (lower than the CT reference image), providing additional out-of-ROI information. The specified ROI for monitoring motion during treatment delivery is confirmed, or modified as required, to ensure accurate SGRT patient motion tracking. The patient position, as monitored using surface imaging, is compared to the CT-based surface image (on the first day of treatment) or to a newly captured reference surface using IDENTIFY, depending on the desired workflow.

When the patient is aligned correctly in the treatment position, submillimeter motion detection[3] can be enabled to track continuously using a one-click command. The information is also displayed on the IDENTIFY monitor at the treatment console.

4.3.3.4 Image-Guided Final Patient Setup

Surface-guided positioning allows the therapist to observe and correct for all 6DOF patient movement. This is facilitated with the use of an

ROI that is designated as an area of interest for surface tracking purposes. Color-coded virtual reality displays translational and rotational
offsets between the reference and the live surface of patients' surface
anatomy onto a live video feed that can be used by the treatment team
in the room to correct patient alignment before verification imaging
and treatment. Once in position, the patient's movement can be tracked
to ensure they remain within the set tolerances during verification
imaging and treatment. After position verification with radiographic
imaging and any required couch shifts are applied, a temporary surface can be captured, creating a new baseline of the patient's position
for subsequent motion that may occur during treatment. When the
patient moves out of tolerance, the handheld controller at the treatment console will provide an audible alarm. The therapist can then
stop the beam. In scenarios where the linac is equipped with a motion
management interface, IDENTIFY can issue a beam hold to automatically pause radiation delivery.

4.3.3.5 Noncoplanar Treatment Delivery

In the normal IDENTIFY workflow, a new reference surface is captured
once cone-beam computed tomography (CBCT) imaging has verified correct patient positioning for treatment. The patient's position is then tracked
against this reference. When using noncoplanar delivery (i.e., couch kicks),
IDENTIFY uses a drop-down menu with prepopulated couch angles
imported from the patient's DICOM RT Plan file. Ad hoc couch angles can
also be entered. An updated surface image with respect to the couch angle
is then recalculated and offsets are displayed. The positions of the three
surface imaging cameras are configured to provide optimal coverage with
minimal occlusion by the gantry. SRS treatments are accomplished with
open-face masks having a maximum FOV.

All IDENTIFY cameras are internally calibrated for SRS to ensure
accuracy during noncoplanar treatments (≤ 0.5 mm and $\leq 0.2°$). A stereotactic accuracy verification and calibration procedure is an integral part of
IDENTIFY and is described later in the chapter.

4.3.3.6 Deep Inspiration Breath Hold

IDENTIFY has a dedicated workflow for performing deep inspiration
breath hold (DIBH) treatments. Once the patient is positioned using
the free-breathing CT DICOM reference, the breath-hold CT DICOM
reference surface is selected from the handheld device, and the patient

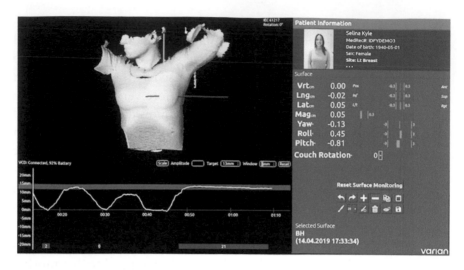

FIGURE 4.10   Monitor view of the patient's breathing trace during DIBH.

is coached into the correct position using the Varian VCD. The target breath-hold is based on the surface of the breath-hold reference image and can be configured to use offsets in the vertical axis or a combination of the vertical and longitudinal axes (default). The DIBH window tolerance is defined in the planning tool or on the handheld controller.

The radiation beam is enabled while the patient's surface is within the breath-hold position threshold, in 6DOF. Whenever the patient moves out of any of the 6D value tolerances, the beam will be paused. An indicator below the respiratory signal indicates the length of the patient's breath-hold (Figure 4.10).

4.4 COMMISSIONING AND QUALITY ASSURANCE

After installation of IDENTIFY in the clinic, Varian staff performs acceptance testing together with the center's physicist user prior to full handover for commissioning and clinical operation. As part of this process, a complete end-to-end test is highly recommended as site-specific components (e.g., OIS, TPS, CT, and linac) will differ from site to site. Chapter 7 provides a guide to the commissioning, end-to-end testing, and routine QA of the IDENTIFY system.

4.4.1 Daily QA

IDENTIFY will automatically start up in the morning in QA mode. QA can only be performed after all cameras have warmed to treatment

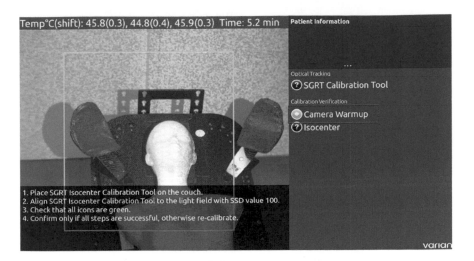

Temp°C(shift): 45.8(0.3), 44.8(0.4), 45.9(0.3) Time: 5.2 min   Patient Information

...

Optical Tracking
(?) SGRT Calibration Tool

Calibration Verification
(●) Camera Warmup
(?) Isocenter

1. Place SGRT Isocenter Calibration Tool on the couch.
2. Align SGRT Isocenter Calibration Tool to the light field with SSD value 100.
3. Check that all icons are green.
4. Confirm only if all steps are successful, otherwise re-calibrate.

varian

FIGURE 4.11   Morning warmup screen, ensuring the cameras are at the right temperature before morning QA.

temperatures to ensure the system is calibrated and verified under the right circumstances to ensure submillimeter accuracy.[3] QA is performed by those users who have appropriate rights in the data manager.

Depending on the modules used, daily morning QA (Figure 4.11) typically consists of two verification tests.

- *Verification of Setup Camera*: The setup camera is used for head-to-toe setup and for accessory navigation. Verification of the setup camera is done with the couch calibrator tool and with the couch in the loading position. The tool is inserted into one predefined indexing slot and then moved to a different slot. IDENTIFY will verify that the distance between both positions is within 3 mm. When the reading is within 3 mm, the verification will be automatically marked as successful. If the value is larger than 3 mm, the option to recalibrate will be presented. Specific user rights to recalibrate are needed.

- *Verification of the Surface Imaging camera isocenter alignment*: The system verifies the correct Isocenter calibration and the correct relative alignment between the three SGRT cameras by tracking a calibration board, which is aligned to the machine isocenter, using lasers or maximum light field (Figure 4.12).

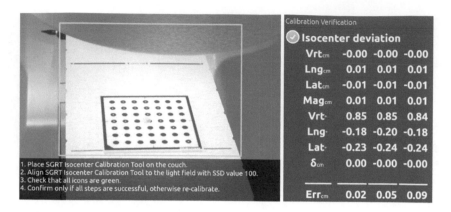

Calibration Verification

✅ **Isocenter deviation**			
Vrtcm	-0.00	-0.00	-0.00
Lngcm	0.01	0.01	0.01
Latcm	-0.01	-0.01	-0.01
Magcm	0.01	0.01	0.01
Vrt·	0.85	0.85	0.84
Lng·	-0.18	-0.20	-0.18
Lat·	-0.23	-0.24	-0.24
δcm	0.00	-0.00	-0.00
Errcm	0.02	0.05	0.09

1. Place SGRT Isocenter Calibration Tool on the couch.
2. Align SGRT Isocenter Calibration Tool to the light field with SSD value 100.
3. Check that all icons are green.
4. Confirm only if all steps are successful, otherwise re-calibrate.

FIGURE 4.12   The IDENTIFY isocenter tool during a calibration verification procedure as seen on the IDENTIFY monitor (left) and the Calibration Verification section of IDENTIFY's SGRT morning QA mode (right).

The reported numbers show the difference between the current board position and the saved isocenter position in the IDENTIFY System. All three cameras are aligned to the same center. A discrepancy between the measurements of each camera to the others (defined by the δ-value) indicates that the relative position between cameras changed (one of the cameras was moved for example).

In addition, each camera verifies its internal calibration (intrinsic camera parameters and the stereo transformation between the left and the right camera eyes, defined by the Err-value). The position of the calibration board is computed for each camera eye separately and the resulting transformation from one eye to the other compared to the stored calibration. If this transformation exceeds a threshold, the 3D camera requires recalibration.

If the calculated error between the calibrated position and the current position is below the defined range, the sign on the wall-mounted screen becomes green. Under deviation, the number of columns with position values corresponds to the number of SGRT cameras used for the setup.

If the calculated error remains outside the defined range, the sign on the wall-mounted screen becomes red and recalibration is necessary. Access to the calibration mode is restricted by user rights.

4.4.2 Stereotactic QA

The SRS accuracy verification and calibration is a streamlined process for verifying machine isocenter and the isocenter of IDENTIFY. This procedure is done at installation and recommended for use prior to any SRS/SBRT treatment. A Modus QA QUASAR™ Penta-Guide phantom is delivered with every surface-guided radiosurgery module for IDENTIFY. It is the responsibility of the physics team to scan the phantom at CT with 1-mm slices and create a simple treatment plan for use as a DICOM reference. As sub millimeter accuracy is paramount for SRS treatments, the Penta-Guide phantom is scanned at an approximately 45° angle to get a usable structure set from the TPS and to facilitate initial phantom setup using the room lasers. This plan forms part of an end-to-end test for stereotactic calibration and is sent to IDENTIFY from the TPS.

It is important to note that the SRS calibration procedure can only be completed for defined "QA Patients," ensuring that accidental recalibrations do not occur.

The SRS calibration procedure is as follows:

1. At the linac, using lasers, align the phantom to isocenter, with the same 45° rotation as during CT simulation.

2. Open the QA patient, which has the imported phantom data and select the reference surface.

3. Using IDENTIFY, the best practice recommended by Varian is to level the phantom with the DICOM reference surface prior to imaging such that axes are within the range of shifts possible with CBCT guidance.

4. Acquire CBCT or MV imaging of the phantom.

5. Optimize window and level to review proper alignment and use automatic registration. Varian recommends several registrations and use of average values as automatic registration values may differ by ±0.2 mm.

6. If the values between the imaging source and IDENTIFY are deemed to be too large, the radiographic isocenter calibration menu

in IDENTIFY can be used to enter the exact shifts calculated by the imaging system.

7. Click the calibrate button and confirm. This will match the IDENTIFY coordinate system to the imaging coordinate system. On the CBCT, apply the calculated couch corrections.

8. All values in the 6DOF display should be close to zero (less than 0.3 mm/0.2°).

9. As a secondary check, perform another CBCT scan and use the automatic registration for verification.

10. The registration values and the values shown at the 6DOF display should match within 0.5 mm/0.2°. If deviations exceed these tolerances, then the user should repeat the procedure.

The system is now calibrated for stereotactic treatments.

4.4.3 Intrinsic and Stereo Camera Calibration

At the time of installation, the cameras always require both an intrinsic and a stereo calibration.

Intrinsic calibration includes the focal length, image center, and distortion parameters of each camera, while stereo calibration is used for the translational and rotational accuracy between cameras in a camera unit and between the three camera units. The IDENTIFY SGRT calibration plate is used for both calibrations. During installation, each camera unit (consisting of two stereoscopic cameras and a projector) is calibrated for intrinsic parameters using an external application. A series of images are captured of the IDENTIFY SGRT calibration board in different positions, and these images are used to calculate the intrinsic camera parameters. Results from this calibration are stored in the camera unit.

After intrinsic calibration, several images of the IDENTIFY SGRT calibration plate are captured with both cameras in an IDENTIFY SGRT camera unit. Both cameras in each camera unit must be calibrated to one another (stereo calibration) and to the room coordinate system.

KEY POINTS

- The IDENTIFY system uses ToF and RGB sensor technology to assist with accessory placement and initial orthopedic patient setup.

- The IDENTIFY system uses stereo vision cameras for surface-based inter-fraction patient positioning and intra-fraction motion monitoring.

- Therapist interaction with the IDENTIFY system is via handheld controllers and augmented reality displays indicating the patient's current position offsets relative to a reference surface overlaid onto a live video feed of the patient.

- The IDENTIFY system has intrinsic camera calibration, and the recommended quality assurance tests for routine daily and stereotactic treatments to ensure proper system performance.

- IDENTIFY is compatible with a variety of linac and OIS platforms from multiple vendors.

REFERENCES

1. Zhao H, Huang Y, Sarkar V, et al. Radiation therapy treatment deviations potentially prevented by a novel combined radio-frequency identification (RFID), biometric and surface matching technology: WE-RAM1-GePD-J(B)-02. *Med Phys*. 2017;44(6):3204.
2. Zhao H. Verification of patient treatment accessories and posture in SAM practical medical physics quality improvement and safety applications of surface imaging: WE-AB-KDBRA2-03. *Med Phys*. 2018;45(6):e576.
3. Fischler MA, Bolles RC. Random sample consensus: A paradigm for model fitting with applications to image analysis and automated cartography. *Commun. ACM*. 1981;24(6):381–395.
4. Zhao H, Sarkar V, Paxton A, et al. Initial phantom evaluation of Varian Identify system SGRT consistency and accuracy: SU-I430-GePD-F6-06. *Med Phys*. 2019;46(6):e166–e167.
5. Besl PJ, McKay ND. A method for registration of 3-D shapes. *IEEE Trans Pattern Anal Mach Intell*. 1992;14(2):239–256.
6. Low KL. Linear least-squares optimization for point-to-plane ICP surface registration. Technical Report TR04-004, Department of Computer Science, University of North Carolina at Chapel Hill, February 2004.

CHAPTER 5

Technical Overview and Features of the Vision RT AlignRT® System

Benjamin Waghorn

CONTENTS

Editor's Note: SGRT vendors have a variety of technologies in their product portfolio and some of those technologies are integrated into their SGRT systems to different degrees. Since this textbook is focused on SGRT, the editors have asked the authors to limit their chapter content specifically to surface image guidance for patient setup and inter- and intra-fraction patient monitoring.

5.1 INTRODUCTION

Founded in August 2001, Vision RT (Vision RT Ltd., London, UK) was established to develop real-time surface tracking technology employing computer vision techniques for radiation therapy (RT). By utilizing video-based stereo photogrammetry, Vision RT has developed a solution called AlignRT for use within the RT environment to generate real-time three-dimensional (3D) renderings of the patient's surface.[1] Utilizing knowledge of the patient's intended treatment position via the treatment planning computed tomography (CT) external body contours, the real-time surface of the patient can be used to position the patient for treatment accurately in all 6 degrees of freedom (6DOF), both with respect to the treatment isocenter,[2] and utilizing postural information.[3,4] The patient's position can then be continually monitored throughout treatment to ensure accurate positioning, with no additional ionizing radiation dose or invasive equipment.[5,6]

The first prototype AlignRT system from Vision RT was installed at the Royal Marsden Hospital, United Kingdom, in 2002. This solution utilized two camera pods rigidly mounted from the ceiling within the treatment room. In 2003, data from the first accuracy validation study were presented at the 45th Annual ASTRO meeting in San Diego, CA.[7] The study found that patient repositioning with errors of less than 1 mm was possible with AlignRT. Moreover, Smith et al. identified several key advantages of high-speed 3D surface imaging for patient setup including not requiring markers and no additional irradiation of the patient. The system was designed to be operated by the radiation therapists both remotely and while inside the treatment room with the patient.

Over the subsequent years, the benefits of utilizing 3D patient surface information for patient setup and intra-fraction monitoring were demonstrated by a growing community and the technique and field of surface-guided radiation therapy (SGRT) progressed. In addition to AlignRT, Vision RT offer two additional products, GateCT and GateRT, designed to assist with the treatment workflow of patients where respiration may cause appreciable tumor motion. GateCT is used to generate a 4DCT data series at the time of treatment planning CT acquisition. GateRT can be used during treatment to provide respiratory gating. Both products directly use the patient's surface information to determine breathing motion. This chapter provides a technical overview of the core SGRT Vision RT technologies and describes some of the key features. In addition to the SGRT solutions, Vision RT also provides SafeRT, a patient identification and accessory verification product that integrates with AlignRT. Using an infrared facial recognition solution, SafeRT provides automated, noncontact patient identification, while optical techniques are used for accessory verification. A full description of SafeRT is beyond the scope of this chapter.

5.2 AlignRT®

AlignRT is an optical imaging system capable of providing submillimeter accuracy[8] for monitoring the precise location and movement of an RT patient for initial setup and during treatment. This is done by projecting an optical speckle pattern onto the patient's surface and utilizing active stereo photogrammetry. The speckle pattern is projected from the AlignRT camera units, with the typical installation consisting of three ceiling-mounted camera units (Figure 5.1).

For use, visible red light with a pseudo-random speckle pattern is projected from each camera pod onto the patient. Corresponding two-dimensional (2D) information from each of the two camera sensors (each calibrated to the treatment room coordinate systems) can be converted into a series of 3D coordinates via "triangulation" and visualized as a real-time surface rendering of the patient. AlignRT combines the surface information generated from combined camera pods and uses this information to perform surface registration. The real-time surface is aligned with either a surface model of the patient that was generated from the treatment planning CT scan or a surface that was previously captured by the AlignRT system itself. Surface matching results are displayed to the users as a series of 6DOF rigid shift values, known as real-time deltas (RTDs) (Figure 5.2).

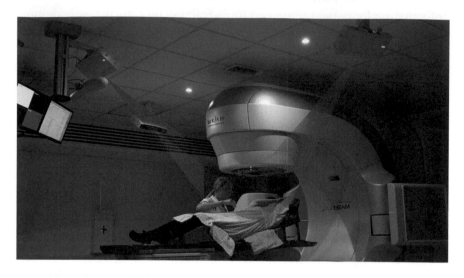

FIGURE 5.1   AlignRT camera pod setup and clinical application in a linear accelerator treatment room.

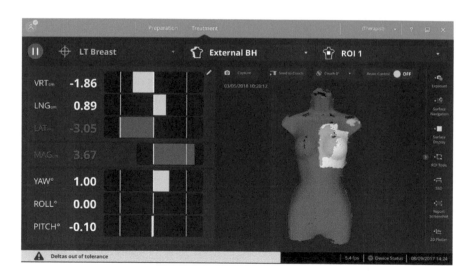

FIGURE 5.2   AlignRT software with surface rendering and real-time delta tracking.

The RTDs quantify the misalignment between the current and the planned patient positions as a series of six unique values, three translational (vertical, lateral, and longitudinal) and three rotational (rotation or yaw, pitch, and roll) values, and a magnitude value (the root-mean-square [RMS] vector magnitude, calculated as the square root of the sum of the squares of the three translational RTD values). This information can be used to adjust the patient's position or to otherwise make clinical decisions based on the patient's real-time position. The complete surface rendering and registration is repeated up to 25 times per second (called the frame rate), allowing for real-time surface tracking.

5.2.1 Algorithms

AlignRT relies on a combination of patented camera technology and real-time 3D reconstruction software to generate high-density 3D models of the patient's external surface, as described below.

5.2.1.1 Surface Reconstruction

Like the human visual system, AlignRT uses stereo vision concepts to generate 3D information, where two cameras within each pod can be used to provide depth information from the corresponding 2D images. The two sensors, with known positions and orientations (as determined during calibration), simultaneously acquire 2D grayscale images of the treatment scene. In the simplest terms, using methods of 3D triangulation, image coordinates from the two images that correspond to the same point in the scene can be projected through the camera focal points.[9] The intersection of these 3D lines generates a 3D coordinate for this one location within the scene. The same 3D triangulation method is repeated for all other matching points that are identified in the two 2D images, and a resultant map of 3D vertices is generated as shown in the AlignRT software (Figure 5.2, Version 6.1 shown).

AlignRT uses a technique known as active stereo, projecting a pseudo-random optical speckle pattern onto the patient's surface, to identify corresponding points between the two images on the patient's surface.

After the stereo matching and triangulation process has taken place, a surface is reconstructed from each camera pod. The generated surfaces (three surfaces for a three camera-unit AlignRT system) will be aligned in space as a result of the system's calibration. A surface merging technique combines the 3D surface data from each of the AlignRT camera pods.

The resultant 3D surface model can be visualized as a wireframe rendering containing the calculated vertices and triangles (comprising three indices to corresponding vertices) or as a smooth rendering. To allow accurate surface reconstruction for a complete range of different skin and surface tones, variable exposure settings are used.

5.2.1.2 Surface Registration

To calculate the RTDs relating the current patient position to the reference surface, a surface matching algorithm is used. The surface matching process employed by AlignRT uses a proprietary algorithm based on the iterative least squared minimization technique to calculate the RTDs.[10] Once the surface registration is complete, AlignRT will compare the RTDs to the user-defined threshold values.

There are several reasons why, in the view of Vision RT, rigid registration techniques are preferred over deformable techniques and are therefore used in the AlignRT surface matching process. Although deformable registrations can play a valuable role in certain areas of radiation oncology, including adaptive RT and automatic segmentation methods, during treatment, it is important to be able to evaluate surface contours without altering the shape and to provide meaningful shift information that can be performed by the 6DOF treatment couch system. Similarly, rigid registration is the means used for patient positioning in other situations, for example, cone-beam CT (CBCT) matching. Moreover, the complexity of deformable algorithms does not make it a favorable option for real-time applications where there is no time for quality assurance of the deformation field that contributes to the final registration. Perhaps most importantly, clinical outcome data for SGRT systems are all based on the rigid registration algorithm.

5.2.2 Hardware

In addition to the software described above, the AlignRT system generally consists of the following hardware components: a computer workstation, three 3D camera units, cables, and a plate that is used for camera calibration. The computer workstation includes a keyboard, monitor, and mouse and is located outside the treatment room at the treatment console area. The workstation is connected to a remote console installed inside the treatment room. Therefore, therapists can interact with the AlignRT software both from within the treatment room for patient setup and during treatment from the treatment console area.

The camera units are connected to a main power supply unit, with an accessible wall switch for convenience when cycling the power. Each camera unit is also connected to a frame grabber, located in the workstation, for high-speed imaging and data transmission. An optional integrated gate controller (IGC) card, or other gating interfaces, also housed within the workstation interfaces the Vision RT equipment to third-party CT and treatment systems, including linacs and proton therapy gating interfaces. To date, Vision RT has system interfaces with several vendors, including Varian, Elekta, Hitachi, IBA, GE, Philips, Siemens, and Toshiba.

As previously described, the optical hardware used to generate patient surface information is housed within camera units (or pods). Within the treatment room, standard AlignRT installations employ three ceiling-mounted camera pods. Pods 1 and 2 are located approximately 30 cm offset from the left and right lateral positions, respectively (as referenced by a head first supine patient), and Pod 3 is located at the foot of the treatment couch (Figure 5.1). As each individual camera pod can be used independently to generate a surface, additional combinations of camera pods and locations are possible and may be used where standard pod configurations are not possible.

While Vision RT hardware and technology has undergone several upgrades since the first installation, the core components remain relatively constant. Each camera pod (Figure 5.3) contains an LED projector, two high-definition image sensors (cameras) and a white LED for calibration and during the acquisition of photorealistic patient images (known as 3D Photo).

The LED projector projects the pseudo-random speckle pattern, via a speckle slide, onto the patient's surface, thereby providing the texture variations required for the reconstruction process. The light projected

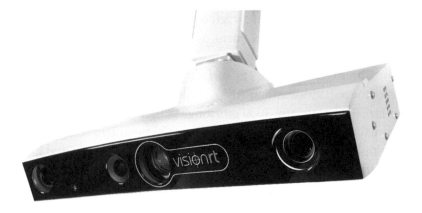

FIGURE 5.3   AlignRT HD camera pod.

onto the patient's surface is in the red visible spectrum. Red light was selected over other wavelengths for several reasons, including (i) red light provides lower skin absorption than other wavelengths, (ii) patient comfort is higher with red light due to the human eye's lower sensitivity perception of brightness in the red wavelength, and (iii) red light falls into the "exemption group" for photobiological safety according to the BS EN 62471:2008 standard.

The two image sensors are separated by a known distance and are used to acquire the raw textured data used for 3D surface reconstruction. The camera sensors have been upgraded over time to reflect advances in the technology, catering to more advanced needs and providing enhanced accuracy.

An additional hardware option is the offline AlignRT workstation (AlignRT Offline), providing users the ability to interact with the AlignRT software to perform certain preparatory tasks that do not require the patient or treatment system, including importing patient information and defining regions of interest (ROIs). The AlignRT Offline solution can be connected to the same database network as the clinical AlignRT systems and can be located in a convenient place such as in the dosimetrist or physicist offices.

5.3 AlignRT® FEATURES

5.3.1 Region of Interest

An aspect of the AlignRT SGRT solution is that the calculated RTDs are based on a user-defined ROI concept. The 6DOF match is constrained to information contained within the ROI, providing more relevant and accurate surface tracking based on the intended application (e.g., tracking an ROI in proximity to the treatment target, or tracking an extremity for postural setup). Using ROIs during treatment, as opposed to whole surface tracking, also increases the system performance. The user can create multiple ROIs to assist with postural setup. The surface data can capture a relatively large region of the patient's anatomy (the AlignRT volume of capture around isocenter is 650 mm laterally, 1000 mm longitudinally, and 350 mm vertically) to ensure the appropriate patient posture on the treatment couch. For example, multiple ROIs can be used to provide accurate arm, chin and hip positions while maintaining the correct isocenter location for breast treatments. During intra-fraction treatment monitoring, it is important to use an ROI where the RTD shifts are calculated around the treatment isocenter (an isocenter ROI). However, during setup, it can be advantageous to perform more localized shifts independent of the treatment isocenter. For this, a centroid, or setup ROI, can be created, generating 6DOF information about the center of mass of the ROI. Moreover, a snapshot of the entire patient surface can be acquired (called a treatment capture), allowing for gross setup errors to be visually identified and corrected.

5.3.2 Treatments

AlignRT can be used to assist with the setup and intra-fraction monitoring of any patient where there is a suitable, exposed area of the patient's surface to use as the tracking ROI. This covers all treatment sites, including intracranial,[11] head and neck,[12] lung,[13] breast,[14] prostate,[15] and extremities,[4] and for pediatric patients.[16] Depending on the treatment site and technique, options exist within AlignRT to vary the surface resolution (density of calculated surface points) and therefore provide a surface accuracy and frame rate appropriate for that treatment application. Stereotactic radiosurgery (SRS) treatments typically require submillimeter positional accuracy, so selecting the SRS treatment site profile in AlignRT can deliver the highest surface tracking accuracy. Although this is computationally more intensive due to the increased data requirements, the frame rates still provide real-time tracking, with the surface generation and 6DOF registration occurring at a rate of approximately 6 frames per second.

5.3.3 Deep Inspiration Breath Hold

Unlike other motion management solutions that measure breath-hold or other respiratory motion in only a single direction (e.g., vertical displacement), AlignRT always uses full 6DOF surface matching to provide more relevant and sensitive motion information. For deep inspiration breath hold (DIBH) applications, AlignRT uses a coaching toolbar to relay 6DOF breathing information to the user and optionally to the patient via a visual feedback device (the Real Time Coach™). Within the coaching toolbar, a static horizontal white bar represents the target area for the patient when they take a deep inspiration. A dynamic, smaller bar provides two pieces of information based on the patient's surface. First, the vertical position shows the vertical RTD such that when it is contained within the white bar, the patient's surface is within the vertical threshold. Second, the color of the smaller bar changes from orange to green when all 6DOF RTDs are within tolerance. Therefore, the full surface information is being utilized to ensure the correct DIBH position is being achieved, enabling incorrect motions, such as back arching, to be successfully identified. As with all treatments utilizing AlignRT, the radiation beam can be controlled based on this movement in 6DOF. More details about DIBH for breast treatments and SBRT can be found in Chapters 10 and 15, respectively.

5.3.4 Stereotactic Module

To achieve more conformal dose coverage or to reduce the dose to organs at risk, treatment plans with nonzero couch rotations are commonly employed for stereotactic and certain other treatments.[17] By selecting the desired couch rotation in the AlignRT software, the reference surface rotates about the calibrated isocenter, allowing AlignRT to continue monitoring the patient's position at all nonzero rotations. Whenever a patient is treated with noncoplanar arcs or fields, additional considerations are required to maintain submillimeter surface tracking accuracy. If there is a calibration offset between the AlignRT isocenter and the treatment isocenter, the surface will "walk-out" at nonzero rotations, thus displaying RTD values with an error proportional to both the calibration offset and couch rotation angle.[18] Therefore, it is important for noncoplanar procedures that the AlignRT isocenter location is accurately co-calibrated to the radiographic isocenter of the treatment machine.

The stereotactic module from AlignRT includes MV Isocenter Calibration software and hardware that enable the user to co-calibrate the AlignRT isocenter to the treatment isocenter using radiographic imaging of a solid calibration cube phantom manufactured by Vision RT (Figure 5.4).

FIGURE 5.4   Vision RT calibration cube.

As the accuracy of surface tracking is dependent on the accuracy of the isocenter location, it is recommended that users who treat patients at nonzero couch rotations use the stereotactic module and perform the MV Isocenter Calibration following monthly calibration. When accurately calibrated, during couch rotations AlignRT RTDs will identify the true motion of the patient relative to radiation isocenter, including a combination of couch walkout and patient motion.

5.3.5 3D Photo Display

The default reconstructed surface in AlignRT can be viewed as either a smooth surface, a wire frame, or a cloud of points per the user's selection. For most applications, these rendered surfaces allow for accurate and effective setup and monitoring. Certain applications, however, can benefit from additional information, and in these cases, AlignRT uses 3D Photo Display, a photorealistic visualization of the patient and immobilization devices (Figure 5.5).

5.3.6 Surface Statistics

To objectively quantify changes in patient anatomy, AlignRT provides a surface statistics tool. Surface statistics provides a surface agreement measurement, quantifying the percentage of the current surface that is within

FIGURE 5.5   3D Photo Display.

a user-defined distance of the reference surface. The user has three options for calculating surface statistics; with the patient in their current position, with best fit translations and yaw registrations virtually applied (primarily for 3DOF couches), and with all translational and rotational RTDs virtually applied. For the latter two cases, if the surface statistics measurement significantly improves following these virtual shifts, repositioning the patient using the calculated RTD values could resolve the setup problems. However, relatively low overlap values in any of the three options could indicate anatomy changes. The visual display within surface statistics can show the anatomical location of the changes and whether the current surface is above (e.g., due to tumor growth) or below (e.g., due to weight loss) the reference surface.

5.3.6.1 Source to Surface Distance Measurements

Although 3D imaging has, in some instances, reduced the need to capture source to surface distance (SSD) measurements, valuable information can still be gained by comparing the treatment planning SSD to the SSD at the time of treatment. Using existing surface and isocenter information, AlignRT can calculate multiple SSD readings from a single treatment captured surface. SSD readings for multiple gantry angles, and offset isocenter locations, can be calculated from either inside or outside the treatment room, even after the patient has been removed from the treatment position.

5.3.7 Gated Capture

Within the AlignRT software, there are two methods for capturing 3D reference surface data. The first is a single snapshot acquisition that reconstructs a record of the surface at that moment. The second, called gated capture, allows a record of the patient to be acquired over several seconds, therefore capturing any respiratory or other cyclical motion. The gated capture method analyses a signal derived from the patient's respiratory motion and then uses this signal to reconstruct a surface that corresponds to a user selectable point in the breathing cycle. For example, data may be repeatedly acquired at maximum exhalation both during simulation and prior to each treatment fraction. This helps to minimize systematic errors that may be otherwise caused by respiratory motion. Alternatively, the location of the captured surface may be selected such that the resultant tracking RTDs will move symmetrically about their zero values. This will minimize any direction bias from the symmetric RTD thresholds.

5.3.8 Treatment Machine Integration

For setup, the AlignRT software contains a function called Move Couch, which enables automatic patient adjustment via treatment couch motions in up to 6DOF.[19] Where enabled, AlignRT can send shift information to the treatment couch via the Move Couch function. The user instructs Move Couch to calculate the current 6DOF rigid shift and to automatically apply the shifts, either applying just the translations, the translations and yaw, or full 6DOF shifts, where possible.

Following the initial setup, AlignRT can continue to monitor the patient, including during any radiographic imaging procedures and during treatment. An IGC can be used to connect AlignRT to the RT treatment systems therefore enabling AlignRT to communicate with the linac or other treatment delivery system. Prior to treatment, motion thresholds can be selected based on the treatment indication (e.g., intracranial SRS, stereotactic body RT or left breast) and then can be edited independently in each degree of freedom to provide patient-specific thresholds. During treatment, if AlignRT detects patient movement exceeding the motion thresholds in any of the 6DOF, it will visually alert the user and either instruct the delivery system to pause the radiation beam (in gating mode) or alert for manual intervention (with gating turned off, or for standalone AlignRT systems). When gating is enabled, AlignRT will send a signal to the IGC, informing the treatment machine to hold the beam.

5.4 RESPIRATORY MOTION MANAGEMENT

5.4.1 GateCT®

GateCT is a surface tracking platform that allows points or regions of a patient's skin surface to be tracked in real-time during treatment planning CT acquisition. Like AlignRT, the system consists of advanced software, a workstation, a 3D camera unit, cables, a calibration plate, and a gate controller hardware interface.

GateCT typically uses a centrally positioned 3D camera unit, equivalent to those used by AlignRT, which is aligned on the central scanning plane of the CT scanner. The goal of GateCT is to track the respiratory motion from the 3D surface of a patient during 4DCT data acquisition. After acquiring a gated 3D surface model of the patient, the user selects tracking points for respiratory motion and patient movement detection. The GateCT system is synchronized with the CT scanner and facilitates 4DCT reconstruction either via prospective signaling to the CT scanner or via retrospective reconstruction, depending on the CT vendor. During the scan, GateCT automatically accounts for couch motion. Phase tags (both inhale and exhale) are computed during real-time tracking. Irregular breathing, or patient movement, are detected and labeled as such in this exported signal. GateCT can also be used during the acquisition of breath-hold CT scans, ensuring the correct breath-hold is maintained throughout the scan.

5.4.2 GateRT®

Using the same camera pods as AlignRT, GateRT is designed to calculate and record the respiratory motion and respiratory state of a patient in the treatment room using real-time 3D surface tracking. GateRT tracks the respiratory motion of a patch on the patient's surface during treatment or imaging and automatically disables the beam when the breathing signal moves outside the user-defined gating window. GateRT supports both phase- and amplitude-based gating. The user can select the relevant tracking point on the patient's surface, determine the type of gating to be used (phase or amplitude) and the associated thresholds and manage abnormal breathing detection handling within the GateRT software.

5.5 CALIBRATION

The quality of calibration information plays a central role in the overall system accuracy and ultimately in the clinical efficacy of SGRT systems. AlignRT systems employ two types of calibration procedures; an intrinsic camera calibration, and an extrinsic isocenter calibration.

The accuracy and stability of surface tracking are dependent on the quality of the camera calibration. Both camera sensors within each AlignRT camera unit acquire real-time images in 2D. The purpose of camera calibration is to define the relationship between an image point (pixel) on the camera sensor and a known position and orientation in the real-world 3D space that the camera is viewing.

The isocenter calibration defines the 3D location of the AlignRT isocenter within the treatment room, and the 6DOF orientation of the camera pods relative to this point. As treatments are delivered to the patient relative to the actual treatment isocenter, it is important that the reconstructed surface references the same isocenter. This will provide a more accurate absolute patient setup at isocenter, and whenever the patient is being treated using couch rotations, it is important that the AlignRT isocenter is coincident with the treatment isocenter so that the rotational axes of the surface and radiation isocenter are common.

5.5.1 Intrinsic Camera Calibration

The intrinsic camera calibration for AlignRT has historically been performed using the Vision RT calibration plate (Figure 5.6) and is completed as part of the monthly (or plate) calibration procedure.

The calibration plate contains 1088 circular blobs that have been accurately positioned on a flat plate. The blob centers are arranged in a symmetric 34 × 32 blob rectangular array centered on the plate. The origin of the plate's coordinate system is implicitly defined at the center of the plate and is indicated with a cross. Prior to calibration in the radiotherapy room, the cross is placed in a well-defined orientation on the treatment couch directly at the isocenter of the machine. The plate is approximately perpendicular to the radiation focal beam with the gantry at 0°.

The determination of the 2D locations of the blob centers is the primary task of the Vision RT calibration software. Once the 3D and 2D blob correspondences have been found, they are passed into an algorithm that computes the associated calibration for each camera image. The calibration is stored in a camera model file, which records the coordinates of the camera focal point, its orientation relative to the origin (i.e., the position and orientation of each camera in 3D space), and a number of intrinsic camera properties. In addition, the algorithm corrects for any errors in the camera center point.

Most treatment plans use isocentric setups, so the patient's surface will not be located at the treatment isocenter (except for some superficial treatments). More commonly, the surface will be raised above the isocenter,

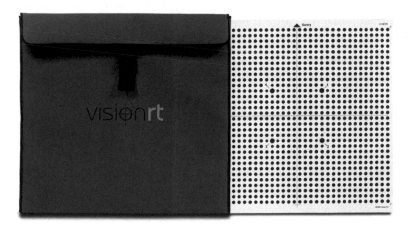

FIGURE 5.6 Calibration plate and storage case.

and therefore it is optimal to generate an optical model for the system in this raised location. As the intrinsic calibration is generally most accurate in the plane in which the system is calibrated, the intrinsic calibration is generally performed with the calibration plate positioned above the treatment isocenter. A raised plate height of 75 mm provides an approximate average surface location for treatments.

5.5.2 Advanced Camera Optimization

In 2018, a new optical setup and intrinsic calibration technique, called advanced camera optimization (ACO), was made clinically available. The optical camera model is determined from static image captures of the calibration plate. To achieve a complete and more accurate model encompassing all possible surface tracking locations, the model requires 3D input data captured over the entire imaging volume thus providing information across a 3D volume.

ACO was introduced to optimize the accuracy and stability of 6DOF tracking over a large field of view (FOV). Using a series of proprietary algorithms and techniques, the ACO procedure includes fine-tuning the optical setup and generating a 3D calibration model by acquiring multiple images of a precision manufactured ACO calibration plate as it is positioned throughout a 3D volume. The images are captured across a large FOV, resulting in 3D calibration data that is designed to encompass all typical clinical surface locations.

5.5.3 Extrinsic Isocenter Calibration—MV Isocenter Calibration

One component of the monthly calibration procedure described above is performed with the calibration plate positioned at isocenter. Information from this procedure is used to determine the isocenter location of the AlignRT system. Due to inaccuracies in the ability to accurately and reproducibly position the calibration plate at the treatment delivery system's radiation isocenter, a minor correction needs to be calculated and applied to the monthly calibration isocenter location to improve the AlignRT isocenter accuracy. This is computed during a procedure known as MV Isocenter Calibration.

Using the Vision RT calibration cube (Figure 5.4), the AlignRT MV Isocenter (or Cube) Calibration procedure requires four orthogonal MV images of the cube when positioned close to the treatment isocenter. Using a sphere detection algorithm, the locations of the known, five internal spheres are calculated and compared to the imaged isocenter location (either the digitally calibrated image center or the image center calculated via field edge detection techniques). Combining this 6DOF radiographic shift information with the surface location from the existing AlignRT calibration whose location with respect to the internal spheres is known, a 6DOF transform matrix is calculated to co-calibrate the AlignRT iso-center to the radiographic isocenter as determined by the electronic portal imaging device (EPID). This corrective transformation is applied to the data after reconstruction to precisely align the AlignRT surface to the linac's treatment isocenter.

5.5.4 Daily QA

A brief daily quality assurance procedure is required for all AlignRT systems. Using the AlignRT calibration plate positioned approximately at isocenter, the location of the blobs as seen from each camera pod is compared. The discrepancy in blob location from pairs of pods is calculated, and the resultant RMS error is returned to the user. An RMS error above a predefined threshold infers that one of the pods has moved since monthly calibration was performed, and that monthly calibration should be repeated. As the calibration plate is only approximately aligned with isocenter, the Daily QA procedure is not designed to alert the user of changes in the absolute isocenter position. However, changes in the AlignRT isocenter calibration can be inferred from the reported RMS values. Furthermore, the plate position relative to the calibrated isocenter location is stored in the Daily QA report.

5.6 AlignRT® ACCURACY DATA

The accuracy of AlignRT with ACO can be measured using recommendations from the American Association of Physicists in Medicine (AAPM) Task Group 147 report on quality assurance for nonradiographic radiotherapy localization and positioning systems.[20] Within TG-147, a series of localization accuracy tests, including end-to-end absolute localization assessment and known shift tracking accuracy tests, are described. Using these recommendations, the accuracy of AlignRT has been assessed via a series of phantom studies, and the results form part of the Vision RT system technical specification and 510(k) claims.[21]

For clinical applications, it is important that the accuracy of any SGRT system is achieved throughout the entire duration of each treatment and not just during the most straightforward configurations and phases of treatment. Therefore, scenarios for AlignRT accuracy testing were chosen to include the most challenging configurations for surface-based patient setup and monitoring. These tests included large couch rotations, deep isocenters (up to 18 cm below the skin surface), clinically relevant ROIs, and camera pod occlusions. The testing also looked at the coplanar tracking accuracy of AlignRT and the ability for AlignRT to accurately position a phantom as part of an end-to-end test.

5.6.1 Absolute Setup Accuracy

The absolute setup accuracy is defined as the accuracy with which AlignRT can position a phantom in the planned treatment position in the treatment room. This study involved measuring the absolute setup accuracy of AlignRT with ACO, using DICOM CT surfaces from the treatment plan (using a CT slice thickness of 1 mm) as the reference surface. End-to-end tests were performed with the Vision RT calibration cube. The maximum error in absolute positioning accuracy of AlignRT, defined as the 6DOF match shifts between the CBCT and treatment planning CT following initial phantom alignment with AlignRT, was 0.2 mm/0.2°.

5.6.2 Coplanar Tracking Accuracy

Tests to measure the accuracy of AlignRT with ACO under coplanar monitoring conditions were performed using the Vision RT calibration cube phantom positioned on a manual stage. Using a FaroArm® (Faro, Lake Mary, FL) to accurately measure introduced translational and rotational shifts, all AlignRT measured RTD errors (difference between FaroArm® measured offset phantom position and RTD value) were below 0.16 mm and 0.05°.

5.6.3 Noncoplanar Tracking Accuracy

Noncoplanar accuracy testing involved tracking a head phantom at various phantom and gantry rotations relative to a reference surface captured at the zero position. MAX-HD (IMT Inc., Troy NY) was positioned on a custom-made rotating platform (0.1 mm/0.1° manufacturing tolerances). The position of the phantom was monitored using AlignRT during various gantry and platform rotations, and RTD information was recorded for each configuration. At each platform rotation (0°, ±45°, and ±90°), gantry positions of 0° and ±30° were used to measure the impact of occluding each lateral camera pod. The sensitivity of accuracy on the isocenter location was also investigated by varying the isocenter location from 3 to 18 cm below the surface.

Across all couch and gantry angles (including pod occlusions), for a mid-depth isocenter (~12 cm below the surface), it was demonstrated that AlignRT with ACO could track the Max-HD phantom with a maximum error of 0.32 mm (MAG RTD) and a maximum rotational error (YAW, ROLL, and PITCH) of <0.2°. The difference in average MAG RTD value between shallow and deep isocenters was ~0.1 mm, and the average change in RTD value with ACO caused by a pod occlusion was 0.06 ± 0.02 mm (maximum change = 0.07 mm). These data are in agreement with work from Wiant et al.[22]

These data suggest that AlignRT with ACO addresses three combinable challenges inherent to any SGRT system: (i) maintaining accuracy across a range of isocenter locations, (ii) providing RTD stability when one of the camera pods is occluded, and (iii) maintaining accuracy for all couch rotations. In the first two instances, changes in RTD values due to varying surface location or pod occlusions were shown to be on the order of 0.1 mm in phantoms. Regarding the third challenge, AlignRT with ACO, even with the largest couch rotations, was shown to provide sub-0.4 mm tracking accuracy.

The AlignRT accuracy measurements above have been independently replicated in phantom studies on several clinical and nonclinical AlignRT systems, and other similar work has been published.[8,19,22] Moreover, in a study of over 900 intracranial stereotactic treatment fractions, early clinical data from the University of Alabama at Birmingham support the phantom data with ACO.[23] Using patient data, Covington et al. conclude that ACO eliminates the dependence of isocenter location on tracking accuracy in patients.

5.7 CONCLUSION

AlignRT is an optical imaging system for delivering SGRT, which is used to image the skin surface of a patient in 3D before and during radiotherapy treatment. During the treatment session, AlignRT is able to continuously monitor the motion of the patient in real-time, and whenever the patient moves away from the intended treatment reference position, AlignRT calculates this motion and presents the information to the user via 6DOF RTD values. AlignRT is designed to be able to operate independently as a standalone system. However, additional benefits can be realized when AlignRT interfaces directly with third-party hardware and software. Interfacing with the treatment couch enables automated couch movement for patient setup. Interfacing with the treatment delivery system allows automatic beam control based on the patient's position during treatment or gating for respiratory-correlated delivery techniques. The combined techniques and features described in this chapter have been demonstrated to provide multiple benefits to the patient and their treatment team, including increased initial setup accuracy, improved treatment throughput, accurate real-time patient position tracking, clinical efficacy for most treatment indications, and enhanced patient comfort.

To further address respiratory motion, GateCT and GateRT use patient surface information to directly measure the patient's motion without the need for surrogate markers or additional devices. All Vision RT systems are noninvasive, do not require the use of body markers, and produce no ionizing radiation during the imaging process.

KEY POINTS

- Vision RT manufactures three surface image-guided products, namely AlignRT, GateCT, and GateRT. All of the products use a projected speckle pattern and active stereo photogrammetry techniques to generate a surface.

- Live surfaces generated by AlignRT can be registered to either the initial CT DICOM surface, or a previously captured surface, to generate 6DOF RTDs using an iterative least squared minimization technique.

- Gating interfaces to treatment delivery machines allow for automated beam hold when the patient motion exceeds a patient specific, predefined motion threshold.

- The generated surface can be used as a respiratory motion management solution during 4DCT acquisition (GateCT) and for respiratory gating (GateRT).

- AlignRT uses calibration techniques that result in sub-0.2 mm tracking accuracy for coplanar configurations and less than 0.1 mm change in tracking accuracy during pod occlusions.

- AlignRT can offer sub-0.4 mm tracking of a head phantom under all clinical configurations, including all noncoplanar couch angles, during pod occlusions, and for all clinically realistic isocenter locations.

REFERENCES

1. Smith N, Meir I, Hale G, Howe R. Image processing system for use with a patient positioning device. 2003; Patent number EP1529264 (Vision RT Ltd.).

2. Chang AJ, Zhao H, Wahab SH, et al. Video surface image guidance for external beam partial breast irradiation. *Pract Radiat Oncol.* 2012;2(2):97–105.

3. Batin E, Depauw N, MacDonald S, Lu HM. Can surface imaging improve the patient setup for proton postmastectomy chest wall irradiation? *Pract Radiat Oncol.* 2016;6(6):e235–e241

4. Gierga DP, Turcotte JC, Tong LW, Chen YL, DeLaney TF. Analysis of setup uncertainties for extremity sarcoma patients using surface imaging. *Pract Radiat Oncol.* 2014;4(4):261–266.

5. Wiant DB, Wentworth S, Maurer JM, Vanderstraeten CL, Terrell JA, Sintay BJ. Surface imaging-based analysis of intrafraction motion for breast radiotherapy patients. *J Appl Clin Med Phys.* 2014;15(6):4957.

6. Li G, Ballangrud A, Kuo LC, et al. Motion monitoring for cranial frameless stereotactic radiosurgery using video-based three-dimensional optical surface imaging. *Med Phys.* 2011;38(7):3981–3994.

7. Smith N, Meir I, Hale G, et al. Real-time 3D surface imaging for patient positioning in radiotherapy. *Int J Radiat Oncol Biol Phys.* 2003;57(2):S187.

8. Wen N, Snyder KC, Scheib SG, et al. Technical Note: evaluation of the systematic accuracy of a frameless, multiple image modality guided, linear accelerator based stereotactic radiosurgery system. *Med Phys.* 2016;43(5):2527.

9. Davies ER. *Machine Vision: Theory, Algorithms, Practicalities.* 3rd ed. Amsterdam, the Netherlands: Elsevier; 2005.

10. Besl PJ, McKay ND. A method for registration of 3-D shapes. *IEEE Trans Pattern Anal Mach Intell.* 1992;14(2):239–256.

11. Pham NL, Reddy PV, Murphy JD, et al. Frameless, real-time, surface imaging-guided radiosurgery: update on clinical outcomes for brain metastases. *Transl Cancer Res.* 2014;3(4):351–357.

12. Gopan O, Wu Q. Evaluation of the accuracy of a 3D surface imaging system for patient setup in head and neck cancer radiotherapy. *Int J Radiat Oncol Biol Phys.* 2012;84(2):547–552.

13. Alderliesten T, Sonke J-J, Betgen A, van Vliet-Vroegindeweij C, Remeijer P. 3D surface imaging for monitoring intrafraction motion in frameless stereotactic body radiotherapy of lung cancer. *Radiother Oncol.* 2012;105(2):155–160.

14. Shah AP, Dvorak T, Curry MS, Buchholz DJ, Meeks SL. Clinical evaluation of interfractional variations for whole breast radiotherapy using 3-dimensional surface imaging. *Pract Radiat Oncol.* 2013;3(1):16–25.

15. Bartoncini S, Fiandra C, Ruo Redda MG, Allis S, Munoz F, Ricardi U. Target registration errors with surface imaging system in conformal radiotherapy for prostate cancer: study on 19 patients. *Radiol Med.* 2012;117(8):1419–1428.

16. Rwigema J-CM, Lamiman K, Reznik RS, Lee NJH, Olch A, Wong KK. Palliative radiation therapy for superior vena cava syndrome in metastatic Wilms tumor using 10XFFF and 3D surface imaging to avoid anesthesia in a pediatric patient; a teaching case. *Adv Radiat Oncol.* 2017;2(1):101–104. doi:10.1016/j.adro.2016.12.007.

17. Thomas EM, Popple RA, Wu X, et al. Comparison of plan quality and delivery time between volumetric arc therapy (RapidArc) and Gamma Knife radiosurgery for multiple cranial metastases. *Neurosurgery.* 2014;75(4):409–417; discussion 417–408.

18. Paxton AB, Manger RP, Pawlicki T, Kim G-Y. Evaluation of a surface imaging system's isocenter calibration methods. *J Appl Clin Med Phys.* 2017;18(2):85–91.

19. Mancosu P, Fogliata A, Stravato A, Tomatis S, Cozzi L, Scorsetti M. Accuracy evaluation of the optical surface monitoring system on EDGE linear accelerator in a phantom study. *Med Dosim.* 2016;41(2):173–179.

20. Willoughby T, Lehmann J, Bencomo JA, et al. Quality assurance for non-radiographic radiotherapy localization and positioning systems: report of Task Group 147. *Med Phys.* 2012;39(4):1728–1747.

21. Waghorn BJ. Advanced Camera Optimization, White Paper Issue 1.0. 2018;1016-0238.

22. Wiant D, Liu H, Hayes TL, Shang Q, Mutic S, Sintay B. Direct comparison between surface imaging and orthogonal radiographic imaging for SRS localization in phantom. *J Appl Clin Med Phys.* 2018.

23. Covington EL, Fiveash JB, Wu X, et al. Optical surface guidance for submillimeter monitoring of patient position during frameless stereotactic radiotherapy. *J Appl Clin Med Phys.* 2019;20(6):91–98.

CHAPTER 6

Commissioning and Routine Quality Assurance of the C-RAD Catalyst™ and Sentinel™ 4D-CT Systems

Alonso N. Gutierrez and Dennis N. Stanley

CONTENTS

Editor's Note: SGRT vendors have a variety of technologies in their product portfolio and some of those technologies are integrated into their SGRT systems to different degrees. Since this textbook is focused on SGRT, the editors have asked the authors to limit their chapter content specifically to surface image guidance for patient setup and inter- and intra-fraction patient monitoring.

6.1 INTRODUCTION

Optical-based, surface imaging is a noninvasive, nonradiographic form of localization that can be used as a tool to perform surface-guided radiotherapy (SGRT) for a variety of disease sites.[1] One of the commercially available, optical-based surface image guidance systems is the Catalyst system (C-RAD AB, Uppsala, Sweden), which is installed in treatment delivery rooms and which works in conjunction with the laser-based, Sentinel system (C-RAD AB, Uppsala, Sweden), which is installed in computed tomography (CT) simulator rooms. Sentinel allows the acquisition of surface information during simulation and also facilitates the acquisition of four-dimensional (4D)-CT for respiratory motion studies. As the advantages of SGRT have been outlined in previous chapters, the Catalyst surface imaging system enables additional capabilities on the treatment unit that help improve initial patient setup, monitor the patient's position during treatment delivery, and track the patient's surface as a breathing surrogate during gated radiotherapy delivery techniques. The Catalyst has been commercially available for a number of years and has been widely used on various radiotherapy treatment units including conventional linear accelerators,[2,3] helical tomotherapy units,[4] and proton therapy delivery units.[5] Moreover, the Catalyst and Sentinel

4D-CT systems have been used to treat a number of disease sites including general breast treatments,[2,4] deep inspiration breath hold (DIBH),[3,6,7] head and neck,[8] pelvis,[2,9,10] and extremities.[2]

6.2 C-RAD CATALYST™ AND SENTINEL™ 4D-CT SYSTEM AND WORKFLOW OVERVIEW

6.2.1 C-RAD Catalyst™ and Sentinel™ 4D-CT System Hardware

The Catalyst patient positioning system is an optical-based imaging system that uses the principals of photogrammetry to generate localized surface images. The complete system is composed of two major hardware components: Catalyst, which is located in the radiation treatment delivery vault, and Sentinel 4D-CT, which is located in the CT simulation room. Within each ceiling-mounted Catalyst camera unit, there is both a projector that projects a striped light pattern and a camera to detect the reflected light from the patient surface—see Figure 6.1. Using the reflected light from the patient, the Catalyst system is able to reconstruct the patient surface by correlating the expected light pattern versus the measured reflected light pattern that has been perturbed by the patient. Typically, treatment rooms are outfitted with one to four units depending on the radiation delivery system geometry. For photon linac-based installations, one unit (centered

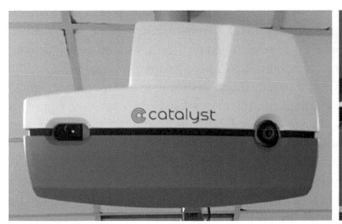

FIGURE 6.1 (*Left*) A ceiling-mounted Catalyst™ camera system is shown. Within the unit housing, a light projector and detecting camera can be seen. (*Right*) A ceiling-mounted Sentinel™ 4D-CT system is shown. Within the unit housing, the scanning laser and CCD camera can be seen.

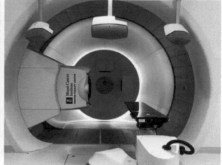

FIGURE 6.2 (*Left*) Three-unit Catalyst™ HD system (C-RAD AB, Uppsala, Sweden) installed on an Elekta Versa™ HD radiotherapy delivery unit. (*Right*) A three-unit Catalyst™ PT system installed on an IBA Proteus Plus proton gantry system.

at the foot of the treatment table) or three units (displaced 120° apart from one another) are commonly installed—the three-unit Catalyst camera configuration is referred to as the Catalyst HD system. For a helical tomotherapy configuration, a single camera is placed at the foot of the treatment table although other configurations are possible (see Chapter 24). For proton therapy applications, it is often a challenge to mount cameras inside the plane of gantry rotation and because of this, units are mounted in front of the gantry with angles between cameras typically less than 90°. Figure 6.2 illustrates a Catalyst HD system installed on a photon delivery unit and a Catalyst PT system installed on a gantry-based proton delivery unit. SGRT for proton therapy is discussed further in Chapter 23.

The Sentinel 4D-CT system functions using a scanning laser light in conjunction with a calibrated camera in a fixed geometry—see Figure 6.1. As the laser scans the patient's surface while on the couch of the CT simulator, a synchronized camera detects the laser reflection on the patient surface and through postprocessing determines the height of the reflected surface. Using this process, the system can generate the surface as the laser scans across the patient with an acquisition time of 3–6 s. Although the complete surface acquisition is relatively slow, the Sentinel 4D-CT may be used to acquire one fixed position at a high temporal frequency (15 Hz) by truncating the scan volume and continuously rescanning a shorter selected area—this region of interest serves as a surrogate for patient breathing and is used to determine the respiratory signal for 4D-CT binning and prospective gated treatment delivery techniques. Overall, the

Sentinel 4D-CT system serves to acquire both surface images to serve as a reference image for gated and nongated treatments and a tool to acquire the breathing trace.

For both systems, the field of view (FOV) for the cameras are sufficiently large to encompass the majority of patients and patient setups. For each camera, the native hardware is able to acquire images at a high acquisition frequency but due to postprocessing of images and image registration, the visualized frame rates range between 8 and 24 Hz depending on certain patient specific parameters. Moreover, the positioning accuracy for system and patient positioning has been shown to be submillimetric.[11]

6.2.2 C-RAD SGRT Workflow Overview

Both the Catalyst and Sentinel 4D-CT systems play a key role in the hardware components of the C-RAD SGRT solution; however, there are other hardware (visual feedback goggles and facial recognition camera) and software components that wholly work together to generate a comprehensive SGRT solution. In particular, the C-RAD SGRT solution is composed of various software modules that aid in patient identification (cPatient™), positioning (cPosition™), monitoring (cMotion™), and gating (cRespiration™).

For patient identification, the cPatient™ module in conjunction with the facial recognition camera can be used to register the patient in the C-RAD c4D™ software environment—which consequently also interfaces with the oncology information system (OIS). After initial patient facial registration, the patient is then identified for each subsequent treatment fraction using the facial recognition camera. The cPatient™ module is designed for patient safety and overall treatment efficiency by automatically opening up the patient in the cPosition™ module and facilitating initial positioning. With the patient on the treatment table and the cPosition™ module activated, the Catalyst cameras are used to acquire live surface images. These images are then processed, and a 6 degrees-of-freedom (6DOF) matching software registers the real-time images to a predefined reference image using a deformable registration algorithm. The deformable registration algorithm calculates the displacement of the isocenter due to the amount of deformation on the patient surface.[12] As an attribute of the registration algorithm, the accuracy of the isocenter registration is dependent on the distance between the isocenter and the patient surface—that is, an isocenter closer to the patient surface is likely to correlate better with the surface deformation due to proximity. Using this methodology, the C-RAD system determines the required 6DOF correction vector to properly align the patient to the radiation delivery unit isocenter. The Catalyst system also projects color-coded light onto the regions of the patient surface that are outside of predetermined, patient-specific tolerances to alert therapists of gross misalignments such as those potentially found in arm positioning during breast treatments.

Once in the correct treatment position, the patient is then actively monitored by the Catalyst cameras and real-time 6DOF correction vectors are displayed in the cMotion™ module. In this module, a new reference image is generated at the module launch, and correction vectors are re-zeroed. This happens each time the cMotion™ module is launched, or the module is refreshed. After which, if the patient moves outside of the patient specified tolerance, the cMotion™ module alerts the treating operator through an audible alarm and can also interrupt the beam delivery if the beam hold interface is engaged.

For gated deliveries, the cRespiration™ module is used to help guide the radiation delivery by synchronizing the beam-on signal of the treatment unit (if the interface is purchased) to the correct surface position of the patient. The cRespiration™ module uses a small, patient-specific area of the patient surface for respiratory amplitude signal tracking. This patient-specific tracking area is first determined at CT simulation using the Sentinel 4D-CT and then propagated to the Catalyst system for patient treatment. Prior to radiation delivery and during CT simulation, the patient is coached using either audio instructions or video via goggles. For DIBH, a reference-free breathing image can be acquired using the Sentinel 4D-CT system and used to initially position the patient in the treatment room in a free breathing state.

6.3 ACCEPTANCE AND COMMISSIONING OF THE CATALYST™ AND SENTINEL™ 4D-CT SYSTEMS

The American Association of Physicists in Medicine (AAPM) released Task Group Report 147: Quality assurance for nonradiographic radiotherapy localization and positioning systems in 2012 and was charged with making recommendations about the use of nonradiographic methods of localization, specifically radiofrequency, infrared, laser, and video-based patient localization and monitoring systems.[13] It did not, however, address specific tests and methodologies needed for the latest generation of surface imaging systems. In keeping with the principals set forth by TG-147, the following sections describe the C-RAD specific acceptance testing requirements as well as a methodology for commissioning of the Catalyst and Sentinel systems.

6.3.1 Acceptance Testing and Integration of Peripheral Equipment

6.3.1.1 Acceptance Testing—Vendor Specific

Acceptance testing of the C-RAD Catalyst and Sentinel 4D-CT imaging systems should quantifiably evaluate the static and dynamic localization accuracy, image reproducibility, proper system functionality, and safe overall operation of each hardware component including any peripheral items such as visualization goggles and facial recognition cameras. C-RAD should be able to demonstrate the correct configuration and functionality of all these systems.

6.3.1.2 Communication and Integration with Record and Verify System, Treatment Unit, and CT Simulator

To test the communication with the OIS, a series of phantom tests should be conducted for both the Catalyst and Sentinel 4D-CT. Phantom patients can be created on the C-RAD c4D™ database to verify direct communication between the OIS and should include the following: transfer of data with patients in different clinical orientations (prone, supine, feet-first, etc.), correct coordinate system, correct isocenter selection, correct isocenter identification with multiple isocenters, and correct field parameters.

The integration with the treatment unit and CT simulator should be performed for both systems using the same phantom patients and should include the following: transfer of data—patients in different clinical orientations (prone, supine, feet-first, etc.), correct coordinate system, correct scanning type (prospective or retrospective), correct export of files, correct field parameters, and correct units. As a clinical example, a Catalyst system paired with an Elekta Versa HD via the Response™ gating control system should be verified for communication by testing the beam hold for several procedures: breath-hold radiation delivery techniques (prospective gating), patient displacement during positioning and motion monitoring (cMotion™) and patient respiratory monitoring (cRespiration™). Similarly, a Sentinel 4D-CT paired with a CT simulator should ensure communication via the appropriate control gating box with the CT simulator. The control gating box is the sole communication method between the Sentinel 4D-CT system and the CT simulator. The control gating box will control the CT scanner for prospective gating (DIBH, etc.) and retrospective gating (4D-CT simulation scan).

6.3.1.3 Determination of Localization FOV

The localization FOV is quantified during the acceptance testing, and the camera positions are adjusted as necessary to ensure minimum specifications. Due to the large FOV of the Catalyst system, the furthest extent of your clinical treatment space (commonly defined by the maximum movement of the couch in each direction) plus an additional margin depending on the limitation of the treatment couch should be evaluated. Doing so should ensure that for any clinical scenario, the system's ability to visualize the patient is known. Practically, this is commonly performed using either the C-RAD Catalyst or Sentinel 4D-CT daily QA phantoms as seen in Figure 6.3. The respective daily QA device is positioned at the isocenter and systemically translated from the origin in six directions until it can no longer be visualized by the imaging system. Table 6.1

FIGURE 6.3 The Sentinel™ 4D-CT daily QA phantom (*left*). The Catalyst daily QA phantom (*right*).

TABLE 6.1 Determination of the Field of View of a Sentinel™ 4D-CT System and a Three-camera Catalyst™ HD Imaging System

	Sentinel™: Distance (mm)	Catalyst™ HD: Distance (mm)
Lateral X_{min}	400	500
Lateral X_{max}	400	550
Longitudinal Y_{min}	850	360
Longitudinal Y_{max}	580	730
Vertical Z_{min}	210	380
Vertical Z_{max}	300	450
Total lateral	800	1050
Total longitudinal	1430	1090
Total vertical	510	830

displays typical FOV values for a Sentinel 4D-CT and three camera Catalyst. It should be noted that differences in vault setup, vault size, couch size, camera placement, and linear accelerator type will determine the FOV.

6.3.2 Commissioning of the Catalyst™ and Sentinel™ 4D-CT Systems

6.3.2.1 System Drift and Static Spatial Reproducibility

The temporal system drift and static spatial reproducibility should be performed as one of the initial evaluation metrics when commission ing both the Catalyst and Sentinel 4D-CT imaging systems. AAPM TG-147 recommends monitoring and recording a test pattern device at initial startup of the equipment for at least 90 min or until sufficient stability is achieved due to system thermal drift vulnerability.[13] To evaluate the thermal system drift, the system can be initialized from a cold start where the entire system has been powered down. After thermal stability is achieved, the static spatial reproducibility can be evaluated by monitoring the position of a static device for a period. These tests are also recommended to be performed at least annually or after equipment changes.

In practice, the evaluation of the system drift and static spatial reproducibility can be quantified using the daily QA phantom. This can be performed by shutting down the system for at least 24 h, so it equilibrates with the ambient room temperature. Before powering up the system, the daily QA phantom can be placed at isocenter using the lasers as reference to ensure daily QA phantom has not moved. Immediately upon powering up the system, isocenter checks can be performed in the Daily QA software interface, and the results are recorded for each camera and in each direction. The process can be repeated every 5 min, recording the absolute positions for each direction until an equilibrium is achieved—see Figure 6.4. Moreover, it might be beneficial to quantify if the amount of ambient room light influences the stability of the imaging systems. This can be performed using a low room light setting and full room light setting as seen in Table 6.2. As an example, performing these measurements, the static spatial reproducibility of the Catalyst HD has been found to be 0.05 ± 0.03 mm. This process is applicable for both to the Catalyst and Sentinel 4D-CT systems.

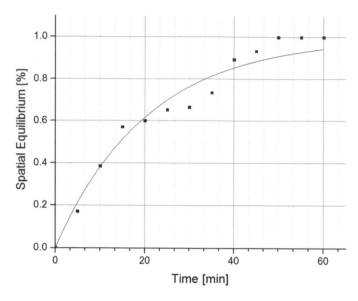

FIGURE 6.4 The temporal 3D spatial system drift due to thermal drift for a three-camera Catalyst™ HD system expressed as % deviation from spatial equilibrium versus time from power-on. The magnitude of displacement was found to be 0.7 mm. The system appears to stabilize within 45 min from being powered on.

TABLE 6.2 Static Spatial Reproducibility of a Catalyst™ HD System in Both Low and Full Room Light Conditions

	Low Light (12.65 lx)			Full Light (236.25 lx)		
	X	Y	Z	X	Y	Z
Lt Camera						
σ (mm)	0.04	0.03	0.03	0.01	0.02	0.03
Range (mm)	0.14	0.10	0.10	0.05	0.05	0.15
Mid. Camera						
σ (mm)	0.02	0.02	0.02	0.02	0.02	0.02
Range (mm)	0.07	0.06	0.06	0.07	0.08	0.06
Rt Camera						
σ (mm)	0.03	0.02	0.03	0.01	0.03	0.01
Range (mm)	0.11	0.05	0.10	0.05	0.12	0.05

6.3.2.2 Static Localization Accuracy

The static localization accuracy evaluation can be performed with any phantom that can be imaged with the Catalyst system and registered for a reference image set using radiographic imaging techniques such as planar kV/MV or volumetric cone-beam CT (CBCT) imaging techniques. Ideally, the phantom should be anthropomorphic as it would mimic patient geometry. The goal of this localization accuracy evaluation is to quantify congruence between the isocenters of the Catalyst cameras and treatment unit imaging systems. The localization accuracy should be within 2 mm of isocenter for standard dose fractionation and 1 mm for stereotactic treatments per TG-147 recommendations. Furthermore, the ability to accurately use the localization system to translate a target a known distance or degree should be evaluated during initial commissioning. Recommended accuracy is within 2 mm over a 10-cm range in all directions for standard dose fractionation and within 1 mm over a 10-cm range for stereotactic treatments. A full end-to-end test, from CT simulation through treatment delivery, is recommended upon initial commissioning and annually thereafter.

In practice, the static localization accuracy of the Catalyst can be quantified using an anthropomorphic head or pelvic phantom, positioned at isocenter with the Catalyst system and verified by the planar/volumetric imaging system of the treatment unit. The phantom can be shifted from isocenter in 1-cm increments (up to 10 cm) using the treatment couch. After each shift, a planar/volumetric and Catalyst image can be acquired, and deviations recorded. Figure 6.5 shows an example of these differences using a pelvic phantom between a Catalyst HD system and Elekta Versa™ HD using a HexaPod™ treatment couch. Figure 6.6 shows example differences using a pelvic phantom between a Catalyst HD system and Elekta Versa™ HD using XVI™ kV-CBCT imaging system.

6.3.2.3 Dynamic Localization Accuracy

Both the Catalyst and Sentinel 4D-CT imaging systems must demonstrate an end-to-end accuracy commensurate with TG-142 recommendation when operated in dynamic mode since they are designed to do dynamic tracking for gated delivery. A full end-to-end test from CT simulation through treatment delivery should be performed at

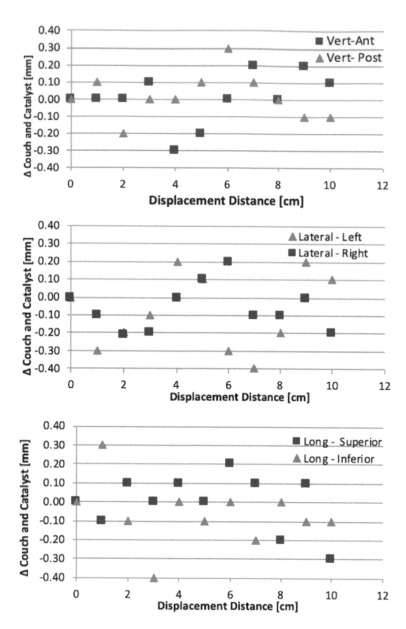

FIGURE 6.5 Static localization displacement accuracy between a Catalyst™ HD and a HexaPod™ treatment couch in the vertical (top), lateral (middle), and longitudinal (bottom) directions.

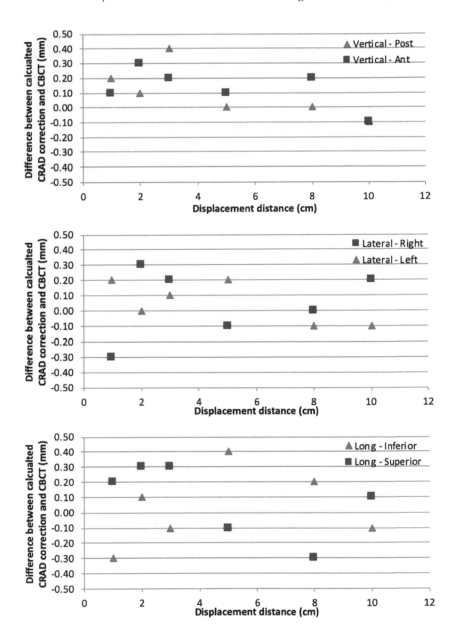

FIGURE 6.6   Static localization displacement accuracy between a Catalyst™ HD and a Elekta XVI™ kV-CBCT in the vertical (top), lateral (middle), and longitudinal (bottom) directions.

commissioning and annually thereafter to assess the dynamic localization accuracy. These tests should utilize phantoms that allow for evaluation of dynamic spatial and temporal accuracy and dynamic radiation delivery capabilities.

In practice, the dynamic spatial and temporal accuracy of the Catalyst system is challenging to quantify as phantoms are not readily available; however, both the dynamic spatial and temporal accuracy can be evaluated using a phantom mounted on the controller arm of a water tank scanning system. Given the good positional accuracy of a water tank controller arm (~0.1 mm), this phantom setup may be used to accurately verify the 3D position of the phantom moving dynamically. Figure 6.7 shows an example of an in-house phantom used to assess the dynamic spatial and temporal accuracy of a Catalyst HD system using

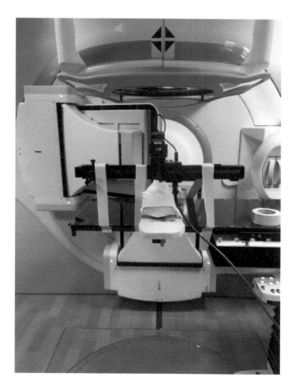

FIGURE 6.7 A foam in-house phantom mounted onto the controller arms of a water tank scanning system used to accurately introduce dynamic shifts for a defined period to evaluate the dynamic spatial and temporal accuracy of the Catalyst™ HD system.

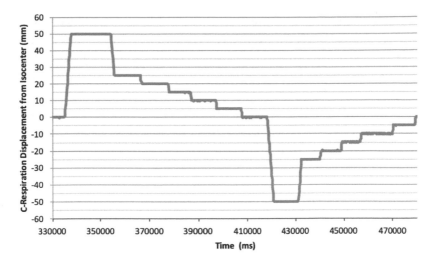

FIGURE 6.8   A sample cRespiration™ breathing signal trace for a dynamic phantom using a foam phantom mounted onto the controller arms of a water tank scanning system.

a water tank controller arm (Standard Imaging, Inc., Middleton, WI). Figure 6.8 shows the temporal breathing signal trace and agreement between the absolute position of the phantom location and the respiration position determined by the cRespiration module of the Catalyst HD system. In this example, the dynamic spatial accuracy of the Catalyst HD system was 0.2 ± 0.1 mm.

6.3.2.4 Dynamic Radiation Delivery—Gating

Similar to assessing the dynamic spatial and temporal accuracy of the surface imaging system, it is also important to validate the dynamic radiation delivery accuracy for the various types of gating techniques to be used with the surface imaging system. As with any system that is capable of automatically controlling the radiation beam of a treatment unit, it is important to have a complete understanding of how the interface functions and the delay or latency in response time of both the beamenable and beam-disable functions of the system. The amount of latency tends to be of a larger concern when free breathing gating is used as opposed to breath-hold gating due to the increased frequency of beam holds with free breathing gating—especially for patient with short breathing periods. Quantifying the exact latency time of a surface imaging system and associated treatment delivery system is difficult in practice, but studies have shown this time to be on the order of 800 ms for a Catalyst/Elekta linear accelerator combination.[14] In clinical practice, the exact latency is not always quantified and will vary based on the surface imaging system and the treatment delivery unit. However, an end-to-end system analysis with a motion phantom should be performed to ensure that the correct dose to a phantom is delivered accurately and precisely for dynamic deliveries of all gating techniques to be used clinically.

6.3.2.5 Sentinel™ Retrospective Gating (4D-CT)

As with the Catalyst system, the Sentinel 4D-CT system must also be assessed for its dynamic temporal and spatial accuracy. Since the Sentinel 4D-CT solely tracking in one dimension (amplitude based), a respiratory motion phantom with a moving platform of variable amplitudes can be used to quantify the spatial accuracy of the Sentinel 4D-CT. Moreover, the motion phantom can be used to quantify the accuracy of the 4D-CT reconstruction, and the target displacement as quantified by a 4D-CT reconstruction can be compared to the actual phantom target excursion. If the Sentinel 4D-CT system is replacing another system such as the Varian RPM (Varian Medical Systems, Palo Alto, CA), it may be prudent to acquire the 4D-CT breathing signal with both systems and reconstruct the 4D-CT using each respiratory breathing file to ensure agreement.

6.4 RECOMMENDED ROUTINE QA

An on-going routine QA program is recommended and crucial to assess the proper operation of any surface imaging system and to ensure the safe and accurate administration of radiation dose to patients, especially those using surface imaging for gating purposes. The AAPM TG-147 report established a schedule of parameter testing that monitors the safety, accuracy, precision, sensitivity, and reliability of surface imaging systems at daily, monthly, and annual intervals. The localization accuracy and reproducibility of a system used for patient positioning must be consistent with the recommendations of AAPM TG-142, that is, within 2 mm for conventional treatments and 1 mm for stereotactic radiosurgery (SRS) and stereotactic body radiation therapy (SBRT). Daily QA should check the integrity of the camera installation(s), geometric and optical calibration of the camera system and static localization accuracy (e.g., a phantom offset test as part of daily image-guided radiation therapy (IGRT testing)). C-RAD provides a daily check device consisting of three spheres for surface detection, which is aligned to the linac light field and in-room lasers. The acquired surface image of the daily check phantom is compared against a reference image to verify static localization accuracy. Monthly QA should verify the gating interlock and system latency, if the SGRT system is used for gated treatments. A localization test should also be performed using a hidden or known target (e.g., using a surface-imaging compatible phantom for a Winston–Lutz radiation-mechanical iso-center test). C-RAD recommends the use of the Penta-Guide phantom (Modus Medical Systems, London, Canada) for conventional and SRS/SBRT localization QA. Dynamic localization accuracy should also be evaluated by shifting the couch by known offsets and comparing against surface imaging-indicated offsets. Annual QA should evaluate camera settings and thermal drift against baseline values and verify that all system functionality and performance are consistent with the baselines established during system commissioning. A test of data transfer capabilities through performance of an end-to-end test is also recommended.

6.5 SUMMARY

An overview of the acceptance testing, clinical commissioning, and on-going quality assurance aspects for the C-RAD surface imaging solution was presented. As described in this chapter, the C-RAD surface imaging solution is composed of multiple hardware components, namely the Catalyst and Sentinel 4D-CT imaging systems, and various software modules (cPosition™, cMotion™, etc.) that interface with different radiation delivery systems and OISs. The use of these components and software modules in various configurations enables different surface imaging work-flows—stemming from a simple patient setup workflow to a respiratory-gated treatment workflow. Clinically, the Catalyst surface imaging system has been shown to exhibit submillimeter accuracy in both phantom and patient positioning studies. However, as with any imaging system used to guide radiation delivery, a comprehensive, robust QA program with on-going QA tasks is necessary to ensure proper functionality and accurate radiation delivery. For the Catalyst and Sentinel 4D-CT systems, recommended testing tasks and frequency are outlined in AAPM TG-147.

KEY POINTS

- C-RAD provides a comprehensive surface imaging solution that is composed of multiple hardware (Catalyst and Sentinel 4D-CT) and software modules (cPosition™, cMotion™, etc.) that interface with different radiation delivery systems and OISs to improve the radiation delivery accuracy and treatment efficiency as well as enable gated delivery capabilities.

- The Catalyst system has been shown to clinically improve patient positioning accuracy and precision for general breast treatments, DIBH treatments, pelvic treatments, and treatment of extremities.

- Both acceptance testing and commissioning of the Catalyst and Sentinel 4D-CT systems can be performed following guidelines recommended by AAPM TG-147 and with the use of commercially available anthropomorphic and dynamic motion phantoms.

- On-going QA tasks and recommended frequencies are outlined in AAPM TG-147 for surface imaging systems such as those available from C-RAD.

REFERENCES

1. Hoisak JDP, Pawlicki T. The role of optical surface imaging systems in radiation therapy. *Semin Radiat Oncol.* 2018;28(3):185–193. doi:10.1016/j.semradonc.2018.02.003.
2. Stanley DN, McConnell KA, Kirby N, Gutiérrez AN, Papanikolaou N, Rasmussen K. Comparison of initial patient setup accuracy between surface imaging and three point localization: a retrospective analysis. *J Appl Clin Med Phys.* 2017;18(6):58–61. doi:10.1002/acm2.12183.
3. Kügele M, Edvardsson A, Berg L, Alkner S, Andersson Ljus C, Ceberg S. Dosimetric effects of intrafractional isocenter variation during deep inspiration breath-hold for breast cancer patients using surface-guided radiotherapy. *J Appl Clin Med Phys.* 2018;19(1):25–38. doi:10.1002/acm2.12214.
4. Crop F, Pasquier D, Baczkiewic A, et al. Surface imaging, laser positioning or volumetric imaging for breast cancer with nodal involvement treated by helical TomoTherapy. *J Appl Clin Med Phys.* 2016;17(5):200–211. doi:10.1120/jacmp.v17i5.6041.
5. Rana S, Bennouna J, Samuel EJJ, Gutierrez AN. Development and long-term stability of a comprehensive daily QA program for a modern pencil beam scanning (PBS) proton therapy delivery system. *J Appl Clin Med Phys.* 2019;20(4):29–44. doi:10.1002/acm2.12556.

6. Kalet AM, Cao N, Smith WP, et al. Accuracy and stability of deep inspiration breath hold in gated breast radiotherapy—a comparison of two tracking and guidance systems. *Phys Med.* 2019;60:174–181. doi:10.1016/j.ejmp.2019.03.025.

7. Schönecker S, Walter F, Freislederer P, et al. Treatment planning and evaluation of gated radiotherapy in left-sided breast cancer patients using the Catalyst™/Sentinel™ system for deep inspiration breath-hold (DIBH). *Radiat Oncol.* 2016;11(1):143. doi:10.1186/s13014-016-0716-5.

8. Cho H-L, Park E-T, Kim J-Y, et al. Evaluation of radiotherapy setup accuracy for head and neck cancer using a 3-D surface imaging system. *J Inst.* 2013;8(11):T11002. doi:10.1088/1748-0221/8/11/T11002.

9. Walter F, Freislederer P, Belka C, Heinz C, Söhn M, Roeder F. Evaluation of daily patient positioning for radiotherapy with a commercial 3D surface-imaging system (Catalyst™). *Radiat Oncol.* 2016;11:154. doi:10.1186/s13014-016-0728-1.

10. Carl G, Reitz D, Schönecker S, et al. Optical surface scanning for patient positioning in radiation therapy: a prospective analysis of 1902 fractions. *Technol Cancer Res Treat.* 2018;17:1533033818806002. doi:10.1177/1533033818806002.

11. Stieler F, Wenz F, Shi M, Lohr F. A novel surface imaging system for patient positioning and surveillance during radiotherapy. *Strahlenther Onkol.* 2013;189(11):938–944. doi:10.1007/s00066-013-0441-z.

12. Li H, Summer RW, Pauly M. Global correspondence optimization for non-rigid registration of depth scans. *Eurographics Symposium on Geometry Processing 2008.* 2008;27(5):1–10.

13. Willoughby T, Lehmann J, Bencomo JA, et al. Quality assurance for non-radiographic radiotherapy localization and positioning systems: report of Task Group 147. *Med Phys.* 2012;39(4):1728–1747. doi:10.1118/1.3681967.

14. Freislederer P, Reiner M, Hoischen W, et al. Characteristics of gated treatment using an optical surface imaging and gating system on an Elekta linac. *Radiat Oncol.* 2015;10:68. doi:10.1186/s13014-015-0376-x.

CHAPTER 7

Commissioning and Routine Quality Assurance of the Varian IDENTIFY™ System

Hui Zhao and Adam B. Paxton

CONTENTS

Editor's Note: SGRT vendors have a variety of technologies in their product portfolio and some of those technologies are integrated into their SGRT systems to different degrees. Since this textbook is focused on SGRT, the editors have asked the authors to limit their chapter content specifically to surface image guidance for patient setup and inter- and intra-fraction patient monitoring.

7.1 INTRODUCTION

IDENTIFY (Varian Medical Systems, Palo Alto, CA) is a surface imaging system intended for surface-guided radiation therapy (SGRT). The IDENTIFY system also combines SGRT capabilities with biometric patient identification and verification of the intended treatment accessories through radio-frequency identification (RFID). IDENTIFY provides two modes of surface guidance in the RT workflow. The first is an orthopedic setup verification mode that utilizes a large field-of-view (FOV) camera to visualize the position and orientation of the entire patient in the loading position of the treatment couch (i.e., the couch is in the position where the patient first lays down, retracted all the way down and out). This mode of SGRT is intended to position patients in the same posture they were in at the time of simulation using patient self-positioning according to augmented reality (i.e., color-coded video feedback of the agreement of the real-time patient surface with a reference surface from simulation). The patient is able to see the reference image via a ceiling-mounted monitor. The second mode of SGRT is patient body surface monitoring that can be used during setup, imaging, and treatment. In this mode, the FOV is smaller and focused at the isocenter of the treatment machine. This mode also provides the user with translational and rotational offsets of the real-time surface relative to a reference surface. The reference surface either can be from the body (or external) surface from a treatment planning system (TPS) or captured by IDENTIFY in the treatment room. Regions of interest (ROIs) can be set to focus on a particular area of the body. Figure 7.1 shows the various components of the IDENTIFY system. A full technical description of the components and features of IDENTIFY is given in Chapter 4 of this book.

The first clinical installation of IDENTIFY was at the University of Utah in 2016. At the time, IDENTIFY was a product of HumediQ Global GmbH and was later acquired by Varian in 2018. This chapter describes the experience of commissioning IDENTIFY for clinical application at the University of Utah. Moreover, clinical workflows that were developed are described. Finally, quality assurance (QA) processes including calibrations and routine QA are described.

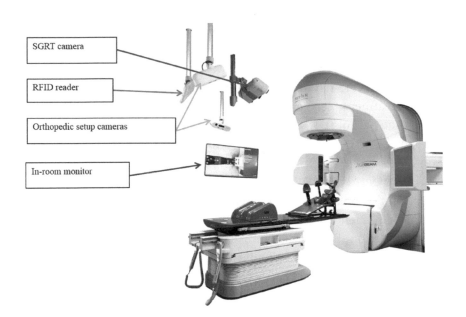

FIGURE 7.1   Some of the hardware components of the IDENTIFY system.

7.2 COMMISSIONING

The clinical commissioning of the IDENTIFY system (Version 2.0) installed at the University of Utah included three major procedures: a functionality test, an SGRT system consistency and accuracy test, and an end-to-end test. Because this was the first clinical installation of IDENTIFY, some of the commissioning tasks overlapped with what may typically be considered part of an acceptance procedure.

7.2.1 Functionality Tests

IDENTIFY system functionality tests were divided into five tests directly related to SGRT: system setup, patient surface information management, handheld device management, system FOV management, and security management.

7.2.1.1 System Setup

At the time of clinical commissioning at the University of Utah, the IDENTIFY system consisted of several SGRT-related hardware components including two setup cameras for patient treatment accessory devices and patient orthopedic setup, three SGRT cameras for surface monitoring, one in-vault monitor for patient self-adjustment, two handheld devices, and a system computer. Some of these components are shown in Figure 7.1. In addition to the hardware components, the IDENTIFY Planning Tool software was installed on a computer other than the IDENTIFY system computer. This software is used for importing patient treatment plans and structure sets and defining ROIs. The integrity of each part of IDENTIFY system was checked, and the whole system was setup.

7.2.1.2 Patient Surface Information Management

The patient orthopedic setup reference image can be taken while patients are in their intended treatment position during computed tomography (CT) simulation. During patient treatment, this orthopedic setup reference is loaded for patient self-adjustment at the couch loading position (Figure 7.2). Again, if an IDENTIFY system is not available in the CT simulation room, the reference image can be acquired in the treatment vault. In practice, this would be completed after the position of the patient has been verified to be correct, such as after imaging with cone-beam CT (CBCT). The verification of the functionality of the patient orthopedic

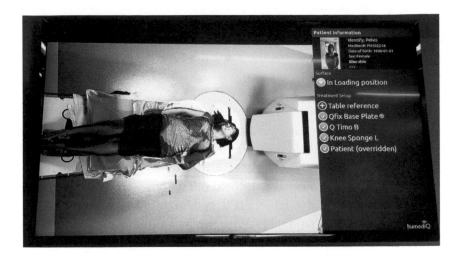

FIGURE 7.2   Photograph of the IDENTIFY orthopedic setup monitor screen with a healthy volunteer on the couch in the loading position. Augmented reality overlays colors on the volunteer indicating the alignment of body regions. White indicates alignment within a preset tolerance, blue indicates the region is too low, and red indicates the region is too high.

setup reference capture and transfer to the treatment vault was completed with a healthy volunteer. The ability to capture a reference in the treatment vault was also confirmed.

Patient surface matching at the treatment position can be performed by matching the real-time patient surface with the external body contour calculated by the TPS from the planning CT or with a reference surface captured by IDENTIFY. The functionality of being able to import the body surface from the TPS and capture a new reference surface with IDENTIFY was confirmed.

Imported and captured reference surfaces can be labeled according to their intended purpose. Four available labels for reference surface options were verified including: free breathing (FB) surface, FB bolus surface, deep inspiration breath hold (DIBH) surface, and DIBH bolus surface.

Intra-fractional patient motion can be monitored during imaging and treatment delivery, and the reference surface can be the external body contour from the TPS, a previously captured patient reference surface, or the current patient surface following IGRT shifts (which is not saved in the database, but only used for that treatment session). All three surface references were verified for intra-fractional motion monitoring.

The tolerance of patient motion can be individually set for all 6 degrees of freedom (6DOF). Once the tolerance is configured for a specific patient and treatment site, it remains the same until it is reset. This functionality was confirmed.

7.2.1.3 Handheld Device Management

The IDENTIFY system is operated at CT simulation and in the treatment vault via handheld devices. There are typically two handheld devices for each treatment vault, one inside the vault and one outside at the treatment console. All handheld operations were verified for their functionality during the IDENTIFY system commissioning. During a DIBH treatment, the handheld device can be mounted on the couch with a holder as a display for patient respiratory coaching. This functionality was also confirmed.

7.2.1.4 System FOV Management

The IDENTIFY system includes two sets of cameras: two orthopedic setup cameras and three SGRT cameras. Camera calibration needs to be completed and their intended FOV confirmed. Camera calibration includes two steps for each set of cameras: camera intrinsic parameter calibration and machine isocenter calibration. The calibration and confirmation of the FOV were performed. The specific processes for camera calibration are discussed later in this chapter.

7.2.1.5 System Security Management

The IDENTIFY system is password protected. One password is assigned to the system computer login. Multiple users can be assigned to the handheld device operation. Each user was assigned an account with a unique user name and passcode during IDENTIFY system commissioning. Two account groups were assigned, therapist and physicists. The different account rights for each group were confirmed.

7.2.2 Surface Imaging Consistency and Accuracy Test

The surface imaging consistency and accuracy was tested in four ways: a consistency check, a shift accuracy check, a rotation accuracy check, and a continuous recording stability check.

All checks were performed using an anthropomorphic pelvis phantom. For the consistency check, the pelvis phantom was set up on the treatment couch, and a reference surface image was captured. Ten separate instant phantom position measurements detected by IDENTIFY system were recorded and compared. The same test was performed at three different couch angles: 0°, 90°, and 270°. The consistency was within 0.1 mm for the three translational axes and 0.05° for the three rotational axes.

For the shift accuracy check, the pelvis phantom was set up on the treatment couch, and a reference surface image was captured. The treatment couch was shifted to eight different positions using a combination of translations. The magnitude of the translations ranged from 1 to 50 mm in each direction (i.e., shifts were 1 mm in vertical, longitudinal, and lateral directions to 50 mm in the vertical, longitudinal, and lateral directions). The direction of the shifts was alternated. The offsets detected by IDENTIFY were compared to the known treatment couch shifts. The same test was performed at three different couch angles: 0°, 45°, and 315°. The shifts reported by IDENTIFY were within 0.6 mm of the known shifts for all tests. The treatment couch used was on a Varian iX linac. The couch motion resolution was 1 mm, so the differences observed with IDENTIFY were within the resolution of the couch motion.

For the rotation accuracy check, the pelvis phantom was set up on the treatment couch, and a reference surface image was captured. The treatment couch was rotated to 15 different positions ranging from 0° to 75° in both the clockwise and counterclockwise directions. For each angle, the yaw detected by IDENTIFY system was recorded and compared to the known couch rotation. The IDENTIFY reported rotation was within 0.1° of the known couch rotation for all tested positions. IDENTIFY also has the ability to rotate the reference image to match the couch rotation. The reference rotation accuracy was verified to be within 0.6° for the full range of couch rotations from 90° to 270°. The treatment couch used was not able to move in 6DOF, so pitch and roll were not investigated.

For the continuous recording stability check, the pelvis phantom was set up on the treatment couch, and a reference surface image was captured. Twenty-five minutes of continuous IDENTIFY reported offsets were recorded after the cameras had reached a thermal equilibrium. The variation of translations was within 0.24 mm, and the variation of rotations was within 0.1°. The standard deviation of translations was within 0.12 mm, and the standard deviation of rotations was within 0.01°. Plots of the IDENTIFY reported translations and rotations versus time are shown in Figure 7.3.

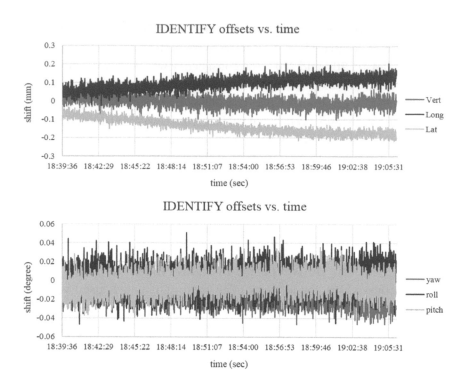

FIGURE 7.3   Plots of the IDENTIFY surface tracking system's indicated translations (top) and rotations (bottom) versus time of a stationary anthropomorphic pelvis phantom.

7.2.3 End-to-End Test

An end-to-end test was performed using an anthropomorphic pelvis phantom. The steps of the end-to-end test were as follows:

1. A mock patient was registered in the IDENTIFY palm reader in the CT simulation room.

2. A pelvis phantom was scanned in the CT simulation room under a mock patient ID.

3. The treatment accessory device list was generated in the CT simulation room, and an orthopedic setup reference image was captured with the phantom at the loading position.

4. CT images were transferred to the TPS.

5. The external body contour of the phantom was generated in the TPS.

6. A treatment plan was created in the TPS using the phantom CT scan.

7. The treatment plan and the structure set were exported to IDENTIFY.

8. The treatment plan and the structure set were imported using the IDENTIFY Planning Tool software.

9. An ROI was created using the Planning Tool software.

10. The same mock patient was verified on the palm reader at the treatment vault.

11. The treatment accessory device list was verified at the treatment vault.

12. The pelvis phantom was set up on the treatment couch in the loading position, and the phantom position was adjusted based on the orthopedic setup reference image captured at the time of CT simulation.

13. The phantom was initially set up to the treatment room lasers.

14. The phantom position was fine-tuned using the IDENTIFY surface imaging, using the phantom external body contour from the TPS as a reference surface.

15. A CBCT was acquired of the phantom.

16. The CBCT images were registered to the CT simulation reference images.

17. The CBCT shifts were compared with the IDENTIFY reported offsets.

The phantom was set up in three different positions: the first was at isocenter, the second was 1 cm offset from isocenter in each of the three translational directions, and the third was a random setup that had a translational offset of 4.3 cm from the intended isocenter position (determined after CBCT image registration). After each setup, CBCT was acquired and the registration was performed. The agreement between CBCT and IDENTIFY was within 1 mm and 1° for all three tests. Once again, the treatment couch used for the end-to-end tests did not have 6DOF capability, so CBCT registrations did not include rotations.

7.3 IDENTIFY™ CLINICAL WORKFLOW DESIGN

Initially, three IDENTIFY clinical workflows were created at the University of Utah, which included an initial setup and intra-fractional tracking workflow, a DIBH breast workflow, and a DIBH CBCT workflow. Another component to implementing workflows, such as these that will be presented, is proper staff training. Simulation with the actual staff who will be completing each of the steps of the entire workflow is helpful in that it allows everyone to understand the process and their individual responsibilities as well as helps to identify any potential problems that may occur.

7.3.1 Initial Setup and Intra-fractional Tracking Workflow

This workflow is able to be used in most radiation therapy treatments besides those involving respiratory management such as DIBH. A typical radiation therapy treatment that does not utilize any surface guidance involves initial patient setup based on the simulation and treatment plan instructions, patient pretreatment setup imaging, image registration, patient shifts based on image guidance, and finally, delivery of the radiation treatment. There are many image-guided radiation therapy (IGRT) modalities, including (but not limited) to megavoltage (MV) and kilovoltage (kV) planar imaging and CBCT. The goal of this workflow was to integrate the IDENTIFY system into the regular radiation treatment procedure that could potential utilize any of the IGRT modalities listed above. The detailed workflow is divided into three sessions: CT simulation workflow, pretreatment patient preparation workflow, and treatment vault workflow.

7.3.1.1 CT-Simulation Workflow

1. The patient is set up on the CT simulation couch.

2. The planning CT scan is acquired.

3. The orthopedic setup reference image is captured with the patient at the loading position.

7.3.1.2 Pretreatment Patient Preparation Workflow

1. After the treatment plan is approved by the physician, the plan and structure set are exported from the TPS to IDENTIFY via DICOM.

2. The plan and structure set are imported into IDENTIFY and patient setup instructions are copied from the treatment plan to the IDENTIFY system using the Planning Tool software.

3. The ROI is drawn on the reference surface using the Planning Tool software. The ROI will vary depending on the treatment site and technique.

4. The external body contour from the treatment plan is assigned to the default reference surface.

7.3.1.3 Treatment Vault Workflow

1. The patient is set up on the treatment couch in the loading position.

2. The patient is coached to self-adjust his or her posture based on the orthopedic setup reference image captured during CT simulation.

3. The patient is initially set up by aligning tattoos (or other skin marks) to treatment room lasers.

4. The IDENTIFY offset (using the external body surface contour from the TPS as a reference surface) agreement is checked.

5. The IDENTIFY system is switched to intra-fractional motion monitoring.

6. The IGRT technique of choice is performed, and patient motion during IGRT imaging is monitored.

7. IGRT image registration is performed, and the patient is shifted based on the IGRT results.

8. IDENTIFY offset agreement with shifts is checked.

9. Patient surface reference is reset to the current position allowing intra-fractional motion monitoring.

10. Treatment begins and the patient is monitored for intra-fractional motion during treatment.

7.3.2 DIBH Breast Workflow

A DIBH technique is regularly applied on left breast radiation therapy patients for sparing dose to the heart. SGRT is a common technique for DIBH breast treatments. A detailed DIBH breast workflow was designed utilizing the SGRT capabilities of IDENTIFY. The workflow is divided into three sessions: CT simulation workflow, pretreatment patient preparation workflow, and treatment vault workflow.

7.3.2.1 CT-Simulation Workflow

1. Breath-hold duration, consistency, and amplitude are checked to determine whether the patient is a DIBH candidate.

2. Both FB and DIBH CT scans are acquired in one CT imaging session.

3. The physician decides if DIBH is necessary by comparing the potential heart involvement in tangent fields on the FB and DIBH CT scans.

4. If DIBH is necessary, treatment isocenter is placed on the DIBH scan.

5. The same isocenter coordinates are copied to the FB CT scan.

6. The patient is marked in the FB position.

7. The patient's FB orthopedic setup reference image at the loading position is captured.

7.3.2.2 Pretreatment Patient Preparation Workflow

1. After the patient treatment plan is approved by physician, both the FB and DIBH treatment plans and structure sets are exported from the TPS to IDENTIFY.

2. The plans and structure sets are imported and patient setup instructions are copied from the treatment plan to the IDENTIFY system using the Planning Tool software.

3. Both FB and DIBH ROIs are drawn using the Planning Tool software.

4. The FB patient body contour from the treatment plan is assigned to the default reference surface.

7.3.2.3 Treatment Vault Workflow

1. The patient is set up on the treatment couch in the loading position.

2. The patient is coached to self-adjust their posture based on the FB orthopedic setup reference image captured during CT simulation.

3. The patient is initially set up by aligning tattoos (or other skin marks) placed during FB to treatment room lasers.

4. The patient is shifted based on the IDENTIFY offset (using the FB external body contour from the TPS as the reference surface).

5. The DIBH reference surface is loaded in IDENTIFY, and DIBH monitoring of the patient is switched on.

6. The patient is guided to perform a DIBH and the offsets between the real-time surface and the reference surface are observed. Patient coaching is performed (if needed) to help guide the patient to the correct breathing amplitude.

7. Treatment portal images are taken when the patient's breast surface is within tolerance (3 mm and 3°). The images are compared to digitally reconstructed radiographs (DRRs) to assess the patient's alignment.

8. A physician reviews and approves the alignment of the portal imaging to the DRRs. This provides a known correlation between the radiographic imaging and the surface imaging.

9. The patient treatment begins when the DIBH is within the specified tolerance.

7.3.3 DIBH with CBCT Workflow

Besides left-sided breast, the DIBH technique can be applied to other treatment sites, such as liver and pancreas, with the intention of reducing the volume of the treatment target due to the reduction of respiratory motion. All DIBH liver and pancreas patient treatments at the University of Utah are localized using daily CBCT. The DIBH with CBCT workflow is also divided into three sessions: CT simulation workflow, pretreatment patient preparation workflow, and treatment vault workflow. The CT simulation and pretreatment patient preparation workflows for DIBH with CBCT are the same as the DIBH breast workflow. The treatment vault workflow for DIBH with CBCT is similar to the DIBH breast workflow, and the only difference is that the IGRT modality is changed to CBCT. The normal length for a full gantry rotation to acquire a CBCT is approximately 1 min, and it usually takes a patient two to three DIBH cycles to complete the acquisition. The CBCT scan is initiated when the patient's DIBH is in tolerance and is paused while the patient is FB. After the patient is comfortable with proceeding, another DIBH is performed and the CBCT is resumed. After the full CBCT scan is acquired, it is registered to the CT simulation scan. If shifts are needed, a new IDENTIFY DIBH reference surface is recaptured after the patient is shifted but still in DIBH. A CBCT is reacquired using the same process as described above to confirm the correlation between the radiographic imaging and the DIBH level indicated by IDENTIFY using the new reference surface.

7.4 QA PROCESSES

The IDENTIFY system includes two sets of cameras: the two orthopedic setup cameras and the three SGRT cameras. Both camera sets need to be calibrated during system installation and setup. If the camera position is moved or exhibits drift, recalibration is necessary. For each set of cameras, two calibrations are required: an intrinsic parameter calibration and an isocenter calibration.

7.4.1 IDENTIFY™ Orthopedic Setup Camera Calibrations
7.4.1.1 Intrinsic Parameter Calibration

Each orthopedic surface camera has one lens and one projector; the intrinsic parameters, such as focal length, image center, and lens distortion, are calibrated using a calibration board with a checkerboard pattern (Figure 7.4). The spatial relationship between the two orthopedic surface cameras is also calibrated using the same calibration board. On the

FIGURE 7.4 The IDENTIFY orthopedic setup camera intrinsic parameter calibration board.

calibration board, there are seven by six black and white squares. The side of each square is 60 mm in length. In the calibration process, 20 different images of the calibration board are acquired. Each image is acquired with the calibration board at a different position. All intrinsic parameters and spatial information for the cameras are calculated from the 20 images.

7.4.1.2 Isocenter Calibration

The orthopedic setup camera allows for patient setup at the initial loading position. The treatment machine isocenter position relative to the table reference first needs to be calibrated. The IDENTIFY orthopedic camera isocenter calibration tool is shown in Figure 7.5. The isocenter calibration tool is a metal plate, which can be mounted on a treatment couch index bar. During the calibration, the center of the metal plate is set up at the machine isocenter, and the plate is then moved to several index positions along the couch. Moving the metal plate toward the couch end allows the effects of couch sag to be included into the daily IDENTIFY surface camera isocenter verification.

7.4.2 IDENTIFY™ SGRT Camera Calibrations

7.4.2.1 Intrinsic Parameter Calibration

Each IDENTIFY SGRT camera has two lenses and one projector; the intrinsic parameters, such as focal length, image center, and lens distortion, are calibrated using a calibration board with a polka-dot pattern (Figure 7.6). The spatial relationship between the three SGRT cameras is also calibrated using the same calibration board. During the calibration process, the calibration board is moved to different spatial locations near the treatment machine isocenter. The intrinsic parameters and spatial information for all cameras are calculated from these combined images.

FIGURE 7.5   The IDENTIFY orthopedic setup camera isocenter calibration tool.

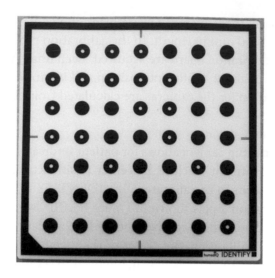

FIGURE 7.6   The IDENTIFY SGRT camera calibration board.

7.4.2.2 Isocenter Calibration

The IDENTIFY SGRT cameras must be calibrated such that the system's isocenter is coincident with the treatment machine's isocenter. The SGRT camera isocenter calibration uses the same calibration board as used for the intrinsic parameter calibration (Figure 7.6). During the SGRT camera isocenter calibration process, the center of the calibration board is set up at the isocenter of the machine with a specific orientation. If the calibration board is aligned to the treatment machine isocenter using surrogates (e.g., room lasers), the accuracy of the surrogates should be verified.

7.5 ROUTINE QA

IDENTIFY daily QA recommended by Varian includes an IDENTIFY system overall check, an orthopedic setup camera isocenter check, and an isocenter check of the SGRT cameras. An overall IDENTIFY system check is performed when the system is turned on at the start of the treatment day. These checks include initialization and connection verification for all cameras, a communication check with the IDENTIFY Data Manager, and a communication check with the oncology information system. After all of the communications and connections are verified, the orthopedic setup camera isocenter verification and SGRT camera isocenter verification can be performed. The tool used for orthopedic setup camera isocenter verification is the orthopedic setup camera isocenter calibration plate

(Figure 7.5). The plate is attached to an index bar, and the index bar is fixed to the same slot on the treatment couch as the calibration process. The plate is then aligned with the treatment machine isocenter. During the verification process, the offset of the orthopedic setup camera isocenter is calculated and displayed on the IDENTIFY monitor. The tolerance for the IDENTIFY orthopedic setup camera isocenter position verification is 2 mm. The tool used for SGRT camera isocenter verification is the SGRT calibration board (Figure 7.6). The center of the board is aligned with the treatment machine isocenter, and the offset of the SGRT camera isocenter is calculated and displayed on the IDENTIFY monitor for each of the three SGRT cameras. The tolerance for the IDENTIFY SGRT camera isocenter position verification is 1 mm. This tolerance criterion is tighter than the 2 mm recommended by the American Association of Physicists in Medicine (AAPM) Task Group (TG) 147 report for QA of nonradiographic localization systems.[1]

Besides the daily QA procedures recommended by Varian, another beneficial QA test is to integrate the IDENTIFY surface tracking imaging with the SGRT cameras into the clinic's daily IGRT QA program. Daily IGRT QA procedures are commonly accomplished using a cube phantom that has a hidden target embedded inside with known offsets from external marks. The cube phantom can be shifted based on an IGRT modality, such as CBCT, and the position of the hidden target can be verified to be aligned to the machine isocenter using MV imaging. Integrating IDENTIFY surface imaging into this task can verify that the IDENTIFY-indicated shifts are correct and cross-validate the system against both another IGRT modality and the MV beam.

Currently there are no recommended monthly QA procedures suggested from Varian, however, thorough daily QA can provide ongoing verification that the system is operating as intended. Camera calibrations may be needed as any issues are discovered or QA results exceed tolerances. Otherwise, camera calibration at set regular intervals may not be needed. If a QA test indicates that recalibration is necessary, it is good practice to repeat the QA test after camera calibration to validate the calibration. Depending on the use of the system, additional monthly QA may include checks that are not performed during daily QA. An example would be checking the system's ability to accurately report couch rotations, which may not be checked during daily QA. This would be important in situation where noncoplanar treatment fields are utilized, such as with stereotactic radiosurgery.

During annual QA of the IDENTIFY system, physicists could perform a thorough IDENTIFY system overall integrity check that follows many of the commissioning tasks. An example of annual QA checks for the IDENTIFY SGRT system is listed below.

1. Functionality check, including the Planning Tool and RFID tagging solution

2. Orthopedic setup camera isocenter calibration check

3. SGRT camera isocenter calibration check

4. Consistency check

5. Shift accuracy check

6. Rotation accuracy check

7. Continuous recording stability check

8. End-to-end test

All system functionality, consistency, and accuracy should be consistent with the baselines established during system commissioning. Some of these checks may be accomplished with ongoing use of the system. As IDENTIFY is a relatively new SGRT system (at the time of this chapter writing), the recommended QA procedures will likely continue to be developed as the system is further developed and clinical use expands.

KEY POINTS

- IDENTIFY is an SGRT system that allows surface tracking of patients to aid in patient setup and monitoring during treatment. The system also includes functionality to allow patients to self-position themselves with the aid of augmented reality technology.

- The IDENTIFY commissioning process from the University of Utah has been described. It included an evaluation of all IDENTIFY components and verified the accuracy of the SGRT abilities of the system.

- Clinical workflows for initial setup and intra-fractional tracking, DIBH breast, and the DIBH with CBCT were described including all steps for simulation, pretreatment, and treatment.

- On-going QA of the IDENTIFY system was described including calibration of the camera systems and suggested daily, monthly, and annual QA.

REFERENCES

1. Willoughby T, Lehmann J, Bencomo JA, Jani SK, Santanam L, Sethi A, Solberg TD, Tomé WA, Waldron TJ. Quality assurance for nonradiographic radiotherapy localization and positioning systems: Report of task group 147. *Med Phys.* 2012;39(4):1728–1747.

Commissioning and Routine Quality Assurance of the Vision RT AlignRT® System

Guang Li

CONTENTS

Editor's Note: SGRT vendors have a variety of technologies in their product portfolio and some of those technologies are integrated into their SGRT systems to different degrees. Since this textbook is focused on SGRT, the editors have asked the authors to limit their chapter content specifically to surface image guidance for patient setup and inter- and intra-fraction patient monitoring.

8.1 INTRODUCTION

The AlignRT system (Vision RT, London, UK) is an optical surface imaging system used for surface-guided radiation therapy (SGRT), with applications for patient setup before treatment and/or patient motion monitoring during treatment.[1-3] Since AlignRT received 510(k) clearance from the US Food and Drug Administration in January 2006, this SGRT system has gone through substantial improvements and has been increasingly utilized in radiation therapy clinics at both academic institutional centers and community hospitals around the world. The two anatomic sites most commonly treated with SGRT are breast cancer and brain cancer,[1,4-11] although other anatomic sites have been described in the literature[12-14] and in other chapters of this textbook. For intracranial cases, the target is typically deep seated within the rigid anatomy of the skull, producing a relatively fixed external-internal position relationship compared with the case of breast cancer where there is a relatively superficial but deformable treatment target. In both cases, the skin surface serves as a surrogate for the internal target position and is tracked in real-time by AlignRT with respect to a reference surface, displaying displacements between live and reference surfaces as real-time deltas (RTDs). The skin surface also serves as a surrogate for respiratory motion for applications such as deep inspiration breath hold (DIBH), which is discussed further in Chapter 10, and four-dimensional computed tomography (4D-CT), which is discussed further in Chapter 15. Vision RT also offers a dedicated product for respiratory motion management called GateCT; however, this chapter will focus only on the commissioning and routine quality assurance (QA) of the AlignRT system for patient setup and motion monitoring in the treatment room.

An AlignRT system consists of two or three ceiling-mounted camera pods in a treatment room: for the three-pod installation, two pods are ceiling mounted laterally to either side of the treatment couch, and a third pod is mounted at the far end of the couch away from the linear accelerator (linac) gantry, as shown in Figure 8.1. Each camera pod contains a light projector and two cameras used to reconstruct a

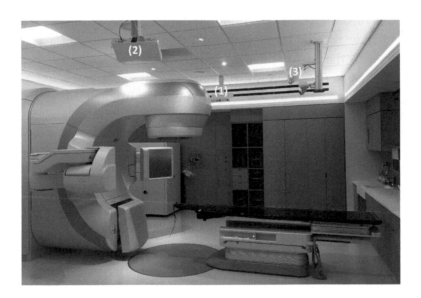

FIGURE 8.1 A typical installation of the AlignRT system with three ceiling-mounted camera pods in a treatment room with a TrueBeam (Varian, Palo Alto, CA) linear accelerator. A 6 degrees-of-freedom (6DOF) couch (PerfectPitch, Varian, Palo Alto, CA) with a couch head extension (CDR Systems, Calgary, Canada) are also shown.

three-dimensional (3D) surface of an object or a human body from its view angle. Depending on the number of camera pods, two or three surfaces are captured, reconstructed, and merged together to form a more complete 3D surface image of an object. The assembly of these 3D surfaces is based on their spatial relationship to the linac isocenter, which is well defined and obtained through AlignRT system calibration. Depending on the calibration technique employed, uncertainty in the surface tracking accuracy may rely on the surface location relative to the isocenter. Clinically, uncertainties can be minimized through a proper calibration and verified via appropriate commissioning and QA procedures. Additional technical details of the AlignRT system are provided in Chapter 5.

This chapter discusses AlignRT system commissioning, calibration, and routine QA processes, which can be customized for particular clinical procedures in a radiation therapy clinic, following the clinical guidelines provided by the report of the American Association of Physicists in Medicine (AAPM) Task Group 147 (TG-147).[2] Other sources of this type of information include a chapter from the "Quality and Safety in Radiotherapy" textbook describing surface imaging systems.[15] As with all radiation therapy equipment, following installation and acceptance testing with the equipment vendor, it is the responsibility of the clinical physicist to perform a full-system commissioning procedure, ensuring the SGRT system meets all requirements for its intended clinical use. This includes quantifying system accuracy, defining procedures for calibration and routine QA, workflows for clinical utilization, and verifying that the SGRT system integrates into the full treatment solution through end-to-end (E2E) testing.

This chapter starts by defining the accuracy requirements of AlignRT for different applications and then describes techniques to perform both the intrinsic (optical), and extrinsic (isocenter and pod position) AlignRT calibrations. Section 8.2 provides an overview and clinical accuracy requirements for surface-guided intracranial stereotactic radiosurgery (SRS) and stereotactic radiotherapy (SRT) and conventional radiotherapy of breast cancer, and Section 8.3 discusses clinical calibration procedures necessary to achieve the highest accuracy (sub-mm). Section 8.4 discusses clinical implementation procedures necessary to prepare for clinical use, with Section 8.5 devoted to routine clinical QA procedures and troubleshooting tips. Finally, Section 8.6 provides some comments and suggestions for future improvements to the AlignRT system. The commissioning, calibration, and QA methods described in this chapter are based on techniques used at Memorial Sloan Kettering Cancer Center (MSKCC) and serve to illustrate how the reader could adapt vendor-specified procedures and AAPM-recommended guidelines to their own clinic's requirements.

8.2 SYSTEM ACCURACY REQUIREMENTS FOR CLINICAL APPLICATIONS

Prior to commissioning an AlignRT system and performing the required calibrations, it is important to define the intended clinical applications of the system and to understand the associated accuracy requirements. These accuracy requirements should depend upon the clinical need in patient setup and motion monitoring, and therefore, different commissioning, calibration, and routine QA processes may be applied. For both intracranial and breast applications, for example, surface imaging provides a good tumor position surrogate. In a rigid anatomy (e.g., the head), the surface position provides an excellent surrogate for the internal tumor position, so the high precision requirements for SRS can be achieved with SGRT. In a deformable anatomy such as the breast, the skin surface is part of, or in close proximity to, the treatment target. Due to the treatment planning techniques commonly employed, and target deformation that varies from fraction to fraction, lower SGRT accuracy requirements (compared to SRS) may be sufficient in breast applications. The following describes accuracy considerations for cranial and breast SGRT applications.

8.2.1 System Accuracy Requirements for Intracranial SRS/SRT

For intracranial SRS, the role of AlignRT is to facilitate patient setup before cone-beam computed tomography (CBCT) as well as patient head motion monitoring throughout the treatment with a set motion tolerance threshold. The motion tolerance threshold can be defined to automatically turn off the radiation beam as soon as patient motion exceeds the threshold in any of the 6DOF, or the operator can do so manually. The radiation will be resumed after the motion is reduced to within the threshold.[16] The overall SRS accuracy, according to the AAPM TG-42[17] and TG-135,[18] is to be within 1.0 mm.

Clinical SRS accuracy is assessed by an E2E test, which is often performed using an anthropomorphic head phantom with two embedded orthogonal radiochromic films across the isocenter inside a hypothetical target for CT simulation, treatment planning, and delivery.[19,20] Both geometric and dosimetric endpoints are used to determine the spatial and dose delivery accuracy. Since there is little motion in the phantom, the E2E test is used to assess the uncertainty of the overall system, producing an ideal result. The ability of the AlignRT system to detect motion has been tested using a head phantom and a high-precision positioning platform.[8] Motion on the order of 0.1 mm is detectable, but the random noise is also on the order 0.1 mm, therefore 0.2 mm is regarded as the motion detection limit, as the mean positions of the phantom with a 0.2 mm shift are statistically separated in the presence of random background noise.

Clinically, the patient's motion is restricted by an immobilization device as shown in Figure 8.2. The head immobilization system should be restrictive, which can be achieved using a system composed of a customized head mold and an open-face mask, so that large patient motion (>1 mm) is unlikely and radiation remains uninterrupted throughout the SRS treatment. This open-face mask not only forms a clam-shell head immobilization with the customized head mold but also provides sufficient exposed facial area for AlignRT motion monitoring.[9,13] Moreover, it is patient friendly, comfortable, and has been shown to be clinically preferred over some other immobilization devices, such as bite blocks.[9] The residual head motion is monitored by AlignRT to ensure uncertainty is below 1.0 mm during beam-on, while smaller head motion will be averaged out within the clinical tolerance.[8,9,21] At MSKCC, the

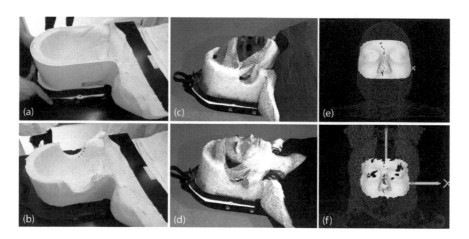

FIGURE 8.2 Illustration of head immobilization for intracranial stereotactic radiosurgery (SRS) patients. A customized head mold (a, b) and open-face mask (c, d) provide firm patient immobilization for SRS. The DICOM reference surface with the region of interest (ROI) (e) and captured reference surface overlaid with the user-specified ROI (f) are shown.

SRS planning target volume is formed by adding a 0–2.0 mm margin to the gross tumor volume, depending on the primary cancer type of the brain metastasis.

For intracranial SRS/SRT patient setup using AlignRT only, the uncertainty is often about 1.0–2.0 mm. This is based on the clinical observation of intracranial SRS patient setups, in which the head is first aligned to achieve sub-mm and sub-degree precision using the AlignRT system for prealignment, and then a CBCT is acquired for final setup. Both the AlignRT and CBCT registration produce shifts in 6DOF, and they often differ by 1.0–2.0 mm, which may be affected by the thresholding of the external contour in the treatment planning system and the CT slice thickness. Real-time head alignment is achieved at MSKCC by using AlignRT RTD information in the treatment room to correct head rotations first, especially roll and pitch rotations, using a commercial couch head extension (CDR Systems, Calgary, Canada), followed by translational couch shifts. This process is typically completed within 60 s inside the room. After leaving the room and acquiring a CBCT, the final residual head rotations and translations are corrected using a 6DOF couch (Perfect Pitch, Varian Medical, Palo Alto, CA). The AlignRT-guided CBCT-based frameless open-face mask SRS procedure has become the clinical standard at MSKCC for several years.[8,9,21]

Using the above considerations, it is important to design a commissioning and routine QA program that can ensure when AlignRT is used for surface-guided SRS, the system has submillimeter tracking accuracy and is capable of tracking motion in an open-face mask throughout the entire noncoplanar treatment.

8.2.2 System Accuracy Requirements for Breast Cancer Treatments

Due to the nature of deformable breast tissue, and the presence of respiratory and other motion, it is not feasible to achieve the same level of setup accuracy as achieved (or required) in SRS cases. The deformation of the breast can make its shape and position variable over daily treatments. At MSKCC, a two-step surface-guided setup procedure has been developed and applied to minimize the deformation of the breast when treating patients with locally advanced breast cancer with intensity modulated radiation therapy (IMRT)[10] or volumetric modulated arc therapy (VMAT).[22]

In this clinical procedure, the patient is first set up using skin tattoos (medial and lateral) and arm position on a breast board (CIVCO Radiotherapy, Orange, IA). The digital imaging and communications in

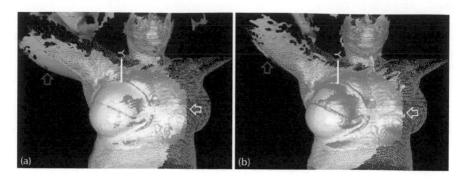

FIGURE 8.3 Illustration of two-step setup (a and b) for patients with locally advanced breast cancer. The alignment of the ipsilateral arm (red arrow) is improved by adjusting the patient's arm to match the DICOM external (body) contour while automatic surface alignment is based on the ROI (solid pink surface, indicated by the yellow arrow).

medicine (DICOM) external body contour derived from the CT simulation is used as the reference (the frontal view is shown in Figure 8.3). Then the first surface treatment capture is acquired in AlignRT. Using the full surface information contained in a treatment capture, the ipsilateral arm and chin are aligned by adjusting the patient with the visual AlignRT guidance in the frontal, lateral, and vertex views.

After confirming the alignment of the arm and chin by capturing one to three additional treatment captures, the impact of breast deformation is minimized when performing breast alignment. The arm position is also important in order to ensure the lymph nodes are in the planned positions when treating locally advanced breast cancer. Using a large, asymmetric region of interest (ROI) for setup by including the entire ipsilateral and one-third of the contralateral breasts, the setup should become more reliable because possible orientation ambiguity due to breast symmetry is eliminated.

Figure 8.3 illustrates a patient with breast cancer setup with the first alignment of the ipsilateral arm followed by breast alignment to simulation CT position (using the DICOM external body contour from CT as the reference surface) with minimal tissue deformation. An optimal patient setup is achieved with reproducible breast and local nodal positions, such as the supraclavicular nodes.[10] Therefore, the treatment targeting accuracy is expected to be improved compared to setups that do not correct for the arm and chin using AlignRT. At MSKCC, conventional setup using skin tattoos is always performed prior to AlignRT setup. If the AlignRT setup requires a shift greater than 5 mm, this suggests a large breast deformation, especially for large patients and requires orthogonal radiographic verification based on bony landmarks for alignment as the final setup, as the skin surrogate may become unreliable in this case.

For patients with left-sided breast cancer, the DIBH technique is often used for treatment to minimize heart irradiation.[23,24] AlignRT has been applied to provide surface image guidance and respiratory monitoring to enable DIBH gating of left-sided breast cancer treatments.[5,25–28] A common tolerance level set for the DIBH positioning in motion monitoring is 3.0 mm. This provides similar accuracy to infrared reflector block-based DIBH (Real-time Position Management, Varian, Palo Alto, CA), but without the need for an external surrogate block, and provides full 6DOF positional information.

In summary, AlignRT commissioning and calibration procedures may be nonuniform and established for different clinical applications of SGRT, such as intracranial SRS and breast DIBH. If a linac is not going to be used for SRS, for example, a less involved isocenter calibration process may be sufficient. However, efforts should be made to perform the optimal system calibration for the intended clinical use, enabling the greatest benefits of SGRT.

8.3 OVERVIEW OF AlignRT® SYSTEM CALIBRATION

8.3.1 AlignRT® Calibration Overview Based on Clinical Accuracy Requirements

Based on the discussions above, the calibration requirements for AlignRT may differ depending on the clinical treatment sites for which it will be used. However, there are some calibration elements that are common for all installations. This section describes the available calibration techniques, along with some details on the workflow for performing them.

The AlignRT system is a stereoscopic surface imaging system based on an active stereo technique, which projects a speckle pattern onto the surface of an object, capturing information from the six cameras in the three pods. The surface images are then reconstructed with a fixed geometric relationship among the two cameras within each pod, and common surface points are identified by the speckle patterns on the object.[29] Similar to human vision, the two captured pictures provide a triangulating relationship between the two cameras and a point on the surface (identified on

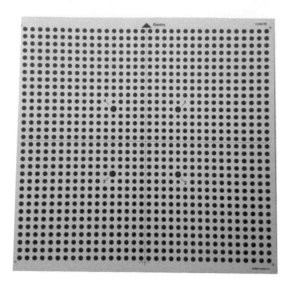

FIGURE 8.4 The AlignRT calibration plate with a crosshair and dot array for monthly calibration and daily QA. The triangle on the edge should point to the gantry, while the four bigger blobs near the center must be labeled on the captured image consistently with the labels shown on the plate (1-->2-->3-->4), determining the plate orientation.

the projected speckle pattern). By applying the law of sines to the triangle, the distances from the cameras to the point can be calculated, and the 3D surface, composed of a point array, can then be reconstructed. The constant parameters in the law of sines are obtained through the calibration process using a known geometric dot array in the AlignRT calibration plate, as shown in Figure 8.4. Therefore, accurate calibration is essential prior to clinical use.

8.3.2 Extrinsic and Intrinsic Calibration Workflow Using the AlignRT® Calibration Plate

The monthly calibration procedure for AlignRT starts by first performing an extrinsic (isocenter) calibration. To perform the extrinsic calibration, the calibration plate, a flat 100×100 cm^2 white-background board with printed black crosshair lines and black dot array,[30] is positioned horizontally at the treatment isocenter by aligning the plate crosshair to the light-projected crosshair from the gantry head. Vertically, the top surface of the plate should be positioned at 100 cm source to surface distance (SSD), using the front pointer, calibrated horizontal room lasers, and/or optical distance indicator (ODI) for setup. The position is then finalized by rotating the gantry head to ±45° to check the alignment of the light projection crosshair with the plate crosshair: they should align throughout the gantry rotation indicating an accurate isocenter position. The accuracy of plate positioning determines the accuracy of this isocenter calibration.

After capturing an image of the calibration plate from each camera sensor, the user can adjust the window/level to make all dots individually visible, manually label the four bigger dots (blobs) in their order (1 to 4), and then check if all visible dots are correctly recognized by the software. The software calibrates one pair of cameras in a pod at a time and progresses in a similar way through all camera pods. The room lighting or illumination condition should be kept constant every day, so that the window/level setting can be tuned, stored, and reused as the calibration condition for future daily QA. This process defines the AlignRT isocenter and pod locations. With the known spatial locations of each dot in the array relative to the isocenter, the AlignRT system is calibrated for reconstruction of a 3D surface image.

The second component of the system calibration, the intrinsic (optical) calibration, can be performed in one of two ways, either using the raised plate calibration as part of monthly calibration, or with advanced camera optimization (ACO), as described in Chapter 5. For systems without ACO, the raised plate calibration is a recommended method that requires moving the plate to a new position, hereafter called the "second plate level." A second plate level at 7.50 cm above isocenter is applied by raising the treatment couch, and a model for the optical properties of each camera is captured at this plate location. Because the isocenter is almost always inside a deep-seated tumor below the skin surface, the elevated plate position provides intrinsic optical calibration information in a plane closer to the expected patient surface measurement conditions.

8.3.3 AlignRT-MV Congruence Calibration Workflow Using a Cube Phantom

Due to the high-precision requirements of frameless surface-guided SRS, and other high accuracy treatments requiring noncoplanar treatment deliveries, a further isocenter calibration procedure beyond the plate calibration may also become necessary. This is the AlignRT-MV (megavoltage) congruence calibration (MV Isocenter Calibration), which uses the Vision RT cube phantom (15 × 15 × 15 cm³) (Figure 8.5) to correct for sub-mm residual error between the AlignRT isocenter and MV

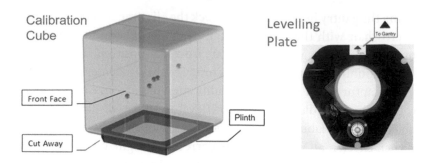

FIGURE 8.5 An AlignRT-MV cube phantom ($15 \times 15 \times 15$ cm³) with five internal ball bearings and its substrate for fine-tuning AlignRT isocenter to align with linac MV isocenter. This additional calibration process is recommended for intracranial SRS/SRT treatments.

radiation isocenter. A small deviation of the two isocenters at couch zero can cause enlarged uncertainties in tracking accuracy at couch rotations up to ±90° due to error propagation, leading to RTD tracking errors proportional to the calibration deviation and couch rotation angle, sometimes exceeding 1.0 mm. The linac radiation isocenter for the MV beam is determined using the electronic portal imaging device (EPID). By correcting the residual difference between the AlignRT and linac MV isocenters using four-directional imaging, similar to the Winston–Lutz test,[31] the cube phantom's five internal ball bearing (BB) markers ($\Phi = 0.75$ cm), and the external surface can be aligned to determine and correct the deviation of the AlignRT isocenter from the radiation treatment isocenter.

Following monthly calibration with the calibration plate, the MV Isocenter Calibration can be performed. The AlignRT system provides software to calculate the isocenters of each system and their difference in 6DOF.

A static surface image of the cube is captured to determine the location of the cube relative to the AlignRT isocenter. Orthogonal MV images are acquired at gantry angles of 180°, 90°, 0°, and 270° to determine the location of the cube relative to the MV isocenter using either the image center (using a calibrated DICOM image with an isocenter location tag) or field center methods (calculating the isocenter on the image based on the centroid of the field edges), as described in Chapter 5. The difference between the radiographic and AlignRT isocenters is then calculated in 6DOF, and a user can either accept or decline to recalibrate the AlignRT isocenter based on the calculated difference. In general, if the difference is greater than 0.2 mm, the workflow at MSKCC is to recalibrate the AlignRT isocenter, and a subsequent verification test is recommended, as described in Section 8.4.5.

The correction for sub-mm deviation of the two isocenters in the two systems (AlignRT vs. linac) may not be required for nonstereotactic treatments, which may not need <1.0 mm accuracy nor noncoplanar beams. The purpose of this cube-phantom calibration is to further reduce the isocenter error at couch zero so that at large couch rotations the AlignRT and treatment isocenters do not rotate about different isocenter locations, which would lead to artificial RTD walkout. Therefore, the cube calibration becomes necessary for SRS treatment since noncoplanar beams are routinely employed. When treating at nonzero couch rotations, the patient's surface needs to be matched to the reference surface that has been rotated around the AlignRT isocenter, via a rotation transformation, thus accounting for the couch rotation. Theoretically, the error of the isocenter difference carried by the reference image will be amplified by up to $\sqrt{2} = 1.4$ at the couch angle of 90°.[32] Generally, the larger the systematic isocenter calibration error and the larger couch rotation (0°–90°), the larger the error will be embedded in the new reference. Practically, couch-walk and patient motion will also contribute to the RTDs observed at a couch rotation. It is essential to first minimize the AlignRT uncertainty at all couch angles through system calibration and to verify the isocenter during commissioning (this will be discussed in Section 8.4), leaving room to detect and tolerate minor patient head motion and reducing the chance of false positives of motion detection. Once a patient is detected to have moved, extra verification steps are necessary, so false-positive indications of head motion will slow down an SRS treatment. In summary, the congruence of the AlignRT and linac MV isocenter can and should be minimized or confirmed using the cube phantom and MV Isocenter Calibration procedure, leading to a smooth clinical SRS treatment.

8.4 COMMISSIONING OF THE ALIGNRT® SYSTEM

8.4.1 Guidelines from AAPM TG-147

AAPM TG-147 provides general guidelines for nonradiographic imaging modalities, which are used in radiotherapy localization and positioning, including SGRT technologies, such as AlignRT.[2] This report describes the general guideline for system commissioning and calibration, including the system integration, spatial reproducibility and drift, static and dynamic accuracy, and recommended assessments and procedures by the vendor. In the following, we provide detailed information on the requirements and procedures for AlignRT system commissioning and calibration.

8.4.2 AlignRT® System Configuration

The AlignRT system with three camera pods is the standard configuration for use, versus older two pod configurations. The 3D surface images from multiple camera pods are combined as a single patient surface image. The more complete the resultant patient body surface, in combination with a well-designed ROI that is visible by the camera pods throughout the treatment, the more accurate and consistent the SGRT monitoring results.

In the AlignRT system, configuration consists of two parts: the technical configuration should be set up as part of the acceptance test and clinical configuration should be set up in the commissioning of the system. The technical configuration may include establishing the IEC (International Electrotechnical Commission) coordinate system, camera sensitivity and communication, and patient data storage arrangement. The clinical configuration includes patient identification settings, setup thresholds for various anatomical sites, calibration procedures, and clinical user accounts with different access rights. The clinical commissioning of the system should be completed prior to clinical use.

8.4.3 System Communication with a Linac

While enabled, AlignRT can communicate with the linac via the motion management interface (MMI), or equivalent, to allow actions such as automatic beam hold. The initial setup of linac communication requires the engineers from Vision RT and the linac vendor, hospital IT service, as well as a clinical physicist. As part of the acceptance test for the beam hold capability, a 3D rigid test object (the "legs phantom") is often employed. This phantom has the shape of two joint cylinders and is provided by Vision RT. This is often used for acceptance testing automatic beam hold triggering by introducing a deliberate couch rotation. While tracking the phantom using AlignRT, a couch rotation is introduced that causes the rotational RTDs to move out of tolerance, therefore initiating a beam hold. Once the surface is rotated back to its original position within the

AlignRT software, the radiation beam should resume. This feature can be used clinically in both intracranial SRS and left-sided breast DIBH, as well as any other treatment indications.

8.4.4 AlignRT® Patient Data Storage and Management

Patient data management is a very important part of AlignRT system acceptance testing and commissioning. Depending on the version of AlignRT deployed, the patient data imported and used by AlignRT can be configured as either a local or a network folder, named PData. AlignRT also offers a database storage solution with the Version 6 software platform, using a central hospital database. This solution is not described within the current chapter.

When configured to use a local PData folder, the patient data will commonly be saved in the F: (VRT) drive of the console workstation. The advantage of the local storage approach is that it supports fast static image capturing and real-time motion monitoring without the possibility of network-related delays. However, for data safety and security concerns, daily PData backup using third-party software is recommended. Saving PData to a network drive is considered the most secure means to store patient data and would be in compliance with data security requirements such as the Health Insurance Portability and Accountability Act, 1996 (HIPAA) in the United States. However, if there is latency in the local area network, some performance may be degraded.

As the number of patients in the PData folder increases, the initialization of the AlignRT software application may slow down because the software must scan through all patients in the PData folder, regardless of which patient data configurations is chosen. Therefore, periodic PData clean-up is recommended to maintain the fast performance of the AlignRT application.

A remote or standalone AlignRT workstation (AlignRT Offline) can be used in the clinic for patient data preparation using the network data storage so that the prepared patient data can be shared by several AlignRT systems in different treatment rooms. Alternatively, a hybrid method can be applied: the local PData is used temporarily to ensure high-performance real-time SGRT and the network PData can be used for offline patient preparation. To connect the two PData folders, a customized script program can be written to distribute the prepared patient data from the network drive to the AlignRT workstation at the treatment console. This hybrid PData setting is valuable for busy clinics where the beam gating feature is

enabled in the AlignRT system for treatment, because the offline workstation configuration allows the patient preparation to be done at any time
without interfering with ongoing treatments. At MSKCC, an in-house
developed "Dispatcher" software was clinically implemented to allow distribution of the prepared patient data from the offline workstation to any
treatment workstation where the patient is scheduled for treatment.

8.4.5 Evaluation of Couch Angle Dependency Using a Head Phantom

At MSKCC, a couch angle dependency test is included as part of the commissioning, in addition to providing verification following the AlignRT
calibration processes. In fact, this test was developed in 2010 at MSKCC
to provide a clinical assessment on the robustness of the AlignRT motion
monitoring when used for frameless SRS treatments.[8] Nowadays, with the
availability of the raised plate calibration and MV cube phantom calibration, MSKCC still performs this couch angle dependency test as part of the
commissioning for any AlignRT system that is newly installed. This test will
provide the calibration baseline for a new AlignRT-linac system, because
the couch angle dependency is not only affected by the isocenter calibration of the AlignRT system but also by the couch walk of the linac system.

In this test, an anthropomorphic head phantom with a customized
mold is placed in the head-first-supine position and the isocenter is placed
in the middle of the brain as shown in Figure 8.6a. A reference image is
captured at couch zero and an ROI is defined from the mid-forehead to
the philtrum above the lips and the lateral cheeks, similar to the area of a
patient's skin that would be exposed in an open-face mask (Figure 8.6b).
Motion tolerance levels are set to 1.0 mm within the AlignRT software.
At 10° couch-rotation intervals, a verification image (treatment capture)
is captured to calculate the residual error between the rotated reference
image and newly captured verification image. The residual errors are
recorded as a function of couch angle (0° to ±90°), as shown in Figure 8.7.

After different calibration procedures, namely the conventional
monthly calibration at isocenter level (or one-level plate calibration) and
raised plate calibration (or two-level plate calibration), the residual errors
of AlignRT as a function of the couch angles were measured and compared. The results from this single test show an error reduction from
±1.0 mm with the one-level plate calibration to ±0.5 mm using the two-
level plate calibration procedure, as shown in Figure 8.7.

This change in tracking accuracy as a function of couch rotation can be
caused by a number of factors, including a systematic offset between the

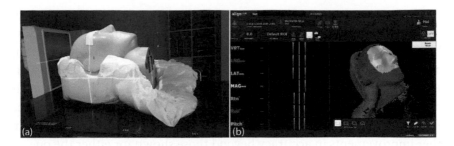

FIGURE 8.6   A head phantom in a customized mold is placed on a treatment couch (zero rotation) (a) for capturing an AlignRT reference image with the region of interest (ROI) in pink (b). The machine isocenter is shown as the laser crosshair (a) and the origin of the coordinate system (b).

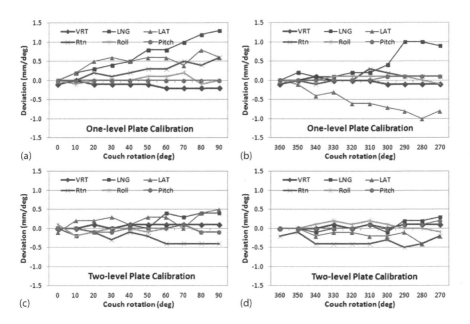

FIGURE 8.7   Residual error of the head phantom as a function of couch rotation. Errors in 1-level calibration procedure (a, b) and errors from raised plate calibration procedure (c, d). These two measurements were performed consecutively before and after the raised plate calibration procedure.

AlignRT and treatment isocenter (which can be minimized using the MV Isocenter Calibration), in the quality and difference of the intrinsic data (including different surface coverage at different couch angles), and extrinsic variables, such as linac couch walk. Assuming the isocenter calibration was the same for the one-level and two-level calibration in this test, the residual uncertainty at increasing couch rotations could be caused by the intrinsic calibration. In this case, the two-level plate calibration provides a better optical calibration than the one-level plate calibration for tracking a nonsuperficial target, which may contribute to the improvements seen in Figure 8.7c and d. Therefore, for non-ACO systems, it is recommended to apply the raised plate calibration as default, regardless of treatment sites. For systems using ACO, which replaces the raised plate calibration, further improvements in accuracy beyond those shown for either the one- or two-level plate calibration are observed.

As noted above, co-calibration errors between the AlignRT and MV radiation isocenters will directly impact the RTD walkout at couch rotations,[32] as discussed previously. This can dramatically impact the accuracy data, such as that quoted above. Therefore, it is essential that on systems that treat patients with couch rotations, for example, SRS treatments, MV isocenter calibration should be performed following monthly calibration to achieve/confirm a low discrepancy and further reduce the couch angle dependency. This will result in more accurate and reliable motion monitoring, providing more confidence over patient positioning throughout the entire treatment.

It is important to note that AlignRT measures the position of the phantom or patient relative to the calibrated isocenter location. Therefore, unless there is no couch walkout, AlignRT will report the location of the phantom relative to the treatment isocenter, and not the couch, thus measuring couch walkout for an otherwise stationary phantom. To correctly measure AlignRT accuracy, couch walkout needs to be subtracted from the observed RTDs.

8.4.6 E2E Test

For frameless SRS treatments, the AAPM TG-135 recommends an E2E test to check the overall accuracy of the entire system.[18] This test can be performed using a head phantom by placing two radiochromic films orthogonally inside and the isocenter is defined by creating a pinhole to record radiation delivery accuracy both spatially and dosimetrically. These dosimetric measurements are beyond the scope of AlignRT QA alone, as it involves the entire SRS workflow, from CT simulation, treatment planning, setup and verification using AlignRT and CBCT, and treatment delivery. However, the E2E test provides an overall assessment of the accuracy of SRS targeting in the absence of intra-fractional motion.

The E2E test can, however, be used to assess the absolute setup accuracy of AlignRT, as well as phantom tracking accuracy during treatment delivery, and general AlignRT workflows as part of the SRS treatment. To perform the E2E test, an entire SRS treatment workflow is followed from CT simulation, planning, setup, and delivery using the head phantom. Following simulation CT using the SRS simulation protocol, the CT dataset is sent to the treatment planning software where the treatment planner generates the external body structure for use as a reference surface and isocenter location (typically located to coincide the with center of a radiopaque marker). A typical SRS treatment plan is generated, and the structure set and plan information (i.e., the isocenter coordinates) are sent in DICOM RT format to the AlignRT system, where the phantom patient is created, along with a clinical ROI. Using AlignRT, the phantom is then positioned in the treatment position using the RTDs, and a CBCT image is acquired to measure any residual setup offset, which is defined as the absolute setup accuracy of AlignRT. Imaging shifts are applied to the phantom, and a new reference surface is captured. The treatment plan is then delivered while recording the tracking accuracy of AlignRT for all couch and gantry angles. Note that for the couch angle dependency test to determine the accuracy of AlignRT at couch rotations, the couch walkout must be accounted for as AlignRT appropriately reports the phantom position relative to the room coordinate system, which may include both patient motion and couch walkout.

As part of the E2E test, known couch shifts can be applied, as confirmed by radiographic imaging, and the resultant discrepancy between the AlignRT RTDs and the applied shift to the phantom can be recorded as the AlignRT static localization accuracy.

8.4.7 System Baseline Drifts

A known characteristic of all optical tracking systems is a thermal drift effect as the camera systems warm up. At the time of writing, AlignRT was known to exhibit a small initial baseline drift on the order of 0.3 mm during the first 5–10 min of real-time motion monitoring. The drift, which mostly occurs in the vertical direction, is stabilized after 10 min, as shown in Figure 8.8. This has been attributed to the thermal stabilization of the cameras. When the system "cools" down, the observed drift will fall back to the original baseline. This small but reproducible drift is a consideration only for SRS treatments, raising the chance of exceeding the threshold due to the baseline drift, causing the false positive for patient motion. This effect can be negated by employing workflows to manage the thermal drift. Clinically, MSKCC manages thermal drift by using AlignRT for continuous motion

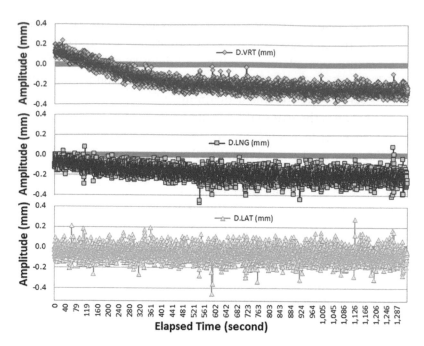

FIGURE 8.8 A small baseline drifts over 21-min continuous imaging using AlignRT real-time motion monitoring. The largest drift is in vertical translation baseline; little drift in the rotational baseline is observed.

monitoring during CBCT after initial AlignRT setup (both monitoring and warming up before treatment). By the time a reference surface has been captured following CBCT imaging and any additional couch shifts, the system will have reached thermal equilibrium and the RTDs will be stable throughout the entire monitoring session. An additional benefit of the AlignRT RTD monitoring during CBCT is that it allows the detection of any large patient motion during image acquisition and registration, affecting patient setup and delivery accuracy if undetected.

8.5 ROUTINE QA OF THE AlignRT® SYSTEM

8.5.1 Daily QA and Troubleshooting

The daily QA procedure for AlignRT is a simple process, placing the calibration plate at the isocenter and performing the software-driven QA procedure. This test provides a consistency check against the calibration file, ensuring that the camera pods have not moved relative to each other, and can be performed daily by radiation therapists or other trained staff. The daily

QA tolerance level is set at 1.0 mm root-mean-square (RMS) error by default, although most daily QA values are in the range of 0.1 to 0.3 mm for stable systems. As relative pod motions are being calculated, the daily QA values are independent of the positioning of the calibration plate relative to isocenter. If the relative pod RMS error exceeds the tolerance level, the AlignRT system cannot be used clinically and this should be reported to a physicist who may repeat the QA process or perform other troubleshooting. If the daily QA consistently fails, the system should be recalibrated. The daily QA results can be saved as a PDF report for record keeping and trending analysis.

In addition to the vendor required daily QA, additional QA procedures can be performed by incorporating AlignRT imaging into the daily QA of the kV and/or MV IGRT modalities. For example, AlignRT can be used for initial placement, and tracking of a daily imaging cube phantom while the cube is aligned with CBCT and/or MV imaging. This provides a simple method for confirming the shifts indicated by the AlignRT are correct and agree with the radiographic imaging system, while also verifying the absolute isocenter location. This test would detect any systematic errors introduced during monthly calibration.

Although AlignRT is a reliable system, technical problems may occur, including corruption of the calibration file, losing communication to the cameras, or losing an active license status. When the calibration file is corrupted, a new calibration is required to re-establish the file. When the communication between the AlignRT system and the camera pods is interrupted and lost, the AlignRT workstation and the camera pods will both require a re-initialization. This procedure can help to re-establish the communication between the hardware and software within the system. Although many routine issues can be resolved by a medical physicist, the last resource is to call the vendor's helpdesk for technical support.

8.5.2 Monthly QA

The vendor recommends that the monthly plate calibration procedure is repeated monthly, or when needed (which may be more or less frequently), as described above. From a clinical point of view, MSKCC experience suggests that the calibration does not need to be repeated on a monthly basis if other ongoing QA is performed and unless a problem occurs that renders the calibration invalid. Instead, other QA tests can be performed to assess the need to repeat monthly calibration, including reviewing historical daily QA log files to assess system stability.

For a system with MMI enabled, the beam hold interface can be tested on a monthly basis, or as needed, by introducing an artificial couch rotation during beam delivery and AlignRT motion monitoring. As the phantom moves out of tolerance based on the AlignRT-defined position, the beam should be held automatically. Returning the couch rotation to the initial position should re-enable the beam automatically.

According to TG-147, the frequency of the QA should be daily, monthly, and annually for nonradiographic imaging systems.[2] However, the frequency of recalibration is not specified (usually annual) and should be specific to an imaging system, depending on its stability and accuracy tolerance. As the AlignRT system becomes more and more robust, and as the stability of the optics is validated, the frequency of calibration could be reduced. Specific monthly QA tests can include a subset of the tests performed during commissioning, with the E2E test providing much of the information required in TG-147.

8.5.3 Annual QA and Preventive Maintenance Service

The AlignRT/OSMS system should receive annual preventive maintenance (PM), covered by the service contract and provided by the vendor. The PM service includes several items, such as a review and adjustment of the camera optics. After the service, a full-system calibration should be performed. The AlignRT calibration should provide the adequate accuracy required for clinical treatment. The tests described above, including the E2E test, can measure the accuracy of AlignRT, and determine whether the system requires recalibration, or if the system is maintaining baseline operation since commissioning. The establishment of the clinical frameless SRS workflow has allowed an increase in the number of SRS treatments by two to three times and extension of the SRS procedure to treat hypofractionated intracranial SRT with a reduced margin and accelerated workflow. For other clinical procedures, such as breast SGRT, that do not use couch rotations nor require sub-mm accuracy, simplified calibrations may be performed accordingly by omitting the AlignRT-MV calibration. The trained physicist should evaluate the expected clinical applications of the AlignRT system and design the periodic QA program accordingly. Since 2010, MSKCC has developed two generations of frameless SRS methods,[8,9] and the raised plate calibration procedure clearly enhances accuracy, with the more recent addition of the MV Isocenter Calibration[32,33] and ACO[34,35] further improving the accuracy of AlignRT.

8.5.4 Other Reported Clinical QA Systems and Procedures

It is worthwhile to mention that other studies have compared the accuracy between different imaging systems. For instance, a study to compare three imaging systems: AlignRT, an infrared system, and CBCT illustrated that there is a sub-mm difference between AlignRT and CBCT alignment.[36] Another study describes a prototype wedge-block phantom made of Styrofoam for monthly QA to check the accuracy before and after a known couch shift.[30]

AlignRT has been used for patient setup in proton therapy, where the tissue thickness from the skin to a potentially deep-seated tumor is critical to accurate dose delivery. In one study, AlignRT was used for chest wall patient setup and both the accuracy and efficiency were improved compared with X-ray imaging-based setup alone.[37] Surface guidance for proton therapy is discussed in more detail in Chapter 23.

8.6 CLINICAL OBSERVATIONS AND FUTURE DIRECTIONS

The AlignRT system has gone through substantial improvements and many upgrades in the past decade, making the system more user friendly and more versatile in handling various clinical situations and procedures. In fact, many academic and community hospitals have made clinical recommendations and contributions that have helped develop the system available today. Many of the latest features are the result of feedback from the clinical forefront. In the following section, several recommendations are made for further improvements.

8.6.1 Frame Rate Enhancement (Current vs. Future)

Depending upon the imaging resolution and size of the surface tracking ROI, the AlignRT frame rate may vary. Whether performing motion monitoring for frameless SRS or left-side breast DIBH treatment, the frame rate is an important factor for accurate positioning.[5,16] For SRS, a high-resolution surface with a small ROI (e.g., the upper face) can be monitored for motion at 3–4 frames per second (fps) with the current software. In the case of DIBH, the ROI is often larger but at lower resolution, so a rate of 2–3 fps is achievable and should be sufficient for beam gating during the procedure. Overall, the AlignRT system is capable of sufficient frame rates for current clinical procedures.

However, to extend the application of SGRT to other anatomic sites, such as the thorax with respiratory motion, the frame rate of AlignRT

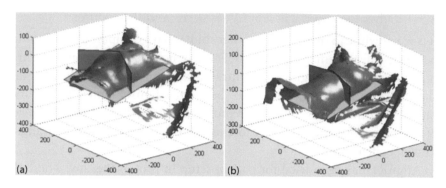

FIGURE 8.9 AlignRT images of the entire male torso (a) and female torso (b) for measuring the respiratory tidal volume and breathing pattern (volume variation between the thorax and the abdomen). The tilted horizontal cut plane is used for calculating the volume under the anterior skin surface to calculate tidal volume using full exhalation as the reference, while the vertical cut planes are separating the thorax from the abdomen for calculating the breathing pattern.

needs to be at least >10 fps, as for other motion monitoring systems, such as RPM (Varian Medical Systems, Palo Alto, CA), Calypso (Varian Medical Systems, Palo Alto, CA), and fluoroscopic systems.[38] The latest version of AlignRT, Version 6, is capable of delivering refresh rates up to 25 fps. Recently, respiratory tidal volume has been measured using AlignRT by imaging the entire torso, providing surface imaging-based spirometry.[39,40] As the entire human torso is used as the ROI (Figure 8.9), real-time image reconstruction and volume calculation is not possible using the current Version 5 system.

Surface image-based spirometry is very promising as a new clinical application because both thoracic and abdominal surfaces are imaged, and the distribution of tidal volume in the thorax (by the costal muscles) and abdomen (by the diaphragm) is shown and quantified to predict the anterior-posterior and superior-inferior motion, respectively. Moreover, as all external surface motions are available, this technique provides the best opportunity to establish a most comprehensive external-internal motion prediction model, either based on statistics or physical laws.[39–41] The breathing pattern, defined as tidal volume distribution between the chest and abdomen, was qualitatively measured using AlignRT.[42,43] However, to implement this technique in the clinic, a frame rate of >10 fps is essential.

8.6.2 Other Recommendations for Improved Clinical Applications

Based on the clinical experience using the AlignRT system at MSKCC, the following recommendations are offered to further facilitate clinical applications. First, it is common for many optical systems to experience occlusions as the rotating linac gantry impedes the line of sight between the camera pods and the patient. This could possibly be avoided by careful positioning of the camera pods, and potentially through the installation of additional pods for certain treatment room settings. More immediate mitigation of the possible accuracy degradation caused by pod occlusions can be realized through the careful design of the ROI, and by implementing calibration techniques such as ACO, where the impact of pod occlusions has been shown to result in a clinically insignificant RTD change of 0.06 mm.[35]

Second, AlignRT research features available to MSKCC enable the user to visualize a color-coded distance map between the reference and verification surfaces. This can be valuable, for instance, when treating a superficial lesion, where any tumor shrinkage will become obvious to the clinician using this feature, thus allowing clinical adaptation. Within the clinical product, AlignRT provides a tool called surface statistics, which provides a surface agreement measurement, and that can be used to provide some information on anatomical changes.

Third, it is desirable that critical actions of AlignRT operation require user sign-off, as currently AlignRT enables regular clinical functions through a group user account, although individual user accounts with role specific rights also exist. User sign-off can create accountability and serves as caution to the clinician when performing a critical procedure. For instance, critical functions requiring user sign-off should include daily QA, the capture of new reference image, and modification of the ROI.

Finally, upcoming software releases (Version 6 and later) will introduce a centralized patient database for improved patient data management and guaranteed real-time patient data saving via temporary cached data buffer, independent of local network performance.

8.7 SUMMARY

In this chapter, we have discussed the clinical commissioning, calibration, and routine QA of the AlignRT system for SGRT in the clinic. Performing the couch-angle dependency test, E2E test, and baseline drift test as part of AlignRT system commissioning is recommended. The necessity of sub-mm calibration of the system depends on the clinical accuracy requirement of existing and future clinical treatment procedures implemented at an individual clinic. For instance, for a dedicated breast treatment machine, sub-mm accuracy may not be necessary, and therefore, the AlignRT MV isocenter calibration is not needed. However, for intracranial SRS treatment, inclusion of the MV Isocenter calibration procedure is necessary as noncoplanar beam fields will likely be used. Submillimeter tracking for coplanar treatment should be possible for all AlignRT users and treatment indication. Monthly plate calibration should be performed as needed, based on the results of other ongoing QA procedures, while daily QA is essential.

KEY POINTS

- Following installation and acceptance testing with the equipment vendor, it is the responsibility of the clinical physicist to commission the system. This includes quantifying system accuracy, defining procedures for calibration and establishing workflows for clinical utilization, and setting up an ongoing QA program.

- It is important to define the intended clinical applications of the system and to understand the associated accuracy requirements. System performance for intracranial SRS is different than for conventional breast radiation therapy, and the calibration and QA requirements should be designed accordingly.

- Daily and monthly QA should follow vendor recommendations as well as AAPM TG-147 guidelines. Additional tests such as couch rotation dependency should be investigated by the user and added to routine QA as necessary.

- E2E testing, particularly for the user's proposed SRS/SRT workflow, is recommended to ensure that the surface guidance system has been fully integrated into the clinic's processes and procedures.

REFERENCES

1. Bert C, Metheany KG, Doppke KP, Taghian AG, Powell SN, Chen GT. Clinical experience with a 3D surface patient setup system for alignment of partial-breast irradiation patients. *Int J Radiat Oncol Biol Phys.* 2006;64(4):1265–1274.
2. Willoughby T, Lehmann J, Bencomo JA, et al. Quality assurance for nonradiographic radiotherapy localization and positioning systems: report of Task Group 147. *Med Phys.* 2012;39(4):1728–1747.

3. Li G, Mageras G, Dong L, Mohan R. Image-guided radiation therapy. In: Khan FM, Gerbi BJ, eds. *Treatment Planning in Radiation Oncology* (pp. 229–258). 4th ed. Philadelphia, PA: Lippincott Williams & Wilkins; 2016.

4. Schoffel PJ, Harms W, Sroka-Perez G, Schlegel W, Karger CP. Accuracy of a commercial optical 3D surface imaging system for realignment of patients for radiotherapy of the thorax. *Phys Med Biol.* 2007;52(13):3949–3963.

5. Cervino LI, Gupta S, Rose MA, Yashar C, Jiang SB. Using surface imaging and visual coaching to improve the reproducibility and stability of deep-inspiration breath hold for left-breast-cancer radiotherapy. *Phys Med Biol.* 2009;54(22):6853–6865.

6. Cervino LI, Pawlicki T, Lawson JD, Jiang SB. Frame-less and mask-less cranial stereotactic radiosurgery: a feasibility study. *Phys Med Biol.* 2010;55(7):1863–1873.

7. Peng JL, Kahler D, Li JG, et al. Characterization of a real-time surface image-guided stereotactic positioning system. *Med Phys.* 2010;37(10):5421–5433.

8. Li G, Ballangrud A, Kuo LC, et al. Motion monitoring for cranial frame-less stereotactic radiosurgery using video-based three-dimensional optical surface imaging. *Med Phys.* 2011;38(7):3981–3994.

9. Li G, Ballangrud A, Chan M, et al. Clinical experience with two frameless stereotactic radiosurgery (fSRS) systems using optical surface imaging for motion monitoring. *J Appl Clin Med Phys.* 2015;16(4):5416.

10. Ho AY, Ballangrud A, Li G, et al. Long-term pulmonary outcomes of a feasibility study of inverse-planned, multibeam intensity-modulated radiation therapy in node-positive breast cancer patients receiving regional nodal irradiation. *Int J Radiat Oncol Biol Phys.* 2018. doi:10.1016/j.ijrobp.2018.11.045.

11. Chang AJ, Zhao H, Wahab SH, Moore K, Taylor M, Zoberi I, Powell SN, Klein EE. Video surface image guidance for external beam partial breast irradiation. *Pract Radiat Oncol.* 2012;2(2):97–105.

12. Gopan O, Wu Q. Evaluation of the accuracy of a 3D surface imaging system for patient setup in head and neck cancer radiotherapy. *Int J Radiat Oncol Biol Phys.* 2012;84(2):547–552.

13. Li G, Lovelock DM, Mechalakos J, et al. Migration from full-head mask to "open-face" mask for immobilization of patients with head and neck cancer. *J Appl Clin Med Phys.* 2013;14(5):243–254.

14. Zhao H, Wang B, Sarkar V, et al. Comparison of surface matching and target matching for image-guided pelvic radiation therapy for both supine and prone patient positions. *J Appl Clin Med Phys.* 2016;17(3):14–24.

15. Cervino L, Pawlicki T. *Stand-Along Localization: Surface Imaging Systems.* Boca Raton, FL: Taylor & Francis Group; 2011.

16. Wiersma RD, McCabe BP, Belcher AH, Jensen PJ, Smith B, Aydogan B. Technical note: high temporal resolution characterization of gating response time. *Med Phys.* 2016;43(6):2802–2806.

17. Stereotactic Radiosurgery, https://www.aapm.org/pubs/reports/RPT_54.pdf. (verified as of December 22, 2018).

18. Dieterich S, Cavedon C, Chuang CF, et al. Report of AAPM TG 135: quality assurance for robotic radiosurgery. *Med Phys.* 2011;38(6):2914–2936.

19. Verellen D, Linthout N, Bel A, et al. Assessment of the uncertainties in dose delivery of a commercial system for linac-based stereotactic radiosurgery. *Int J Radiat Oncol Biol Phys.* 1999;44(2):421–433.

20. Klein EE, Hanley J, Bayouth J, et al. Task Group 142 report: quality assurance of medical accelerators. *Med Phys.* 2009;36(9):4197–4212.

21. Ballangrud A, Kuo LC, Happersett L, et al. Institutional experience with SRS VMAT planning for multiple cranial metastases. *J Appl Clinl Med Phys.* 2018;19(2):176–183.

22. Kuo L, Ballangrud AM, Ho AY, Mechalakos JG, Li G, Hong L. A VMAT planning technique for locally advanced breast cancer patients with expander or implant reconstructions requiring comprehensive postmastectomy radiation therapy. *Med Dosim.* 2018. doi:10.1016/j.meddos.2018.04.006.

23. Mageras GS, Yorke E. Deep inspiration breath hold and respiratory gating strategies for reducing organ motion in radiation treatment *Semin Radiat Oncol.* 2004;14(1):65–75.

24. Sixel KE, Aznar MC, Ung YC. Deep inspiration breath hold to reduce irradiated heart volume in breast cancer patients. *Int J Radiat Oncol Biol Phys.* 2001;49(1):199–204.

25. Tang X, Zagar TM, Bair E, et al. Clinical experience with 3-dimensional surface matching-based deep inspiration breath hold for left-sided breast cancer radiation therapy. *Pract Radiat Oncol.* 2014;4(3):e151–e158.

26. Rong Y, Walston S, Welliver MX, Chakravarti A, Quick AM. Improving intra-fractional target position accuracy using a 3D surface surrogate for left breast irradiation using the respiratory-gated deep-inspiration breath-hold technique. *PLoS One.* 2014;9(5):e97933.

27. Tanguturi SK, Lyatskaya Y, Chen Y, et al. Prospective assessment of deep inspiration breath-hold using 3-dimensional surface tracking for irradiation of left-sided breast cancer. *Pract Radiat Oncol.* 2015;5(6):358–365.

28. Zagar TM, Kaidar-Person O, Tang X, et al. Utility of deep inspiration breath hold for left-sided breast radiation therapy in preventing early cardiac perfusion defects: a prospective study. *Int J Radiat Oncol Biol Phys.* 2017;97(5):903–909.

29. Djajaputra D, Li S. Real-time 3D surface-image-guided beam setup in radiotherapy of breast cancer. *Med Phys.* 2005;32(1):65–75.

30. Wooten HO, Klein EE, Gokhroo G, Santanam L. A monthly quality assurance procedure for 3D surface imaging. *J Appl Clin Med Phys.* 2011;12(1):3338.

31. Winston KR, Lutz W. Linear accelerator as a neurosurgical tool for stereotactic radiosurgery. *Neurosurgery.* 1988;22(3):454–464.

32. Paxton AB, Manger RP, Pawlicki T, Kim GY. Evaluation of a surface imaging system's isocenter calibration methods. *J Appl Clin Med Phys.* 2017;18(2):85–91.

33. Sarkar V, Paxton A, Szegedi MW, et al. An evaluation of the consistency of shifts reported by three different systems for non-coplanar treatments. *J Radiosurg SBRT.* 2018;5(4):323–330.

34. Covington EL, Fiveash JB, Wu X, et al. Optical surface guidance for submillimeter monitoring of patient position during frameless stereotactic radiotherapy. *J Appl Clin Med Phys.* 2019;20(6):91–98.
35. Wiant D, Liu H, Hayes TL, Shang Q, Mutic S, Sintay B. Direct comparison between surface imaging and orthogonal radiographic imaging for SRS localization in phantom. *J Appl Clin Med Phys.* 2019;20(1):137–144.
36. Peng JL, Kahler D, Li JG, Amdur RJ, Vanek KN, Liu C. Feasibility study of performing IGRT system daily QA using a commercial QA device. *J Appl Clin Med Phys.* 2011;12(3):3535.
37. Batin E, Depauw N, MacDonald S, Lu HM. Can surface imaging improve the patient setup for proton postmastectomy chest wall irradiation? *Pract Radiat Oncol.* 2016;6(6):e235–e241.
38. Keall PJ, Mageras GS, Balter JM, et al. The management of respiratory motion in radiation oncology report of AAPM Task Group 76 *Med Phys.* 2006;33(10):3874–3900.
39. Li G, Huang H, Wei J, et al. Novel spirometry based on optical surface imaging. *Med Phys.* 2015;42(4):1690.
40. Li G, Wei J, Huang H, et al. Characterization of optical-surface-imaging-based spirometry for respiratory surrogating in radiotherapy. *Med Phys.* 2016;43(3):1348.
41. Yuan A, Wei J, Gaebler CP, Huang H, Olek D, Li G. A novel respiratory motion perturbation model adaptable to patient breathing irregularities. *Int J Radiat Oncol Biol Phys.* 2016;96(5):1087–1096.
42. Li G, Xie H, Ning H, et al. A novel analytical approach to the prediction of respiratory diaphragm motion based on external torso volume change. *Phys Med Biol.* 2009;54(13):4113–4130.
43. Hughes S, McClelland J, Tarte S, et al. Assessment of two novel ventilatory surrogates for use in the delivery of gated/tracked radiotherapy for non-small cell lung cancer. *Radiother Oncol.* 2009;91(3):336–341.

CHAPTER 9

Surface Guidance for Whole Breast and Partial Breast

Technical Aspects

Laura Padilla

CONTENTS

9.1 INTRODUCTION

Radiotherapy has long been a part of the standard care treatment for breast cancer patients. Thanks to increased access to early detection and improvements to treatment, breast cancer patients typically have high survival rates and long life expectancies. This leads to an increased concern over secondary effects from radiotherapy, even those with long latency periods, because patients may live long enough to experience them. Cardiac morbidity[1,2] and secondary malignancies[2,3] are of highest concern.

In both cases, there is a correlation between increased dose to the organs at risk and higher probability of occurrence of side effects. Incidence of secondary primary malignancies is higher in patients less than 40 years of age.[4] Side effects from radiotherapy related to normal tissues within the treatment area (i.e., surrounding breast tissue, ribcage, and muscle) that do not increase mortality but decrease quality of life and increase long-term morbidity have also been reported. These include breast shrinkage, breast hardness, changes in breast appearance, and chest wall discomfort.[2] These facts have prompted the implementation of newer techniques and approaches to breast cancer treatments beyond the conventional tangential beam geometry. Concerns over heart dose have led to deep inspiration breath hold (DIBH) treatments, which will be discussed further in Chapter 10. Concerns over unnecessary irradiation of surrounding breast tissue and other normal organs for lower-risk breast cancer patients have led to partial breast irradiation (PBI) treatments, which will be discussed in the current chapter along with non-DIBH whole breast irradiation.

PBI only irradiates a planning target volume (PTV) that encompasses the surgical cavity plus a margin. This decreases the total volume of normal breast tissue that is irradiated to reduce side effects while still covering the target. The nature of this treatment implicitly increases the importance of proper setup since the treatment margins are tighter; it does not allow the magnitude of uncertainty permitted by traditional tangential whole breast treatments. Furthermore, in accelerated partial breast irradiation (APBI) treatments, the higher dose per fraction and fewer fractions per treatment, typically 38.5 Gy in 10 fractions, exacerbate this issue.

The importance of more refined patient setup is also true for any breast treatments using intensity modulated radiation therapy (IMRT) or volumetric modulated arc therapy (VMAT). These more complex techniques are being utilized to increase dose homogeneity in the breast tissue during whole breast irradiation and to minimize dose to normal structures such as the lungs and contralateral breast. Traditional patient setup verification protocols utilizing megavoltage (MV) portal images acquired with an electronic portal imaging device (EPID) are not sufficient to ensure proper positioning in these cases.[5] These complex delivery techniques require more careful positioning of the patient due to their reduced margins and more conformal dose distributions; any discrepancies, shifts, or drifting of the patient from the expected treatment position will have a larger impact for these techniques than for the more robust and simplistic tangential breast treatments. As a result, increased image guidance is used. Although the rise in risk of secondary malignancies from the inclusion of image guidance is within the uncertainty levels of the risk prediction models,[6] it is ideal to strive for a patient positioning technique that does not require the use of additional ionizing radiation to avoid any potential issues in these long-term survival patients. Aside from additional patient dose, one should also take into account other practical considerations when designing an imaging sequence for patient position verification, especially if it includes cone-beam CT (CBCT). This type of volumetric imaging can be used to verify the patient's posture through visualization of the internal anatomy. Hence, this option provides not only the possibility of checking bone alignment but also the ability to confirm soft tissue positioning. On the other hand, since image acquisition and review takes time, its implementation can inadvertently increase the chance of patient motion, as it has been shown that longer times on the table lead to higher intra-fractional variations.[7-9] Also, CBCT requires the machine to rotate around the patient for imaging acquisition which can lead to clearance issues with the patient's elbows if setup in the supine position with arms above the head and/or if the patient is positioned on an angled breast board. If the treatment team attempts to adjust the patient's position to avoid collisions, the overall treatment time will be prolonged, negatively impacting clinical workflow, and positional discrepancies will potentially arise in the treatment area from the altered arm placement to ensure clearance. Lastly, because IMRT and VMAT use non-tangential fields, they are more susceptible to breast shape changes.[10] Unfortunately, these are difficult to assess with standard image guidance techniques.

Surface imaging can alleviate some of the concerns raised by current radiographic imaging methods. As will be discussed in this chapter, surface imaging is a very useful tool for whole breast and PBI treatments. It can provide the radiotherapy team valuable information on the patient's position without the use of ionizing radiation. As the breast tissue is a superficial target, assessing how its surface compares to the expected geometry from the treatment plan can guide therapists on how to manipulate the patient to achieve the desired position before starting the radiographic imaging protocol. In certain situations, it might even be able to replace standard imaging techniques.[11] The additional ability to perform intra-fraction monitoring of the patient may also allow for smaller clinical target volume (CTV) to PTV margin expansions, and consequently decrease the amount of normal tissue being irradiated. When used correctly, surface imaging can assist and improve patient setup, provide means to perform intra-fraction monitoring without any additional dose to the patient, and help ensure the delivered dose matches more closely with the planned dose.

9.2 CURRENT PATIENT POSITIONING AND VERIFICATION IMAGING PROTOCOLS

Immobilization of breast cancer patients varies by institution. Supine patients are typically set up with one or both arms raised above the head, sometimes on a breast board, with arm supports to help achieve a more stable and reproducible arm position. This arrangement may or may not include the use of a vacuum bag or cradle. Another approach is to position these patients flat on the table with a wing board and vacuum bag or cradle molded to the shoulders and arms. The extent of immobilization used and how it is implemented at each institution affects the inter- and intra-fraction variability of the treatment.[12,13] This can become more important for PBI or more complex treatment techniques like IMRT or VMAT.

Typically, the amount and type of imaging performed for patient position verification depends on the treatment technique, with frequency and complexity increasing for more conformal treatments. As with immobilization approaches, the variation in imaging protocols across institutions is also large. For conventional tangential breast radiotherapy delivered with or without field-in-field (FiF) or an irregular surface compensator/sliding window technique, MV portal imaging of the tangential fields is standard on the first day of treatment. This is done to ensure the patient's position is acceptable for radiation delivery and that the field apertures are correct with respect to the anatomy. When using FiF or sliding-window, patients are still treated with tangential beams, but not all the dose is delivered through open tangents. Part of the daily dose, typically less than 20%, is delivered through segments designed to improve dose homogeneity. Planar MV portal images acquired at the treatment angles can be compared to digitally reconstructed radiographs (DRRs) and provide information on the position of the chest wall relative to the planned position. They can also show the breast outline by modifying the window and level accordingly. After the first day of treatment, MV portal imaging acquired daily or weekly (depending on the institution) is the standard imaging protocol with patients being initially set up based on skin marks, lasers, and light field projections on non-imaging treatment days. Although this is a very simple imaging protocol, tangential breast radiotherapy does not require very precise patient positioning for successful delivery due to its large built-in margins. The differences between bony setup based on MV portal or CBCT images have been investigated for this type of treatments, and although significant, they have a small impact on dose delivery due to the robustness of the tangential beam geometry. Topolnjak et al. found that using the MV EPID for bony alignment underestimated setup errors compared to CBCT by 20%–50%.[14] The 2D versus 3D nature of the images, the differences in image quality which lead to diminished visibility of anatomical landmarks on the EPID images, and the more challenging registration of these images to the DRRs, especially when little bony anatomy is seen in the field, resulted in more unreliable bony alignment compared to CBCT. Even though this difference was not deemed clinically impactful in this study for tangential breast radiotherapy, it can be relevant for more complex techniques, especially since patient positioning with EPID based on bony anatomy does not ensure proper breast tissue alignment.[15] Despite the robustness of tangential breast radiotherapy, there is still literature showing that positional discrepancies can in fact affect the delivered dose distribution of both open and FiF/sliding-window tangent treatments in a way that could influence the expected outcomes.[5] Even when this is the case, the inclusion of CBCT in the imaging protocol to improve positioning is still not favorably viewed for tangential radiotherapy.[16] For these more simple treatments, the benefits derived from volumetric imaging often do not outweigh the downfalls it can introduce, including additional radiation dose, acquisition time, and increased clearance considerations. In general, the addition of CBCT is only viewed as potentially beneficial for larger patients or patients with larger breast volumes.[12,16] Because the treatment target for breast patients is fairly superficial, the implementation of surface imaging for whole breast tangential radiotherapy patients can provide similar advantages to CBCT without the hindrances.

For more complex treatments of whole breast with IMRT and VMAT, or for PBI, patients are often imaged on a daily basis with a kilovoltage (kV) orthogonal image pair to check the patient's position based on bony alignment, and with weekly to daily CBCTs depending on institutional policy. When verifying the position of PBI patients with internal imaging, clips are often used as they have been shown to be a better surrogate for the cavity than the chest wall.[17–22] Aside from the better image quality and contrast of planar kV imaging over MV imaging, orthogonal kV imaging deposits less dose to the patient, but it still does not provide information about the breast tissue itself. This can be inadequate for certain treatments as it has been shown that changes in breast shape impact the dose delivery of non-tangential treatment geometries, like VMAT.[10] A recent study investigating the changes of breast shape over the course of a radiotherapy treatment found that changes of at least 4 mm occurred in 30% of the fractions analyzed, and were more prevalent in patients with breast volumes exceeding 800 cc.[23] Due to the sensitivity of more conformal, non-tangential treatments to these changes, there have been several studies investigating the dosimetric effect of setup errors and inter-fractional anatomical changes on breast VMAT.[24,25] These have shown that volumetric imaging is important to avoid setup errors, minimize the effect of inter-fractional variations, and identify patients that may need re-planning due to soft tissue changes. This can be especially useful for PBI as information about cavity changes can potentially help further reduce PTV margins and spare more normal tissue than when using skin marks or 2D imaging alone for setup.[26] As was discussed previously, however, the implementation of CBCT is not always simple, and even though more complex techniques can indicate dosimetric advantages for patients during planning, the more difficult and stringent setup can make their implementation cumbersome and practically inferior to more robust techniques. Furthermore, even though it is straightforward to gauge the impact of breast shape changes on the dose distribution during offline analysis, it is hard to discern these effects based on CBCT while the patient is on the table. It is challenging to tell if shape differences detected are due to setup and can be adjusted, or if they are anatomical changes considerable enough to necessitate replanning. When using CBCT images only, it can be unclear how, or if, the tissue can be manipulated to better match the planned position. Additionally, the field of view of the CBCT can be limited beyond what the treatment field covers, so not all the relevant anatomy will be visible in the volumetric imaging set.

Depending on the patient's anatomy, and how pendulous their breast tissue is, patients might be treated in the prone position instead of supine. For patients with large, pendulous breast tissue, standard supine positioning often leads to their breast tissue falling laterally. This makes treatment planning extremely difficult as it leads to more inhomogeneous dose distributions and the inclusion of a larger portion of normal tissue in the field.[2] An alternative is to treat these patients prone,[27] where they are positioned on top of a prone breast board with an opening through which the breast tissue hangs. This allows for a better geometry for treatment of the breast however it may not be a good option if nodes need to be included in the treatment volume. Moreover, prone setup reproducibility can be difficult to achieve and can lead to increased setup times.[28] Prone setups require thorough training of the radiotherapy staff and more extensive checks, especially upon initial implementation.[27] Although prone positioning may reduce intra-fractional motion from respiration, it can lead to larger inter-fractional variations.[29,30] This will require increased margins and more internal imaging if PBI is being performed.

Although complex treatment techniques may theoretically benefit patients during the treatment planning stage, the uncertainties inherent to the setup and delivery of the technique need to be considered when determining their true value. The need for more accurate setup, and the additional time this requires, needs to be contrasted with the fact that longer treatment times lead to larger intra-fractional motion.[7,8] Unless intra-fraction monitoring is implemented and an action plan is enforced, extended treatment times due to extensive imaging verification may require larger margins. Hence, this could diminish the value of using these more complicated treatment delivery techniques.

9.3 SURFACE IMAGING FOR WHOLE BREAST RADIOTHERAPY

The use of surface imaging in the setup of tangential whole breast radiotherapy treatments has been investigated by several groups. As previously discussed, typical setup procedures for whole breast radiotherapy patients are relatively simple since delivered doses are robust to inter-fractional changes due to the large margins of these treatments. Therefore, the addition of surface imaging to this process might not lead to large clinically significant improvements in treatment delivery or patient setup, but it will help improve the overall safety and quality of treatments and will detect larger motions that may be clinically significant.

For instance, the presence of surface imaging can help avoid large setup errors that could occur when aligning patients to skin marks and lasers only, especially in fractions where no imaging verification is performed. Without the use of an independent system to check the position, these errors could go unnoticed for 4–5 fractions. Studies consistently show that surface imaging alignment is superior to laser and skin marks.[9,31–33] Since more information about the patient's overall position is easily available to the therapists, the use of surface imaging can allow an assessment of the patient's external position in real-time to refine the setup prior to leaving the room. Surface imaging will display and highlight tissue and breast shape differences so they can be adjusted, if possible, before radiologic imaging, and provides a more complete evaluation of the patient's isocentric and overall postural setup information. The benefit of having real-time in-room guidance is twofold: (1) it has been reported to motivate therapists to actively scrutinize the position achieved with skin marks and lasers if surface imaging shows a discrepancy,[31] and (2) it will highlight any gross anatomical changes the patient might experience (like extreme swelling or a deflated expander) whether the patient notifies the radiotherapy staff or not.[15]

Shah et al. conducted a study to compare the positioning of whole breast patients with surface imaging to skin marks and lasers, and EPID.[31] Portal imaging was considered the gold standard in this study. They found that the position dictated by portal imaging differed from surface imaging by more than 3 mm in only 14% of the fractions, and of those, 37% occurred within the first 5 days of treatment. When comparing skin marks and lasers to surface imaging, they found that the former led to larger discrepancies in the anterior-posterior direction that had the potential to result in excess dose to the heart and lungs. Since this study focused on tangential breast treatments and compared surface imaging setup to bony alignment on MV imaging, the surface imaging region of interest (ROI) selected by the authors included the region around the treated breast, excluding the actual breast tissue (see Figure 9.1).

This type of ROI selection obviously prioritizes chest wall alignment and does not detect any changes in the breast tissue shape, which is acceptable for the tangential breast treatments included in this study. However, it may lead to incorrect positioning when the breast shape can influence the quality of the dose delivery. In a study by Wei et al., where whole breast treatments were delivered using arcs, the surface monitored included the breast tissue.[34]

In this study they found that surface imaging shifts agreed with those determined by CBCT to within 1 mm and 0.5°, when using a 6 degree-of-freedom couch to apply the shifts. They measured residual errors from positioning by capturing a surface after CBCT-based position correction and determined that a PTV margin of less than 5 mm in all directions would be sufficient to account for these remaining uncertainties. Based on their data, they concluded that positioning with surface imaging could replace the use of CBCT for their patient cohort. Other groups have also reported close agreement between surface imaging and CBCT alignment even with

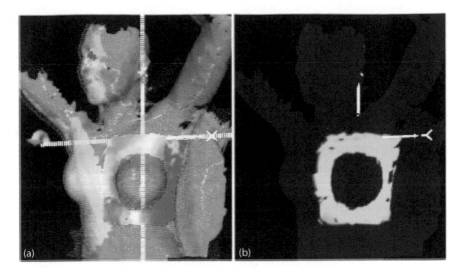

FIGURE 9.1 Examples of AlignRT reference and verification surface images: Panel (a) shows the region of interest from the reference surface in pink and the verification surface acquired with the AlignRT cameras in green. Panel (b) shows the registered region of interest within 2 mm of the reference surface after registration. (Reprinted with permission from Shah, A.P. et al., *Pract. Radiat. Oncol.*, 3, 16–25, 2013.)

the use of a standard treatment table.[35] The potential practical equivalency of volumetric imaging and surface imaging for these treatments was also discussed by Crop et al. when treating whole breast patients with nodal involvement using helical tomographic delivery.[9] In their work they found that when taking into account the additional time needed to acquire and register the megavoltage CT (MVCT) and the inter-user variability of the match, there was no significant difference between patient setup with surface imaging or with MVCT for the patients included in this study.

For post-mastectomy patients receiving radiation to the chest wall, the target area is well correlated with the surface, so the applicability of surface imaging in this scenario is clear, although different considerations apply than for whole breast patients. For patients who have not undergone reconstruction, reliable registration of the real-time patient surface with the reference can be challenging depending on ROI selection. In order for surface imaging systems to find a unique match between the reference surface and the "live" surface acquired during setup, topographically salient features that can anchor the match are necessary. This can be challenging for post-mastectomy patients, and the user should visually ensure that the ROI is acceptable before clinical implementation. This is not an issue for reconstructed patients and surface imaging is an especially useful tool for these setups as the treatment target is superficial and bony anatomy is further away from it.[11] Surface imaging has successfully been used to position post-mastectomy radiotherapy patients being treated with proton beams.[11,36] Positioning with surface imaging proved faster and more accurate than with an orthogonal kV pair, bringing the residual errors within the robustness limits of their pencil beam scanning treatment.[11] Surface image guidance for proton therapy is covered in more detail in Chapter 23.

While surface imaging can be of great use in supine breast treatments, its applicability to prone breast treatments is narrower. Surface imaging can still help position prone patients, but the ceiling-mounted cameras that these systems utilize only achieve limited direct visualization of the breast tissue when the patient is positioned in the prone orientation. Therefore, the utility of surface imaging for detection of breast shape changes or intra-fraction monitoring is more restricted.

Establishing an SGRT program for whole breast and chest wall radiotherapy can aid in efficient patient positioning, reduce the need for repeat verification imaging, and provide intra-fraction motion monitoring, particularly when using more complex techniques that have smaller treatment margins. Current SGRT systems allow flexibility and versatility in the delivery of breast services, and thus can prove highly beneficial to radiotherapy clinics.

9.4 SURFACE IMAGING FOR PARTIAL BREAST IRRADIATION WITH EXTERNAL BEAM RADIOTHERAPY

It has been established in the literature that surgical clips are a reliable surrogate for the resection cavity in partial breast irradiation (PBI) patient alignment. However, this is an internal landmark, so when using surface imaging for PBI patient positioning, the assumption is that the patient's breast surface is a good surrogate for the cavity. The validity of this assumption has been studied by several groups analyzing the patient's surface obtained from the external contour of a volumetric imaging scan or using surface captures from surface imaging systems directly. Hasan et al. compared APBI patient alignment based on surface, clips, and bones by registering the planning CT scan to a second CT scan acquired an average of 27 days apart for 27 patients.[18] All registrations were done manually, and the correct surface match was decided based on the shifts that provided the best average breast surface alignment as determined by the user. The differences between the surrogates were quantified based on the distance between the center of mass (COM) of the cavity in the two image sets. They found that clips provided the best alignment, followed by surface, and bones (the magnitude of difference of the cavity's COM location was 3 ± 2 mm, 5 ± 2 mm, and 7 ± 2 mm, respectively). Surface alignment was increasingly advantageous over bony alignment for patients with large breast volumes. They also found that registration to clips was significantly better than the surface only for cavities located superiorly of the breast COM. This indicates that the reliability of the skin surface as a surrogate for the cavity may be dependent on the cavity location. They concluded that either clips or breast surface can be good surrogates for lumpectomy cavity localization in APBI, although they did warn that non-isotropic shrinkage of the cavity might change its position relative to the whole breast surface.

These findings are in congruence with studies using surface imaging systems to determine the feasibility of surface alignment for APBI patient positioning.[17,37,38] In a study conducted by Gierga et al., the performance of surface-guided alignment for patients undergoing APBI was investigated and compared to positioning with skin marks and lasers, and bony anatomy on portal imaging.[17] The ground truth for the cavity location was taken to be the alignment to surgical clips based on orthogonal kV imaging. The quality of each alignment was quantified using Target Registration Error (TRE) which provides information on the residual setup error of the treatment target after positioning with the technique of choice. The patients included in this study were simulated in free breathing and aligned for treatment based on the implanted clips. TREs for other surrogates were calculated retrospectively. Surface image acquisition for alignment was gated in 8 out of the 12 patients included in this study to suppress the effects of respiratory motion on the alignment after the extent of this was found to be larger than expected for the first 4 patients. In this study, the ROI selected was the entire breast (see Figure 9.2). The TRE for surface imaging was calculated by registering the surface captured at the treatment position with both the surface from the external contour of the planning CT and the treatment surface acquired on the first fraction. This study found that the TRE for gated surface imaging-based alignment using the reference surface acquired during the first fraction was 3.2 mm, in comparison to 5.4 mm and 7.1 mm for the chest wall alignment with kV imaging and laser alignment, respectively. However, the TRE for surface

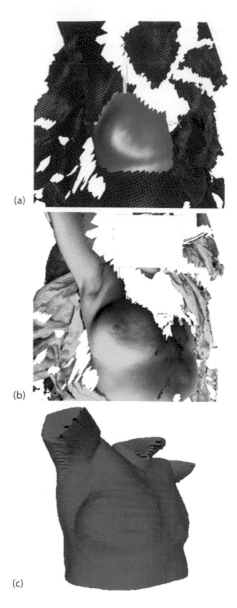

FIGURE 9.2 Example of a reference surface acquired with the AlignRT system with the Region of Interest (ROI) selected for the entire breast (a), a sample video texture image of the same patient (b), and the surface generated from the external body contour of the planning CT (c). Note that the missing medial portion of the surfaces seen in panels A and B is due to the former two-camera installation of this system. Current clinical systems contain a three-camera arrangement that alleviates this artifact. (Reprinted with permission from Gierga, D.P. et al., *Int. J. Radiat. Oncol.*, 70, 1239–1246, 2008.)

imaging when using the planning CT external surface as the reference was 4.9 mm, and the value for non-gated surface captures was 6.2 mm. This is a clear indication that the quality of the setup achieved with surface imaging can be highly variable depending on how it is used. It should be noted that this study was performed when only two camera pods were installed for SGRT, in contrast to modern three pod installations. This may have contributed to some of the variability seen in the setup accuracy results.

Chang et al. reached similar results when they compared surface imaging alignment to bony alignment and skin marks and lasers for 23 patients over 207 fractions.[37] They found that surface imaging yielded the smallest residual setup error out of the three options. They also used a surface imaging ROI encompassing the whole breast. They used the COM of the clips on orthogonal kV imaging as the ground truth for the position of the cavity. Through this study, the vector spatial deviation found with surface imaging was 4.0 ± 2.3 mm, versus 8.3 ± 3.8 mm for bony alignment on orthogonal imaging, and 8.8 ± 4.2 mm when using skin marks and lasers. Therefore, surface imaging was clearly superior to the other positioning techniques. Unlike the previous group, they did not find a need to gate the surface imaging acquisition, and they only used the external contour of the planning CT as the reference surface.

It is unsurprising that the skin surface would provide a better surrogate for the cavity position than bony anatomy. The cavity is located in the breast tissue, which can move independently of the chest wall. Although surgical clips are the most reliable indicator of cavity position, the surface can be used as an acceptable surrogate over bony alignment. In addition, better overall position can be achieved using surface imaging simply due to the fact that a larger field of view of the patient's posture is available. The position of the patient's arm, which can affect the breast tissue shape, is more visible and can be adjusted to achieve a better match between the real-time breast surface and the planned breast surface. Figure 9.3 illustrates the differences in arm position recorded by surface imaging when skin marks and lasers, and portal imaging are used for patient position verification.[38] However, the performance of surface imaging systems can be highly dependent on user input and utilization. The choice of reference surface, ROI delineation, respiratory motion management, planning CT acquisition, etc. can all affect its performance. The effects of some of these factors are discussed next, but oftentimes their true magnitude is very much dependent on institutional protocols and procedures.

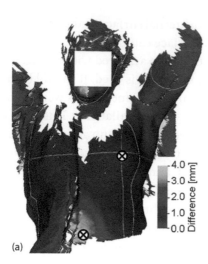

(a) (b)

FIGURE 9.3 Panel (a) illustrates the extent of breathing motion for this patient by displaying the mean distance to the first frame of a typical breathing sequence. Panel (b) displays the inter-fraction difference in the arm position of a patient by overlaying the reference surfaces acquired during different fractions. The surface for each fraction is represented by a different color. (Reprinted with permission from Bert, C. et al., *Int. J. Radiat. Oncol.*, 64, 1265–1274, 2006.)

9.5 PRACTICAL FACTORS THAT MAY AFFECT QUALITY AND PERFORMANCE OF SURFACE IMAGING GUIDANCE

As with any other complex tool used in the clinic, the quality of surface imaging performance is heavily dependent on the user's input and expertise. There are two main categories that one must consider when investigating the factors that affect system's overall performance: user input for the creation of the patient's treatment profile, and proficiency in interpreting the real-time feedback of the system. The selection of reference surface, ROI, and tolerance levels can all have a strong effect on how successfully surface imaging is implemented into the clinic. These parameters are all typically specified prior to treatment during the creation of the patient's treatment profile. How well the user interprets the feedback provided by the system during initial positioning or intra-fraction monitoring also affects how useful and reliable surface imaging ultimately will be in the clinic. While the first category will be expanded upon in this section, the latter is harder to elucidate in a book chapter and is largely developed from hands-on experience; it will not be explored further here. Even though both aspects are important for optimal use of surface imaging, the factors outlined in this section will help create a solid foundation from which to develop the surface imaging program as user proficiency grows. The user should also be well-versed on how their particular surface imaging system works and what type of registration algorithm it uses in order to maximize performance. Specifics of how each currently commercially available systems work have already been discussed in previous chapters and will not all be repeated in this section.

ROI definition for breast patient positioning can greatly affect the results achieved with surface imaging. The anatomy encompassed by this region must both be relevant for the type of treatment being delivered and display enough topographically salient features to ensure the quality of the registration. The anatomy relevant to the treatment will be that which can affect the accuracy of the dose delivery most. Hence, it is important to know what anatomical landmarks should be prioritized during patient positioning to ensure appropriate dose delivery for a given treatment type. As we have seen, for tangential treatments where the dose is delivered largely through open fields, the literature indicates that chest wall alignment, not breast shape, should drive patient positioning as discrepancies in bony alignment lead to the greatest dose uncertainty.[10] An ROI delineation excluding the breast tissue but including the chest wall, like the one shown in Figure 9.1, should be selected in this case. On the other hand, if breast tissue shape and position are the driving factor for proper dose delivery, as is the case for more modulated treatments with non-tangential fields[10] and PBI treatments,[17,37,38] the breast tissue should be included in the ROI definition, as shown in Figure 9.2. If more than one ROI can be created in a given patient's profile, it can be helpful to design a protocol that includes several standardized ROIs for different setup and monitoring purposes. For instance, if both bony alignment and soft tissue shape are of interest, two ROIs can be created. A larger one, encompassing the patient's torso bilaterally but excluding breast tissue, can help better detect overall yaw and roll rotations of the patient's body without including the effects of breast shape differences into the shifts calculated by the surface imaging system. A smaller one, focused on the treated breast tissue, can then be utilized to refine the position and check the breast shape. Extra ROIs beyond the standard protocol set can always be added to address patient-specific situations if applicable. For example, if the contralateral breast dose is of concern for a specific patient's treatment, and the position of that breast needs to be checked, an ROI encompassing the bilateral breasts can also be created to ensure the relative position of the two breasts is consistent with what is expected. The same thing can be done for the arm or chin if their position is deemed important. Although the utility of additional ROIs can be great, one should be cautious not to crowd the patient's profile with too many options. The more ROIs that are available for setup and monitoring, the more potential can exist for error during treatment, through factors such as increased mental workload by therapy staff or simply selecting the wrong ROI.

Another important consideration for ROI delineation is the topography it encompasses. Registration between the real-time patient position and the reference surface can sometimes reach a non-unique solution if there are no distinct features to anchor the registration. As stated earlier in the chapter, this can occur when creating ROIs for non-reconstructed chest wall patients. If the ROI only includes the flat portion of the chest wall, the system may have reduced performance calculating reliable longitudinal and lateral translational shifts. Lateral shifts can be resolved by extending the ROI laterally over the patient's side along the ribcage. Anatomical characteristics to more accurately define the longitudinal match can be more challenging to find in patients with no longitudinal topography. This same issue could be encountered when using surface imaging for prone treatments, especially if the lateral aspect of the pendulous breast tissue is not captured by the system. Since in that case the ROI would have to be drawn on the patient's back, it is possible that it will lack differentiating features. One approach that can be taken to decrease some of the surface registration uncertainties from ROI definition is to constrain the registration using pre-identified and distinctive anatomical landmarks.[39] This can help when the ROI contains insufficient unique topography or when it encompasses deformable tissues where rigid registration might lead to a good overall surface match at the expense of misaligning the anatomical landmark known to correlate better with the target position. Riboldi et al. present a constrained surface fitting algorithm where a photographic-like image of the patient's surface, which is often available in the current commercially available systems, is merged with the 3D surface to create a textured map where landmarks, like the nipple, can be identified.[39] These structures are then used to assist the registration algorithm in reaching a solution that prioritizes the alignment of the identified structure, at the possible expense of decreased overall surface congruence. Although this workflow was not tested against internal imaging to assess its clinical effects, it appears to resolve some of the clinical limitations present in surface imaging systems utilizing rigid registration algorithms. However, this feature is currently not commercially available.

Reference surface selection can also influence the performance of surface imaging systems. As previously explained, one can select either the surface created from the external contour of the planning CT or a surface acquired in the treatment room with the system cameras as the reference (captured after confirming the adequacy of the position with radiographic internal imaging). Since the expected dose delivery is based on the patient's position from simulation, the use of the external contour of the planning CT as the reference surface will aim to replicate this position as closely as possible. Although this would ideally be achievable for all patients and all fractions, the reality is that the simulation position may not be perfectly attained in the treatment room for a wide range of reasons. The surface derived from the planning CT can be affected by image artifacts arising from a free breathing acquisition or could be impacted from using a suboptimal HU value for thresholding the external body contour. Furthermore, the external body contour may only include a limited range of anatomy which may be insufficient for proper positioning. Under any of those conditions, the reference surface from the planning CT can be suboptimal and it may be desirable to acquire a new reference surface with the in-room cameras after radiographic imaging. The downfall of this acquisition, if done too often or carelessly, is that its use as a reference for initial positioning might introduce systematic errors in the setup if the position captured is not ideal, or if it masks anatomical changes that could eventually affect dose delivery. On the other hand, using a camera-acquired reference for intra-fraction monitoring has the advantage over the use of the planning CT reference that it will zero-out all discrepancies between the patient's current position and the planned position. This will allow for the detection of any motion that occurs during treatment. If the external contour of the planning CT is used as the reference for intra-fraction monitoring instead, tolerances will likely have to be increased to keep these baseline discrepancies from reaching the action levels that might prompt beam hold. A more extensive account of the detailed advantages and disadvantages of the reference surface selection have been published elsewhere.[40]

Tolerance levels selected as action levels for initial positioning, but especially for intra-fraction monitoring, should be carefully defined based on system performance, treatment margins, delivery technique, known correlations between the magnitude of surface movements and the extent of motion of the target, and reference surface selection. Since the breast treatments encompassed in this chapter are delivered while the patient is free breathing, periodic respiratory motion of the patient's breast surface is expected and should be taken into account when deciding tolerance values for beam gating, if enabled. There are several publications studying the degree of motion of the breast surface with breathing, and these indicate that the motion is generally below 3 mm.[8,12,38,41–43] Figure 9.3 shows an example of measured breathing motion on a patient's surface. If the intra-fraction monitoring surface is based on a gated capture acquisition, and the mid-ventilation position is selected, tolerance levels of 1.5–2.0 mm may be sufficient to encompass normal breathing motion without interrupting beam delivery, assuming that camera occlusion from the gantry during treatment does not affect the shifts calculated by the surface imaging system. If the intra-fraction monitoring reference surface is not acquired in gated mode, then it could represent any point of the breathing cycle. This can lead to larger discrepancies detected by the system biased in one direction, depending on whether the acquired surface is closer to exhalation or inhalation. This will either force the user to allow for larger tolerances during treatment, or if the tolerances are kept the same, it could cause the beam to be interrupted more often. Hence, the adjustment of tolerance levels to allow for respiratory motion should be determined by taking both respiratory motion amplitude and reference surface selection into account. For reference, both Shah et al.[31] and Gierga et al.[17] used translational threshold levels of 3 mm for their whole breast and partial breast studies, respectively.

Lastly, while more extensive, properly crafted immobilization devices can help reduce the magnitude of inter- and intra-fraction motion uncertainties when using standard patient setup and verification approaches (laser and skin marks, and internal imaging), their effects on the surface imaging system performance should be taken into account. Having immobilization accessories that occlude the patient's surface and restrict the ability of the therapists to adjust the patient's position through surface imaging might hinder the benefits of this technology. If the inclusion of surface imaging is established, the immobilization of patients should be revised and optimized to ensure it does not interfere with the system's performance and is compatible with how surface imaging is used at that institution.

KEY POINTS

- The use of surface imaging for whole breast radiotherapy provides a more complete evaluation of the patient's overall posture, and studies have shown SGRT to be more accurate than a traditional three-point/laser-based setup.

- Surface imaging allows the patient's external position and breast tissue shape to be assessed in real-time to refine the setup prior to leaving the room and can reduce the need for repeat imaging.

- For conventional tangential treatment fields, surface imaging complements current daily or weekly image verification protocols without adding additional dose and can facilitate the accurate setup required for complex treatment techniques such as IMRT or VMAT.

- In addition to whole breast radiotherapy, surface imaging can be of benefit when setting up patients for PBI or chest wall treatments.

REFERENCES

1. Darby SC, Ewertz M, McGale P, et al. Risk of ischemic heart disease in women after radiotherapy for breast cancer. *N Engl J Med*. 2013;368(11):987–998. doi:10.1056/NEJMoa1209825.
2. Kirby A. Whole-breast irradiation following breast-conserving surgery for invasive breast cancer. In: Veronesi U, Goldhirsch A, Veronesi P, Gentilini OD, Leonardi MC, eds. *Breast Cancer: Innovations in Research and Management* (pp. 621–630). Cham, Switzerland: Springer International Publishing; 2017. doi:10.1007/978-3-319-48848-6_51.
3. Grantzau T, Overgaard J. Risk of second non-breast cancer among patients treated with and without postoperative radiotherapy for primary breast cancer: a systematic review and meta-analysis of population-based studies including 522,739 patients. *Radiother Oncol*. 2016;121(3):402–413. doi:10.1016/j.radonc.2016.08.017.
4. Brown LM, Chen BE, Pfeiffer RM, et al. Risk of second non-hematological malignancies among 376,825 breast cancer survivors. *Breast Cancer Res Treat*. 2007;106(3):439–451. doi:10.1007/s10549-007-9509-8.
5. Jain P, Marchant T, Green M, et al. Inter-fraction motion and dosimetric consequences during breast intensity-modulated radiotherapy (IMRT). *Radiother Oncol*. 2009;90(1):93–98. doi:10.1016/j.radonc.2008.10.010.
6. Donovan EM, James H, Bonora M, Yarnold J, Evans P. Second cancer incidence risk estimates using BEIR VII models for standard and complex external beam radiotherapy for early breast cancer. *Med Phys*. 2012;39(10):5814–5824. doi:10.1118/1.4748332.

7. Wiant DB, Wentworth S, Maurer JM, Vanderstraeten CL, Terrell JA, Sintay BJ. Surface imaging-based analysis of intra-fraction motion for breast radiotherapy patients. *J Appl Clin Med Phys.* 2014;15(6):147–159. doi:10.1120/jacmp.v15i6.4957.

8. Ricotti R, Ciardo D, Fattori G, et al. Intra-fraction respiratory motion and baseline drift during breast Helical Tomotherapy. *Radiother Oncol.* 2017;122(1):79–86. doi:10.1016/j.radonc.2016.07.019.

9. Crop F, Pasquier D, Baczkiewic A, et al. Surface imaging, laser positioning or volumetric imaging for breast cancer with nodal involvement treated by helical TomoTherapy. *J Appl Clin Med Phys.* 2016;17(5):200–211. doi:10.1120/jacmp.v17i5.6041.

10. van Mourik A, van Kranen S, den Hollander S, Sonke J-J, van Herk M, van Vliet-Vroegindeweij C. Effects of setup errors and shape changes on breast radiotherapy. *Int J Radiat Oncol.* 2011;79(5):1557–1564. doi:10.1016/j.ijrobp.2010.07.032.

11. Batin E, Depauw N, MacDonald S, Lu H-M. Can surface imaging improve the patient setup for proton postmastectomy chest wall irradiation? *Pract Radiat Oncol.* 2016;6(6):e235–e241. doi:10.1016/j.prro.2016.02.001.

12. Fatunase T, Wang Z, Yoo S, et al. Assessment of the residual error in soft tissue setup in patients undergoing partial breast irradiation: results of a prospective study using cone-beam computed tomography. *Int J Radiat Oncol.* 2008;70(4):1025–1034. doi:10.1016/j.ijrobp.2007.07.2344.

13. Michalski A, Atyeo J, Cox J, Rinks M. Inter- and intra-fraction motion during radiation therapy to the whole breast in the supine position: a systematic review. *J Med Imaging Radiat Oncol.* 2012;56(5):499–509. doi:10.1111/j.1754-9485.2012.02434.x.

14. Topolnjak R, Sonke J-J, Nijkamp J, et al. Breast patient setup error assessment: comparison of electronic portal image devices and cone-beam computed tomography matching results. *Int J Radiat Oncol.* 2010;78(4):1235–1243. doi:10.1016/j.ijrobp.2009.12.021.

15. Padilla L, Kang H, Washington M, Hasan Y, Chmura SJ, Al-Hallaq H. Assessment of inter-fractional variation of the breast surface following conventional patient positioning for whole-breast radiotherapy. *J Appl Clin Med Phys.* 2014;15(5):177–189. doi:10.1120/jacmp.v15i5.4921.

16. Batumalai V, Phan P, Choong C, Holloway L, Delaney GP. Comparison of setup accuracy of three different image assessment methods for tangential breast radiotherapy. *J Med Radiat Sci.* 2016;63(4):224–231. doi:10.1002/jmrs.180.

17. Gierga DP, Riboldi M, Turcotte JC, et al. Comparison of target registration errors for multiple image-guided techniques in accelerated partial breast irradiation. *Int J Radiat Oncol.* 2008;70(4):1239–1246. doi:10.1016/j.ijrobp.2007.11.020.

18. Hasan Y, Kim L, Martinez A, Vicini F, Yan D. Image guidance in external beam accelerated partial breast irradiation: comparison of surrogates for the lumpectomy cavity. *Int J Radiat Oncol.* 2008;70(2):619–625. doi:10.1016/j.ijrobp.2007.08.079.

19. Park CK, Pritz J, Zhang GG, Forster KM, Harris EER. Validating fiducial markers for image-guided radiation therapy for accelerated partial breast irradiation in early-stage breast cancer. *Int J Radiat Oncol.* 2012;82(3):e425–e431. doi:10.1016/j.ijrobp.2011.07.027.

20. Topolnjak R, de Ruiter P, Remeijer P, van Vliet-Vroegindeweij C, Rasch C, Sonke J-J. Image-guided radiotherapy for breast cancer patients: surgical clips as surrogate for breast excision cavity. *Int J Radiat Oncol.* 2011;81(3):e187–e195. doi:10.1016/j.ijrobp.2010.12.027.

21. Kim LH, Wong J, Yan D. On-line localization of the lumpectomy cavity using surgical clips. *Int J Radiat Oncol.* 2007;69(4):1305–1309. doi:10.1016/j.ijrobp.2007.07.2365.

22. Weed DW, Yan D, Martinez AA, Vicini FA, Wilkinson TJ, Wong J. The validity of surgical clips as a radiographic surrogate for the lumpectomy cavity in image-guided accelerated partial breast irradiation. *Int J Radiat Oncol Biol Phys.* 2004;60(2):484–492. doi:10.1016/j.ijrobp.2004.03.012.

23. Alderliesten T, Heemsbergen WD, Betgen A, et al. Breast-shape changes during radiation therapy after breast-conserving surgery. *Phys Imaging Radiat Oncol.* 2018;6:71–76. doi:10.1016/j.phro.2018.05.006.

24. van der Veen GJ, Janssen T, Duijn A, et al. A robust volumetric arc therapy planning approach for breast cancer involving the axillary nodes. *Med Dosim.* 2018. doi:10.1016/j.meddos.2018.06.001.

25. Rossi M, Boman E, Skyttä T, Haltamo M, Laaksomaa M, Kapanen M. Dosimetric effects of anatomical deformations and positioning errors in VMAT breast radiotherapy. *J Appl Clin Med Phys.* 2018;19(5):506–516. doi:10.1002/acm2.12409.

26. Ahunbay EE, Robbins J, Christian R, Godley A, White J, Li XA. Interfractional target variations for partial breast irradiation. *Int J Radiat Oncol.* 2012;82(5):1594–1604. doi:10.1016/j.ijrobp.2011.01.041.

27. Krengli M, Masini L, Caltavuturo T, et al. Prone versus supine position for adjuvant breast radiotherapy: a prospective study in patients with pendulous breasts. *Radiat Oncol Lond Engl.* 2013;8:232. doi:10.1186/1748-717X-8-232.

28. Veldeman L, Gersem WD, Speleers B, et al. Alternated prone and supine whole-breast irradiation using IMRT: setup precision, respiratory movement and treatment time. *Int J Radiat Oncol Biol Phys.* 2012;82(5):2055–2064. doi:10.1016/j.ijrobp.2010.10.070.

29. Kirby AM, Evans PM, Helyer SJ, Donovan EM, Convery HM, Yarnold JR. A randomised trial of Supine versus Prone breast radiotherapy (SuPr study): comparing set-up errors and respiratory motion. *Radiother Oncol.* 2011;100(2):221–226. doi:10.1016/j.radonc.2010.11.005.

30. Morrow NV, Stepaniak C, White J, Wilson JF, Li XA. Intra- and interfractional variations for prone breast irradiation: an indication for image-guided radiotherapy. *Int J Radiat Oncol Biol Phys.* 2007;69(3):910–917. doi:10.1016/j.ijrobp.2007.06.056.

31. Shah AP, Dvorak T, Curry MS, Buchholz DJ, Meeks SL. Clinical evaluation of inter-fractional variations for whole breast radiotherapy using 3-dimensional surface imaging. *Pract Radiat Oncol.* 2013;3(1):16–25. doi:10.1016/j.prro.2012.03.002.

32. Li S, DeWeese T, Movsas B, et al. Initial validation and clinical experience with 3D optical-surface-guided whole breast irradiation of breast cancer. *Technol Cancer Res Treat.* 2012;11(1):57–68. doi:10.7785/tcrt.2012.500235.

33. Cravo Sá A, Fermento A, Neves D, et al. Radiotherapy setup displacements in breast cancer patients: 3D surface imaging experience. *Rep Pract Oncol Radiother.* 2018;23(1):61–67. doi:10.1016/j.rpor.2017.12.007.

34. Wei X, Liu M, Ding Y, et al. Setup errors and effectiveness of Optical Laser 3D Surface imaging system (Sentinel) in postoperative radiotherapy of breast cancer. *Sci Rep.* 2018;8(1):7270. doi:10.1038/s41598-018-25644-w.

35. Ma Z, Zhang W, Su Y, et al. Optical surface management system for patient positioning in inter-fractional breast cancer radiotherapy. *BioMed Res Int.* 2018. doi:10.1155/2018/6415497.

36. Depauw N, Batin E, Daartz J, et al. A novel approach to postmastectomy radiation therapy using scanned proton beams. *Int J Radiat Oncol Biol Phys.* 2015;91(2):427–434. doi:10.1016/j.ijrobp.2014.10.039.

37. Chang AJ, Zhao H, Wahab SH, et al. Video surface image guidance for external beam partial breast irradiation. *Pract Radiat Oncol.* 2012;2(2):97–105. doi:10.1016/j.prro.2011.06.013.

38. Bert C, Metheany KG, Doppke KP, Taghian AG, Powell SN, Chen GTY. Clinical experience with a 3D surface patient setup system for alignment of partial-breast irradiation patients. *Int J Radiat Oncol.* 2006;64(4):1265–1274. doi:10.1016/j.ijrobp.2005.11.008.

39. Riboldi M, Gierga DP, Chen GT, Baroni G. Accuracy in breast shape alignment with 3D surface fitting algorithms. *Med Phys.* 2009;36(4):1193–1198. doi:10.1118/1.3086079.

40. Al-Hallaq HA, Gutiérrez AN, Padilla L. Surface image-guided radiotherapy: clinical applications and motion management. In: Alaei P, Ding GX, eds. *Image Guidance in Radiation Therapy: Techniques, Accuracy, and Limitations* (pp. 309–345). Madison, WI: Medical Physics Publishing, 2018.

41. Yue NJ, Goyal S, Zhou J, Khan AJ, Haffty BG. Intra-fractional target motions and uncertainties of treatment setup reference systems in accelerated partial breast irradiation. *Int J Radiat Oncol.* 2011;79(5):1549–1556. doi:10.1016/j.ijrobp.2010.05.034.

42. Reitz D, Carl G, Schönecker S, et al. Real-time intra-fraction motion management in breast cancer radiotherapy: analysis of 2028 treatment sessions. *Radiat Oncol Lond Engl.* 2018;13(1):128. doi:10.1186/s13014-018-1072-4.

43. Glide-Hurst CK, Shah MM, Price RG, et al. Intra-fraction variability and deformation quantification in the breast. *Int J Radiat Oncol.* 2015;91(3):604–611. doi:10.1016/j.ijrobp.2014.11.003.

CHAPTER 10

Surface Guidance for Breast

Deep Inspiration Breath Hold

Xiaoli Tang

CONTENTS

10.1 INTRODUCTION

The risk of radiation therapy-associated cardiovascular disease in women with breast cancer has been a concern for decades.[1,2] One approach to reduce incidental cardiac irradiation is to treat patients specifically during a deep inspiration breath hold (DIBH) maneuver, where for most patients, a deep inspiration displaces the heart medially, inferiorly, and posteriorly

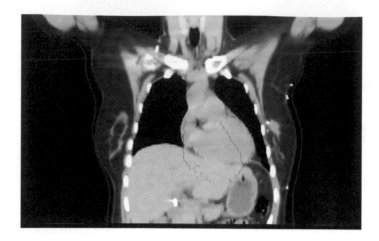

FIGURE 10.1 Comparison of the heart position of normal free breathing (*yellow line*) and breath hold (*purple line*) positions. By taking a deep inspiration, the heart is moved interior, posterior, and medial, increasing separation from the left breast.

(i.e., away from the left breast and chest wall; Figure 10.1). Using a variety of methods, the clinician can monitor the patient's respiration, instruct them to take in a deep breath, and when the patient achieves the required level of deep inspiration, the breath is then held for as long as possible while the radiation dose is delivered. This approach allows a treatment plan to maintain coverage of the target tissues while markedly reducing the degree of incidental cardiac irradiation.[3,4]

The success of the DIBH technique requires an accurate and reliable method for monitoring the level and duration of the breath-hold. A variety of systems have been implemented to deliver radiation during specific portions of the normal respiratory cycle, that is, gating, or during a breath-hold. One approach is to monitor the respiratory phase directly via spirometry (i.e., how much air is inhaled) and either actively hold the breath, with the active breathing coordinator (ABC) system (Elekta, Stockholm, Sweden)[3,5] or voluntarily, with the SDX Spirometric Motion Management System (Qfix, Avondale, PA).[6] Another approach is through systems that monitor respiration via a surrogate of respiration, such as the motion of the chest or abdomen. Such monitoring systems include the real-time position management (RPM) system (Varian Medical Systems, Palo Alto, CA) and the Anzai system (Anzai Medical Co Ltd, Tokyo, Japan). The RPM system monitors the movement of an infrared marker block placed at a single point on the patient's surface,[7] while the Anzai system uses a strain gauge to measure changes in pressure induced in a belt placed around patient's chest or abdomen.[8] Most recently, DIBH can be performed using surface-guided radiation therapy (SGRT) systems such as AlignRT (Vision RT Ltd, London, UK), Catalyst (C-Rad AB, Uppsala, Sweden), and IDENTIFY (Varian Medical Systems, Palo Alto, CA). The AlignRT, Catalyst, and IDENTIFY systems directly track the skin surface using optical techniques via ceiling-mounted cameras inside the treatment vault.

Of these options, the SGRT systems are potentially more accurate as they directly measure patient chest wall and abdominal motion without an additional sensor device or reflector block, and they consider multiple points simultaneously within a region of interest (ROI) on the patient's surface (rather than just one or a few points), and thus can assess for variations in both translational and rotational position. These features are particularly useful for patients being treated for breast cancer where the tracked surface is part of the target itself, rather than just a surrogate for internal target motion.

In Chapter 9, the general use of SGRT for guiding delivery of whole breast, chest wall, and partial breast radiation therapy was presented. In this chapter, a clinical workflow is presented for using SGRT to deliver radiation therapy to patients with left-sided breast cancer using the DIBH technique for cardiac sparing. The clinical workflow includes clinical evaluation and presimulation activities, computed tomography (CT) simulation, treatment planning, pretreatment data processing for the SGRT system, and treatment delivery. Clinical data on SGRT-based DIBH setup accuracy as well as the dosimetric effect of residual motion during DIBH will also be reviewed. A more detailed discussion of cardiac toxicity and clinical outcomes with DIBH is presented in Chapter 11.

10.2 OVERVIEW OF THE SGRT-BASED DIBH WORKFLOW

Figure 10.2 shows a sample workflow for using SGRT to deliver DIBH radiation therapy, using AlignRT as an example.

10.2.1 Clinical Evaluation and Patient Selection

Patient selection is one of the most important steps of any DIBH program. The success of the treatment delivery largely relies on whether the patient is capable of performing consistent breath-holds. Patient selection can start at the initial consult. Physicians prescreen left-sided breast cancer patients who might derive the most clinical benefit from cardiac sparing techniques.

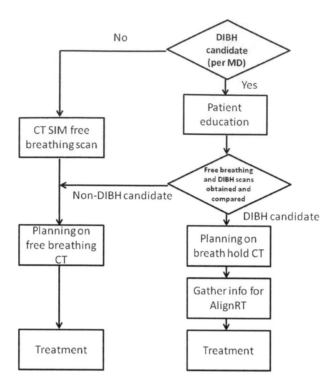

FIGURE 10.2 Overall procedure for DIBH patient selection, simulation, and treatment. Two decision-making steps are in *diamonds*, and others are in *rectangles*.

Patients are typically deemed good candidates for consideration of DIBH if the target tissues are believed to be in close proximity to the heart (e.g., tumor bed in the inferior medial breast) although this usually cannot be confirmed until CT planning images are available. Furthermore, patients must be capable of achieving and maintaining the DIBH position (e.g., can understand instructions and are cooperative and lack respiratory issues such as severe pulmonary disease). Prior to the CT simulation, therapists may coach the patient in the DIBH technique. At some institutions, this includes reviewing a DIBH educational information sheet that describes the procedures to be used at the CT simulation and during treatment delivery. It is important to ensure patients that their breathing will be monitored and the beam will be turned off if they can no longer hold their breath as patients can become upset or stressed if they feel responsible for compromising treatment delivery accuracy. Providing patients with a brief overview of the associated cardiac benefit and risk can also be helpful.

As mentioned, patient selection is one of the most important elements in a successful DIBH program and the patient's suitability for DIBH can usually be accurately assessed at the time of consult and simulation. This is based on a study that found that treatment durations are clinically acceptable and appear not to change significantly over time on an intra-patient basis, and improve over time on an interpatient basis.[11] This indicates that for an individual patient, if they find it challenging to perform a consistent breath-hold at simulation or the beginning of treatment, it is unlikely that their breath-hold will become more consistent after further treatments. However, this study also found that the treatment duration per patient, in the chronologic order of being treated, displayed a decreasing trend.

10.2.2 CT Simulation Activities

Typically, patients are set up on the CT couch with a positioning aid such as a breast board, with both arms above the head. When the patient is on the CT couch and before commencing any scans, a breathing coaching and evaluation session are conducted. The patient is coached to take two to three deep breaths in and to hold for approximately 20 s each time. To increase the reproducibility of the depth of inspiration and maintenance of this level during the breath-hold, patients are instructed to try and hold their breath at a comfortable level for CT simulation and for daily treatments. The consistency of the breath-hold level is then measured using an available device in the CT room (e.g., GateCT [Vision RT Ltd, London, UK], Sentinel [C-Rad AB, Uppsala, Sweden] or RPM).[9] Patients who can perform two to three breath-holds with reasonable consistency (e.g., ±5 mm of the same level with each DIBH) and who demonstrate a measurable surface excursion between the free breathing and breath-hold positions are considered good candidates and can continue with the DIBH workflow. If using an infrared market block such as with RPM, it is important to document the location of the block on the patient's abdomen, for example, relative to the xyphoid or umbilicus. Variable placement of the block can affect the measured breathing amplitude, which in turn affects the reproducibility of the DIBH chest wall excursion.

During the CT simulation, both a free-breathing CT and a breath-hold CT scan are obtained. The attending physician can then compare them to assess whether there is sufficient cardiac displacement to warrant treating patients with DIBH. Typically, physicians review paired axial free-breathing and DIBH images from the same approximate location within the breast and compare the location of the heart relative to the chest wall and tumor bed, thus assessing the degree of cardiac displacement with DIBH. By visualizing where the deep border of a tangent field would need to be placed on each of the CT data sets, with complete cardiac blocking, the physician also assesses if treatment using the DIBH position maintains coverage of the target. For patients where DIBH treatment is deemed beneficial, the breath-hold CT will be used for treatment planning. Moreover, the free-breathing CT can be sent to the treatment planning system for defining the external body contour used for initial patient setup prior to treatment, in clinics where initial setup is always performed while the patient is free-breathing.

10.2.3 Treatment Planning

The heart should be segmented on the CT typically following the Radiation Therapy Oncology Group atlas,[10] extending superiorly from an axial level just inferior to the pulmonary artery to the cardiac apex inferiorly. Whole-breast radiation therapy patients are typically treated with tangent fields. Techniques such as mixed-energy, field in field (FinF), and irregular surface compensators can be used to improve dose homogeneity. Different institutions might have slightly different dose/volume constraints for the heart and other critical organs. During treatment planning, the physician defines the borders of the treatment field while balancing the competing goals of target coverage and normal tissue sparing, which are patient specific. For most cases, the heart is totally excluded from the primary treatment beam unless this compromises coverage of the glandular breast tissue within several centimeters of the tumor bed (usually, marked by clips).

For some patients, tangent fields might not be able to spare enough contralateral breast, for instance, when internal mammary nodes (IMNs) are involved, or when both sides of the breast have reconstructions. In these cases, intensity modulated radiation therapy (IMRT) or volumetric modulated arc therapy (VMAT) planning techniques can be used instead. However, the increased modulation possible with these techniques also comes with increased treatment time. IMRT or VMAT plans for DIBH patients likely will require a greater number of breath-holds for the patient. Patients may become tired and unable to perform consistently after multiple breath-holds. Therefore, compromises may need to be made between the ability to shape the dose in the plan with IMRT or VMAT and limiting the treatment time to prevent the patient from getting too fatigued to efficiently complete the treatment.

10.2.4 Plan Preparation and Pretreatment Quality Assurance for the SGRT System

Depending on the SGRT system being used, the plan preparation and pre-treatment quality assurance (QA) steps might be different. The following workflow is specific to AlignRT but is representative of the general DIBH workflow and can be modified as necessary for other systems. After the treatment plan is reviewed and approved, the external body contours from the breath-hold planning CT, and from the free breathing CT (used during initial setup), are reviewed to ensure a consistent and smooth transition from one slice to another, and the DICOM RT structure sets containing the external body contour are sent to the AlignRT workstation. If used for setup, the user should ensure that the free-breathing body structure is defined in coordinates relative to the treatment isocenter of the DIBH plan and body structure. This can be achieved by copying the free-breathing body structure into the DIBH structure set prior to export or by transferring the DIBH plan onto the free-breathing image set. The treatment isocenter coordinates in the DIBH plan are then sent from the treatment planning system to the AlignRT workstation, along with other treatment plan information such as patient name and identification number. The accuracy of the transfer is verified by comparing the isocenter coordinates to those in the DIBH planning CT. Visual inspection is also done by comparing the 3D location of the isocenter in the AlignRT workstation and in the planning CT side by side. Using the AlignRT workstation, an ROI is defined on the free-breathing and DIBH external body contours related to or including the treatment area for use in setup. When irradiating the supraclavicular nodes, some clinics find it useful to include a part of the neck or chin in the ROI to ensure the head and neck is turned away from the supraclavicular field. Similarly, when arm position has a strong influence on breast shape or position, it can be useful to include some portion of the arm or arm pit in the ROI. However, parts of the body that might have large day-to-day positioning variation and do not affect breast shape or position should not be included in the ROI.

10.2.5 Treatment Delivery

As with plan preparation and pretreatment QA, treatment delivery processes may vary between different SGRT systems. We again use AlignRT as an example to review the principal of the SGRT-based DIBH treatment. As previously described in this example, the patient's first day at the treatment machine is dedicated to patient setup verification and QA, and an actual treatment is typically not delivered. A patient is considered to be in an acceptable setup position if the real-time displacements of the current breath-hold ROI surface relative to the planned surface are within predetermined tolerances. A common workflow is to initially set up the patient while free breathing, using the free breathing CT DICOM surface as the reference. Once the current surface displacements are within tolerance of the reference surface, the clinician switches monitoring to be relative to the breath-hold CT DICOM surface and the patient is instructed to take a deep breath. Here small positional changes can be made, but the vertical position should not be changed as this can have the effect of compensating for an incorrect chest wall position due to insufficient (or too deep) inspiration and introduce a systematic error to the setup. Both the free breathing and DIBH position should be in tolerance relative to their respective DICOM reference surface. The physicist is encouraged to be present at the verification/QA session to ensure the proper selection of the ROI. No skin markers are needed for real-time surface tracking. During setup monitoring, the SGRT system will automatically generate a real-time patient surface image and registers it to the imported reference surface to calculate the real-time displacements. Both medial and lateral tangential portal images are also taken on the verification/QA day.

For all subsequent treatment days, the same procedure is followed for patient setup with verification that the skin surface during DIBH matches the reference image. For treatment, the patient is instructed to take a deep breath and hold it, and the radiation beam is turned on while the patient is holding their breath, and all real-time displacements are within the predefined tolerance. During treatment, the therapists can provide audio coaching to adjust the patient's breathing level (e.g., "breathe in a little more") through the intercom as needed. If a visual coaching device is available, the patient will have visual feedback of breathing. Usually a breathing window is displayed to inform the patient of their current breathing level with respect to the desired breathing level. Portal images are typically taken at least once per week alternating between the medial and lateral tangent fields. New reference surfaces can be captured based on the feedback from the radiographic imaging.

With the typical 3 mm threshold (±3 mm allowance from the breath-hold CT external body contour-based reference surface), most patients can hold their breath within the desired range long enough for delivery of the field with the most monitor units (typically the tangential base or open field), that is, in a single breath-hold. The modulated tangential fields (usually one to three smaller fields if using FinF or a single irregular surface compensator field) can be treated in another breath-hold. A typical VMAT breast plan has four to five partial arcs. The full arc is broken up into partial arcs so that one field can be treated in a single breath-hold.

10.3 CHALLENGES IN USING SGRT FOR DIBH

SGRT for left-sided breast cancer irradiation has been successfully implemented and widely adopted in recent years. There are, however, some practical issues that occasionally limit the clinical utility of the technique unless carefully addressed. Bolus material is often used in breast and chest wall irradiation to increase superficial dose. However, the effective use of a bolus together with surface imaging can be difficult as the bolus needs to be placed perfectly on the patient's skin each day in order to agree with the reference surface. If there are daily placement variations, a new reference surface image needs to be taken and verified. In that scenario, longer treatment times might also be needed. It can also be challenging for the camera system to generate images of the surface of the conventional bolus, as it has different reflective properties compared to skin. For these reasons, some clinics continue to find it challenging to perform SGRT-based DIBH for patients in need of bolus. However, there are commercially available nonreflective conformal bolus materials such as Elasto-Gel (Southwest Technologies Inc, Kansas City, MO) and off-white brass mesh bolus (Radiation Products Design, Albertville, MN), and many clinics have successfully implemented SGRT-based DIBH with bolus.

Field matching of a supraclavicular and tangential field with a breath-hold is another challenge for SGRT. In a free-breathing delivery, the match line is feathered by the normal variation in chest wall position during breathing. However, when delivered with DIBH, subtle differences in the level of the breath-hold between the fields might render the dose at the match line either higher or lower than intended. When treating three-field breast patients, special care is needed to ensure the breath-holds of the fields are reasonably consistent and fall onto the match-line. In patients where the regional nodes are being treated, it is common to use a separate medial electron field to cover the IMNs or the medial breast, matched to shallow "off heart" tangent fields all without DIBH. Finally, when simple materials are used to displace tissue (e.g., bubble wrap in a skin fold or tape on the skin to displace redundant tissue out of the radiation field), care must be taken to ensure that these materials are not in the ROI used for the skin surface matching, nor that they block the SGRT cameras' views of the ROI.

10.3.1 Reproducibility of SGRT-Based DIBH Setup

The setup reproducibility of SGRT-based DIBH was evaluated in a study of 50 breast patients treated with tangential plans.[11] Two megavoltage (MV) portal images were acquired on the verification/QA day, and one image was acquired per tangent field alternating each following week. The MV portal images were analyzed with respect to the planned digitally reconstructed radiographs (DRRs) to assess whether the heart was displaced away from the breast in a manner consistent with the treatment plan.

On the planning DRRs and MV portal images, the distance between the field edge and the anterior pericardial shadow was measured (Figure 10.3).

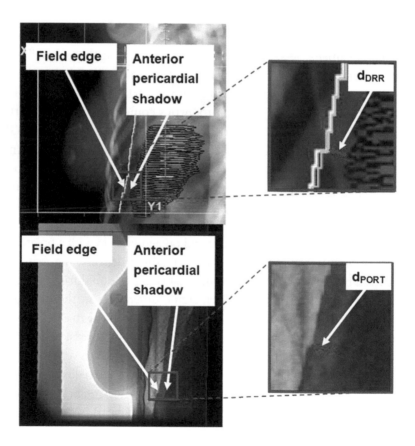

FIGURE 10.3 The digitally reconstructed radiographs (DRR) d_{DRR} and portal (d_{PORT}) images of a patient during deep inspiration breath hold. The d_{DRR} and d_{PORT} were measured on the DRR and portal images as shown. An enlarged view of the boxed area is shown on the right.

This distance is defined as the shortest straight line between the field edge and the anterior pericardial shadow, in the plane of the image. Comparisons of the distance measured on the DRR (d_{DRR}) and the portal image (d_{PORT}) were taken as an indication for the accuracy and reproducibility of the entire DIBH system.

Differences between the d_{DRR} versus d_{PORT} are shown in Figure 10.4 for all patients. Descriptive statistics are summarized in Table 10.1.

The mean random uncertainty of 0.19 cm reflects the reproducibility of the SGRT-based DIBH technique. Nevertheless, the range and standard deviation illustrate that there were instances with poorer reproducibility. There could be other factors that might influence this metric (e.g., time of the day, therapists treating the patient, and day of the week). Similar results were obtained by Alderliesten et al. in a study of 20 patients treated with SGRT-DIBH. Retrospective registration of captured surfaces to cone-beam CT (CBCT) found a systematic setup error of less than 0.17 cm.[12]

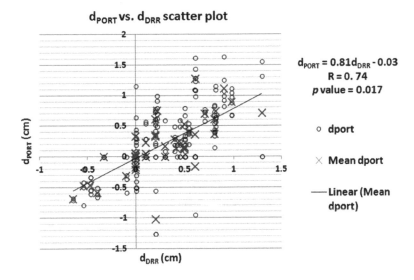

FIGURE 10.4 The plot of digitally reconstructed radiographs (d_{DRR}) versus portal image (d_{PORT}) for all patients. Each *open circle* represents 1 comparison of a portal image to its corresponding reference DRR. The crosses are computed 1 per patient as the mean of their d_{PORT} values. Because for each patient there is only 1 d_{DRR}; the data appear clustered with groups of data sharing the same x axis value and thus are in a vertical line. The best-fit line via linear regression for the d_{DRR} versus mean d_{PORT} data and associated statistics are shown.

TABLE 10.1   Comparison between the d_{PORT} and d_{DRR} and Associated Statistics

	|Systematic Uncertainties| (cm)			Systematic Uncertainties (cm)				Random Uncertainties (cm)			
Measurements	Mean	Max	Min	σ_s	Mean	Max	Min	σs	Mean	Max	Min	σ_r
All patients	0.20	1.22	0	0.23	−0.07	0.67	−1.22	0.30	0.19	0.84	0	0.17

Note: σ_s, standard deviation of the systematic uncertainties; σ_r, standard deviation of the random uncertainties.

10.3.2 Heart Position Variability during DIBH

In addition to overall setup accuracy and chest wall position during DIBH, another source of uncertainty when delivering this technique is variability in heart position. Numerous studies have investigated heart position during DIBH, including SGRT-based DIBH. One study of 50 patients found that heart position on CBCT during DIBH exhibited daily variability up to 10 mm.[13] A study of 20 patients using surface imaging and CBCT found a moderate correlation between surface position and heart position.[14] They used systematic and geometric uncertainties to compute recommended heart margins of 0.11, 0.67, and 0.25 cm in the left-right, cranio-caudal, and anterior-posterior directions, respectively.

10.3.3 Dosimetric Impact of Residual Motion during DIBH

During the DIBH technique, residual motion of the chest wall is possible as the patient gradually loses the ability to maintain the breath-hold or other factors related to breathing. One concern when implementing DIBH is if the dosimetric effect of this residual motion during the breath-hold is acceptable. Surface guidance systems are well suited to evaluating this problem as they continuously monitor surface position, and these data can be stored in log files and studied offline. One such study was conducted to address the concern about residual motion.[15] The analysis used the stored SGRT log files containing the translational and rotational real-time displacement data over time of 30 left-sided breast cancer patients treated with tangential fields and DIBH, using a variety of fractionation schedules.

Data from the beam-on portions of the surface displacement measurements were used to calculate an estimate of the actual delivered dose (compared to the planned dose using CT without considering any breath-hold motion). For each position or set of recorded real-time displacements, the dose was recomputed, and the equally weighted average of the multiple resultant doses was computed. For each patient, the delivered mean heart and total lung V20 doses were compared to the plan that assumed no breath-hold motion. For patients where the IMN were treated, the percentage of the IMN receiving 40 Gy (coverage and mean dose) was also considered.

Comparisons of the planned and delivered dose statistics for heart, lung, and IMN are shown in Figure 10.5. Descriptive statistics are summarized in Table 10.2.

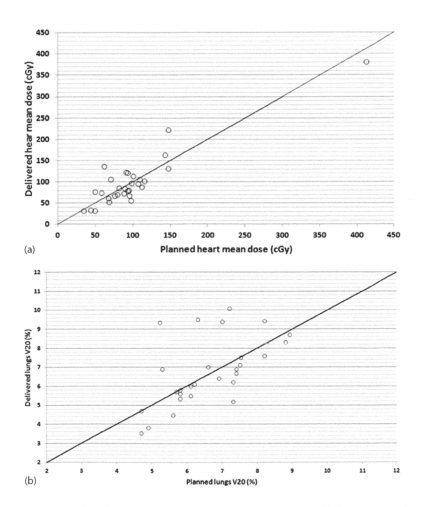

(a)

(b)

FIGURE 10.5 The planned and delivered dose comparison: (a) heart mean dose, (b) lung V20. (*Continued*)

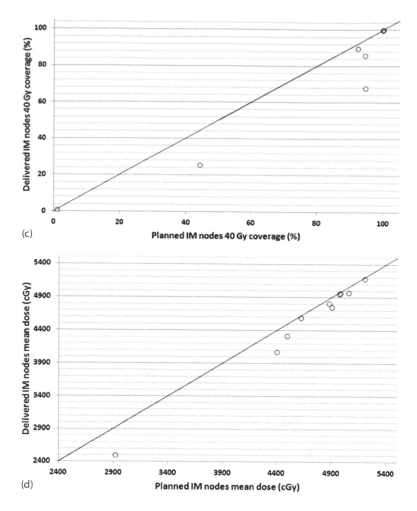

FIGURE 10.5 (Continued)   (c) internal mammary node (IMN) 40 Gy coverage, and (d) IMN mean dose.

An example of the planned and delivered dose distributions is shown in Figure 10.6, where it can be seen that the 50% isodose line has been pushed closer to the heart and the 80% line has been pushed slightly away from the heart. The delivered 100% and 98% isodose lines are also different than in the planned distribution.

However, there is less than 1% difference between the delivered and planned heart and lung dose volume histograms (DVHs) as shown in Figure 10.7.

TABLE 10.2   The Planned and Delivered Heart and Internal Mammary Node (IMN) Dose Coverage

	Heart Mean Dose (cGy)		IMN 40 Gy Coverage (%)		IMN Mean Dose (cGy)	
	Planned	Delivered	Planned	Delivered	Planned	Delivered
Average	99	101	83	77	4642	4518
Standard deviation	66	66	33	36	658	781
Max	412	381	100	100	5203	5186
Min	35	32	1	1	2915	2503
Average difference	2		−6		−123	
Average absolute difference	20		6		123	
Range of the difference	[−41, 76]		[−26, 0.1]		[−411, −11]	
Standard deviation of the difference	28		9		140	

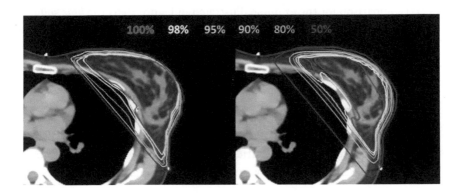

FIGURE 10.6   An example of the planned (left) and delivered (right) dose distribution on an axial CT image. The prescription was 267 cGy × 16 fractions.

Since both the patient setup real-time displacements and the treatment beam-on/off real-time displacements were absolute values calculated against the same reference surface image throughout the treatment course, the dosimetric analysis of the beam-on portions of the surface displacements provide a reasonable estimate of the breath-hold motion effect compared to the planned dose. The study found that dose variations due to setup uncertainties do not propagate to the breath-hold motion-induced uncertainties. On average, the mean heart dose and lung V20 were reasonably close to what has been planned. However, the IMN 40 Gy coverage may be modestly reduced for certain cases.

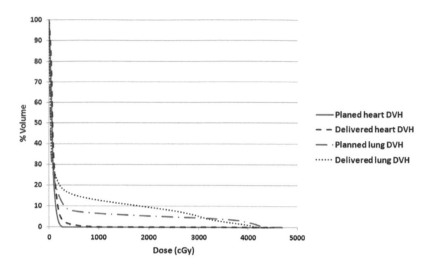

FIGURE 10.7 The planned and delivered heart and lung dose volume histogram (DVH) comparison for the example illustrated in Figure 10.6. The *blue solid line* is the planned heart DVH, the *red dashed line* is the delivered heart DVH, *the blue dashed and dotted line* is the planned lung DVH, and the *red dotted line* is the delivered lung DVH.

KEY POINTS

- Patient selection is an important element of a successful DIBH program and requires that patients be able to follow instructions, be compliant, and have no serious respiratory issues.

- Patient setup with SGRT employs free-breathing and DIBH surface information. A dedicated setup QA and verification imaging session using MV portal imaging is used to ensure the patient is correctly set up and that the DIBH technique is adequately sparing the heart as planned.

- The reproducibility of an SGRT-based DIBH setup including heart and chest wall position has been shown to be clinically reasonable when evaluated with MV portal imaging.

- Residual motion when performing DIBH did not significantly impact cardiac sparing or target dose coverage. Minor variations can occur with IMN coverage. Caution is necessary when evaluating dose at field match lines when using DIBH.

REFERENCES

1. Early Breast Cancer Trialists' Collaborative Group (EBCTCG), Clarke M, Coates AS, Darby SC, Davies C, Gelber RD, et al. Adjuvant chemotherapy in oestrogen-receptor-poor breast cancer: patient-level meta-analysis of randomised trials. *Lancet.* 2008;371:29–40.
2. Paszat LF, Mackillop WJ, Groome PA, Schulze K, Holowaty E. Mortality from myocardial infarction following postlumpectomy radiotherapy for breast cancer: a population-based study in Ontario, Canada. *Int J Radiat Oncol Biol Phys.* 1999;1:755–762.
3. Sixel KE, Aznar MC, Ung YC. Deep inspiration breath hold to reduce irradiated heart volume in breast cancer patients. *Int J Radiat Oncol Biol Phys.* 2001;49:199–204.
4. Zagar T, Kaidar O, Tang X, et al. Utility of deep inspiration breath-hold for left sided breast radiation therapy in preventing early cardiac perfusion defects: a prospective study. *Int J Radiat Oncol Biol Phys.* 2017;97(5):903–909.
5. Remouchamps VM, Letts N, Vicini FA, et al. Initial clinical experience with moderate deep-inspiration breath hold using an active breathing control device in the treatment of patients with left- sided breast cancer using external beam radiation therapy. *Int J Radiat Oncol Biol Phys.* 2003;56:704–715.
6. Kim MM, Kennedy C, Scheuermann R, Freedman G, Li T. Whole breast and lymph node irradiation using Halcyon™ 2.0 utilizing automatic multi-isocenter treatment delivery and daily kilovoltage cone-beam computed tomography. *Cureus.* 2019;11(5):e4744.
7. Jiang SB. Technical aspects of image-guided respiration-gated radiation therapy. *Med Dosim.* 2006;31(2):141–151.
8. Li XA, Stepaniak C, Gore E. Technical and dosimetric aspects of respiratory gating using a pressure-sensor motion monitoring system. *Med Phys.* 2006;33(1):145–154.
9. Cervino LI, Gupta S, Rose MA, Yashar C, Jiang SB. Using surface imaging and visual coaching to improve the reproducibility and stability of deep-inspiration breath hold for left-breast-cancer radiotherapy. *Phys Med Biol.* 2009;54(22):6853.
10. White J, Tai A, Arthur D, Buchholz T, et al. Breast cancer atlas for radiation therapy planning: consensus definitions. *RTOG.* Available at: www.rtog.org/LinkClick.aspx?fileticket=vzJFhPaBipE=. Accessed June 10, 2013.
11. Tang X, Zagar T, Bair E, et al. Clinical experience with 3D surface matching-based Deep Inspiration Breath Hold (DIBH) for left-sided breast cancer radiotherapy. *Pract. Radiat Oncol.* 2013;4(3):e151–e158.
12. Alderliesten T, Sonke JJ, Betgen A, Honnef J, van Vliet-Vroegindeweij C, Remeijer P. Accuracy evaluation of a 3-dimensional surface imaging system for guidance in deep-inspiration breath-hold radiation therapy. *Int J Radiat Oncol Biol Phys.* 2013;85(2):536–542.

13. van Haaren P, Claassen-Janssen F, van de Sande I, Boersma L, van der Sangen M, Hurkmans C. Heart position variability during voluntary moderate deep inspiration breath-hold radiotherapy for breast cancer determined by repeat CBCT scans. *Phys Med*. 2017;40:88–94.

14. Alderliesten T, Betgen A, Elkhuizen PH, van Vliet-Vroegindeweij C, Remeijer P. Estimation of heart-position variability in 3D-surface-image-guided deep-inspiration breath-hold radiation therapy for left-sided breast cancer. *Radiother Oncol*. 2013;109(3):442–447.

15. Tang X, Cullip T, Zagar T, et al. Motion effect during deep inspiration breath hold for left sided breast cancer radiotherapy. *J Appl Clin Med Phys*. 2014;16(4):91–99.

CHAPTER 11

Breast Cancer Outcomes and Toxicity Reduction with SGRT

Orit Kaidar-Person, Icro Meattini, Marianne C. Aznar, and Philip Poortmans

CONTENTS

11.1 INTRODUCTION

Breast cancer is the most commonly diagnosed cancer and the leading cause of cancer death in women.[1] Importantly, breast cancer is the leading overall cause of death in younger women between the ages of 40 to 49 years.[1] The treatment of breast cancer, especially in non-metastatic patients, is multimodal and usually includes surgery, systemic treatments, and radiation therapy (RT). In the curative setting, RT is mostly given post-operatively with target volumes comprising either the residual breast (in case of conservative surgery), chest wall or reconstructed breast (in case of mastectomy) with or without regional lymph nodes. Commonly, breast and chest wall irradiation with or without lymphatic coverage can be accomplished safely with conventional three-dimensional conformal RT (3D-CRT) and patient localization is verified using dedicated localization systems mounted on-board the linear accelerator (linac) or by ancillary systems within the treatment room.

Advances in technology, including both imaging and radiation techniques, have permitted a wide selection of approaches to be used for delineation, immobilization, and verification of localization as well as a selection of a wide variety of treatment planning techniques including conventional field arrangements or volume-based techniques. Considering the best treatment option for a given patient should be a part of standard practice.[2] This can be achieved by understanding both the nature of the disease (the target and potential spread) and the tools available to achieve the best target coverage and minimal exposure of organs-at-risk (OARs).[3] Interpretation of the treatment plan characteristics for the selection of the best treatment plan should take into account the RT technique used as the radiation exposure to OARs can significantly vary from one technique to another such as differences in lung doses for 3D-CRT versus intensity modulated RT (IMRT).

There is increasing attention to potential long-term RT-related toxicity in breast cancer patients. Radiation-related cardiac injury in breast cancer patients was described for the first time more than three decades ago, and continues to be discussed in the literature.[4-7] Toxicity is especially of major concern as these patients have an increasingly good prognosis and thereby many years to develop treatment-related toxicity. This has become even more challenging with the introduction of newer RT techniques resulting in different dose distributions and the integration of RT with new systemic therapies. This requires a full understanding of all factors contributing to toxicity, including the potential means to reduce or avoid risks. Numerous treatment-related, disease-related, and patient-related factors contribute to late side effects, including: systemic therapy (e.g., anthracyclines contribute to cardiac toxicity), radiation technique (3D-CRT versus IMRT), fractionation (conventional fractionation, hypofractionation, accelerated fractionation), use of cardiac-sparing techniques (e.g., breath-hold, active breathing control), extent of disease (e.g., target volumes, proximity to OARs), medical comorbidities (e.g., severe lung disease, preexisting cardiac risk factors), smoking, and body habitus. All should be taken into consideration when selecting the most appropriate RT technique and treatment plan. Therefore, it is of utmost importance that the treating team (i.e., radiation oncologist, radiation therapist, physicist) is aware of these factors to be able to discuss the optimal treatment approach within the multidisciplinary team (breast surgeon, pathologist, radiologist, clinical oncologist) and to preferably apply techniques that are best adapted for a given patient to reduce toxicity.

Constant changes in clinical practice mandate our awareness to minimize potential errors to create a safer patient environment.[3] Each new technique should be validated and constantly re-assessed in clinic to avoid potential harm for future patients.[8]

Surface image guidance is a recently applied technology for guiding lung and cardiac sparing treatment techniques such as deep inspiration breath hold (DIBH), which was described in Chapter 10. This chapter will discuss the use of surface-guided radiation therapy (SGRT), focusing on the evaluation of dosimetric benefits from DIBH and relevant clinical outcome studies including measures of lung and cardiac toxicity.

11.2 METHODS OF DIBH IN BREAST CANCER RT

In DIBH, the inflation of the lungs moves the heart in a dorsal, medial, and caudal direction away from the breast and chest wall, resulting in a more favorable anatomical arrangement[9] relative to the radiation fields. The patient usually lies supine for breast RT with DIBH, although some applications of DIBH in the prone position have been reported.[10,11] During CT simulation, the patient is instructed to take in a deep breath and then hold their breath at a "comfortable" level (e.g., approximately 80% of lung vital capacity) for approximately 20–30 seconds at a time, as many times as necessary to complete the CT scan. The DIBH process is then repeated during every day of treatment delivery. This method can be performed using various approaches such as voluntary or active breathing control, and have been shown to be reproducible and easy to apply.[8,12–15]

The treatment is only delivered when the DIBH position of the chest wall is reached and the "beam on/beam off" can be controlled either automatically or manually by the radiation therapist. Not all breathhold methods require special or sophisticated equipment. DIBH can be achieved by relying only on staff training and patient compliance or ability to perform breath-hold.[12,13] In the UK HeartSpare study Stage IB, a free breathing prone position was compared with a DIBH supine position in patients with large-sized breasts (volume over 750 cm³). The DIBH supine position was superior in terms of reproducibility and cardiac sparing.[16]

11.3 EVALUATION OF DOSIMETRIC BENEFIT FROM SURFACE-GUIDED DIBH

DIBH has been shown to reduce the dose to the heart[17] and lungs,[18] although the magnitude of dose reduction depends both on the individual patient anatomy and on the target volumes. Larger reductions are observed for treatments including nodal regions. SGRT has generated considerable interest as a non-ionizing image guidance modality as opposed to traditional radiographic imaging, i.e., planar kV and cone-beam CT (CBCT). At least one study has suggested that setup errors can be reliably observed and corrected using SGRT alone.[19] Another study compared SGRT to CBCT and suggested that SGRT is more suitable as a complement rather than a "stand alone" positioning approach[20] because of the difficulty distinguishing between errors in patient positioning and changes in the DIBH position of the patient's surface. In addition, SGRT alone does not enable the radiation therapist to verify the position of the heart and other internal OARs. There is, however, general agreement that SGRT provides an excellent option for intra-fraction DIBH position monitoring and it is a nonionizing, noninvasive method for visualization that can co-exist with other methods such as CBCT.[21,22]

11.4 TREATMENT-RELATED TOXICITY AND SGRT

A main priority of RT, especially in the curative setting, is to simultaneously maintain coverage of the target volumes (the remaining breast and chest wall with or without regional lymph nodes) while minimizing the radiation dose to adjacent OARs such as lungs, heart, and contralateral breast. Reproducibility of the treatment position and the patient's comfort in the treatment position are important for assuring that the RT is given as planned.[15]

11.4.1 Cardiac Toxicity

The risk of cardiovascular disease after postoperative RT for breast cancer is well documented and is derived from direct exposure of the heart to radiation, mostly in cases of left-sided disease or internal mammary nodal (IMN) irradiation.[6,23,24] Moreover, breast cancer patients are often treated with adjuvant targeted and systemic treatments (such as anthracyclines and trastuzumab) that may contribute to treatment-related cardiac toxicity.[6,25]

Exposure of the heart to radiation is often inevitable due to anatomical reasons. The left ventricle and the lateral descending artery are often included within the tangential fields in the case of left-sided RT.[26] Cardiac perfusion changes detected via single photon emission computed tomography (SPECT) can be noted within six months from radiation, thus providing an objective quantitative imaging tool to show radiation-associated cardiac injury occurring early after RT and not as previously thought after a long latency period of 10 years or more.[27-30]

There are a number of non-DIBH methods to reduce the dose to the heart. For example, "blocking the heart" is achieved by shaping the radiation fields (originally using shielding blocks and later by closing the multileaf collimator) to reduce cardiac exposure. However, depending on the patient's body habitus and the target volumes (including extent of initial disease), heart shielding may inadvertently shield a portion of the target tissues (e.g., breast with or without IMN) that might result in inferior oncologic outcomes.[31] If the primary tumor bed is located medially or centrally within the breast, cardiac shielding may unintentionally reduce coverage of the high recurrence risk region. IMN irradiation can be significant for disease free survival, and cardiac shielding may compromise coverage in the case of left-sided IMN irradiation.[32-34] Regardless of the technique, it should be kept in mind that reproducibility of patient positioning is a key factor both to reducing the required uncertainty margins and to avoid underestimation of cardiac doses by setup errors.[35]

Respiratory-gated RT techniques, such as DIBH, are being more widely adopted to reduce cardiac exposure.[36,37] These approaches increase the distance between the heart and the target volumes. Thus, DIBH is a means to exclude the heart while minimizing potential under-dosage of the target volumes. DIBH especially provides an advantage in cases where IMN RT is indicated, since the IMN are moved away from the heart during breathhold.[38] However, DIBH techniques have not yet been applied in large studies evaluating cardiac toxicity, resulting in a lack of data on the exact risk reduction. The SAVE-HEART study[37] estimated the individual radiation-induced risk for death due to ischemic heart disease comparing free-breathing RT with DIBH RT for left-sided early stage breast cancer using a probability model based on 3D plans and corresponding individual of any mortality (i.e., cancer, ischemic heart disease, age). The study showed that by using DIBH RT, the mean heart doses were reduced by 35% (interquartile range: 23%–46%) compared to free-breathing. Mean "expected years of life lost" due to radiation-induced ischemic heart disease were 0.11 years in free-breathing, and 0.07 years in DIBH. Moreover, the main benefits from DIBH were in patients with favorable tumor prognosis, high mean heart dose, or high baseline ischemic heart disease risk, independent of their age. In another study, DIBH was reported to reduce the cardiac dose by 26% to 80% compared to free-breathing in left-sided radiation for breast cancer, depending on the patient's anatomy, target volumes, and treatment plan.[36]

Importantly, even if cardiac sparing techniques are employed, the heart and coronary arteries are often exposed to some degree of radiation dose.[17,27] A large case-control study in the setting of post-operative RT for breast cancer[4] demonstrated that there is a linear increase in the *relative rate* of major coronary events by 7.4% per Gy in the mean heart dose received (95% confidence interval, 2.9 to 14.5; $p < 0.001$). The *absolute rate* of major coronary events was increased in patients with cardiac-risk factors, signifying that radiation-associated risk depends on baseline risk for coronary artery disease, supporting the findings of SAVE-HEART study.[37] These findings support the view that efforts should be made to reduce the cardiac dose as much as possible, even more so in patients with preexisting cardiac morbidity risk factors.[4] RT-associated cardiovascular morbidity and mortality is a major concern in this curative population.[24] The latency period for developing radiation-induced cardiovascular disease is shorter than previously assumed in early studies (less than 10–15 years),[4,7] with 44% of the excess major coronary events attributed to radiation occurring less than 10 years post-radiation, further increasing over time from radiation therapy.[4,39] The absolute magnitude of the excess events in the early post-radiation years is modest; therefore, the uncertainties in the associated risk estimates remain large with wide confidence intervals.[4] Current data supporting the use of regional nodal irradiation will make this an ongoing concern as IMN RT increases the risk for incidental irradiation of the heart.[17,33,34,40]

A prospective single arm study of 20 patients who were treated for left breast and chest wall (±IMN) using 3D-CRT with SGRT-DIBH evaluated the RT-induced cardiac perfusion and cardiac wall-motion changes using pre-RT and 6 month post-RT SPECT scans.[29] In that study, the median heart dose was 0.94 Gy (range, 0.56–2.0), and median cardiac V_{25Gy} was zero (range, 0–0.1) when using SGRT-DIBH. None of these patients had post-RT perfusion or wall motion abnormalities. Compared to a similar study without SGRT-DIBH,[27,28] these results suggest that this RT-technique for patients receiving 3D-CRT for left-sided breast cancer may be an effective means to avoid early RT-associated cardiac perfusion defects.[29] The full clinical magnitude of SGRT-DIBH and reducing exposure will only be seen in years from its application.

11.4.2 Lung Toxicity

Factors that were suggested by some to contribute to lung toxicity associated with RT include smoking, chemotherapy, tamoxifen, older age, preexisting lung disease, low lung capacity, and low performance status.[39,41-43] RT-related factors that contribute to toxicity are the lung volume exposed and RT dose. The lung volume included in the RT field should be considered at the time of treatment planning to minimize potential toxicity.[41] It is important to keep in mind that the lung doses are increased with more extensive regions irradiated: each nodal region adds about 3 Gy of ipsilateral mean lung dose (MLD).[18] Importantly, the larger the volume irradiated, the more potential exists for lung dose reduction using DIBH.[18] It is estimated that compared to free breathing, the ipsilateral MLD can be *decreased* by 1 Gy for breast or chest wall only; by 2 Gy for breast or chest wall and supraclavicular nodal fields and by 3 Gy for breast or chest wall, supraclavicular and IMN fields.[18] These reductions do not take into account patients with a unique body habitus.

Acute lung toxicity, presenting as radiation pneumonitis, can occur within 1 to 6 months after completion of RT.[42] The rate and severity of clinical radiation pneumonitis tends to be lower than 2%–4% even when regional nodal irradiation is performed,[42] mostly grade 1–2 toxicity.[41,44] These studies reported the rates of radiation pneumonitis using conventional radiation; therefore, great attention should be given to irradiated lung volumes when using new radiation techniques such as IMRT, helical tomotherapy, or volumetric modulated arc therapy (VMAT) especially in patients with large target volumes (e.g., breast or chest wall and lymphatic drainage, bilateral breast or chest wall). Using these techniques will result in large volumes of the lungs receiving a low dose of radiation (e.g., lung V_{5Gy}) which can lead to significant toxicity.[45] Moreover, most of the reports of toxicity and the constraints recommendations are based on conventional fractionation (1.8–2.0 Gy).

Long-term lung toxicity should also be taken into consideration, including lung fibrosis (about 4% in the EORTC 22922/10925)[33] and increased risks (approximately double) of secondary lung cancer.[39,46,47] In a case-control study of secondary lung cancer in a population-based cohort of 23,627 early breast cancer patients treated with post-operative RT, the median time from breast cancer treatment to secondary lung cancer diagnosis was 12 years (range, 1–26 years).[47]

Smoking is a pivotal contributor for secondary lung cancer in this population, more than 90% of the cases that developed secondary lung cancer were categorized as ever-smokers (versus 40% among the controls).[47] The MLD is also significant for the risk of secondary lung cancer, the rate of which increased linearly at 8.5% per Gy (95% confidence interval, 3.1%–23.3%; $p < 0.001$). This rate was enhanced for ever-smokers with an excess rate of 17.3% per Gy (95% confidence interval, 4.5%–54.0%; $p < 0.005$).[47] Therefore, efforts should be made to reduce the lung dose using advanced techniques as well as advising the patient for smoking cessation.

11.5 SUMMARY

Advanced RT techniques such as SGRT-DIBH have been shown to reduce the dose to OARs without compromising the dose to the target; therefore, it is expected that the rates of late complications, such as cardiac disease or lung cancer, will be reduced in the long term.

KEY POINTS

- Breast cancer is the most common diagnosed cancer and the leading cause of cancer death in women.

- Toxicity is especially of major concern as these patients have an increasingly good prognosis and thereby many years to develop treatment-related toxicity.

- DIBH, including SGRT, has been shown to reduce the dose to the heart and lungs, with anticipated larger benefit of dose reductions in patients who are planned to nodal irradiation.

- It is anticipated that normal tissue sparing with techniques such as SGRT-DIBH will result in reduced rate of late toxicity in breast and chest wall patients.

REFERENCES

1. Bray F, Ferlay J, Soerjomataram I, Siegel RL, Torre LA, Jemal A. Global cancer statistics 2018: GLOBOCAN estimates of incidence and mortality worldwide for 36 cancers in 185 countries. *CA Cancer J Clin.* 2018;68:394–424.
2. Poortmans P, Aznar M, Bartelink H. Quality indicators for breast cancer: revisiting historical evidence in the context of technology changes. *Semin Radiat Oncol.* 2012;22(1):29–39.
3. Marks LB, Jackson M, Xie L, et al. The challenge of maximizing safety in radiation oncology. *Pract Radiat Oncol.* 2011;1(1):2–14.

4. Darby SC, Ewertz M, McGale P, et al. Risk of ischemic heart disease in women after radiotherapy for breast cancer. *N Engl J Med.* 2013;368(11):987–998.

5. McGale P, Darby SC, Hall P, et al. Incidence of heart disease in 35,000 women treated with radiotherapy for breast cancer in Denmark and Sweden. *Radiother. Oncol.* 2011;100(2):167–175.

6. Boekel NB, Schaapveld M, Gietema JA, et al. Cardiovascular disease risk in a large, population-based cohort of breast cancer survivors. *Int J Radiat Oncol Biol Phys.* 2016;94(5):1061–1072.

7. Cuzick J, Stewart H, Rutqvist L, et al. Cause-specific mortality in long-term survivors of breast cancer who participated in trials of radiotherapy. *J Clin Oncol.* 1994;12(3):447–453.

8. Bartlett FR, Donovan EM, McNair HA, et al. The UK HeartSpare Study (Stage II): multicentre evaluation of a voluntary breath-hold technique in patients receiving breast radiotherapy. *Clin Oncol (R Coll Radiol).* 2017;29(3):e51–e56.

9. Pedersen AN, Korreman S, Nystrom H, Specht L. Breathing adapted radiotherapy of breast cancer: reduction of cardiac and pulmonary doses using voluntary inspiration breath-hold. *Radiother Oncol.* 2004;72(1):53–60.

10. Mulliez T, Van de Velde J, Veldeman L, et al. Deep inspiration breath hold in the prone position retracts the heart from the breast and internal mammary lymph node region. *Radiother Oncol.* 2015;117(3):473–476.

11. Mulliez T, Veldeman L, Speleers B, et al. Heart dose reduction by prone deep inspiration breath hold in left-sided breast irradiation. *Radiother Oncol.* 2015;114(1):79–84.

12. Bartlett FR, Colgan RM, Donovan EM, et al. Voluntary breath-hold technique for reducing heart dose in left breast radiotherapy. *J Vis Exp.* 2014;(89):e51578.

13. Bartlett FR, Colgan RM, Carr K, et al. The UK HeartSpare Study: randomised evaluation of voluntary deep-inspiratory breath-hold in women undergoing breast radiotherapy. *Radiother Oncol.* 2013;108(2):242–247.

14. Bergom C, Currey A, Desai N, Tai A, Strauss JB. Deep inspiration breath hold: techniques and advantages for cardiac sparing during breast cancer irradiation. *Front Oncol.* 2018;8:87.

15. Mulliez T, Veldeman L, Vercauteren T, et al. Reproducibility of deep inspiration breath hold for prone left-sided whole breast irradiation. *Radiat Oncol.* 2015;10:9.

16. Bartlett FR, Colgan RM, Donovan EM, et al. The UK HeartSpare Study (Stage IB): randomised comparison of a voluntary breath-hold technique and prone radiotherapy after breast conserving surgery. *Radiother Oncol.* 2015;114(1):66–72.

17. Taylor CW, Wang Z, Macaulay E, Jagsi R, Duane F, Darby SC. Exposure of the heart in breast cancer radiation therapy: a systematic review of heart doses published during 2003 to 2013. *Int J Radiat Oncol Biol Phys.* 2015;93(4):845–853.

18. Aznar MC, Duane FK, Darby SC, Wang Z, Taylor CW. Exposure of the lungs in breast cancer radiotherapy: a systematic review of lung doses published 2010–2015. *Radiother Oncol.* 2018;126(1):148–154.

19. Wei X, Liu M, Ding Y, Li Q, Cheng C, Zong X, Yin W, Chen J, Gu W. Setup errors and effectiveness of Optical Laser 3D Surface imaging system (Sentinel) in postoperative radiotherapy of breast cancer. *Sci Rep.* 2018;8(1):7270.

20. Betgen A, Alderliesten T, Sonke JJ, van Vliet-Vroegindeweij C, Bartelink H, Remeijer P. Assessment of set-up variability during deep inspiration breath hold radiotherapy for breast cancer patients by 3D-surface imaging. *Radiother Oncol.* 2013;106(2):225–230.

21. Tang X, Cullip T, Dooley J, Zagar T, Jones E, Chang S, Zhu X, Lian J, Marks L. Dosimetric effect due to the motion during deep inspiration breath hold for left-sided breast cancer radiotherapy. *J Appl Clin Med Phys.* 2015;16(4):91–99.

22. Xiao A, Crosby J, Malin M, Kang H, Washington M, Hasan Y, Chmura SJ, Al-Hallaq HA. Single-institution report of setup margins of voluntary deep-inspiration breath-hold (DIBH) whole breast radiotherapy implemented with real-time surface imaging. *J Appl Clin Med Phys.* 2018;19(4):205–213.

23. Darby SC, Cutter DJ, Boerma M, et al. Radiation-related heart disease: current knowledge and future prospects. *Int. J Radiat Oncol Biol Phys.* 2010;76(3):656–665.

24. Gagliardi G, Constine LS, Moiseenko V, Correa C, Pierce LJ, Allen AM, Marks LB. Radiation dose-volume effects in the heart. *Int J Radiat Oncol Biol Phys.* 2010;76(3 Suppl):S77–S85.

25. Zagar TM, Cardinale DM, Marks LB. Breast cancer therapy-associated cardiovascular disease. *Nat Rev Clin Oncol.* 2016;13(3):172–184.

26. Taylor CW, Nisbet A, McGale P, Darby SC. Cardiac exposures in breast cancer radiotherapy: 1950s–1990s. *Int J Radiat Oncol Biol Phys.* 2007;69(5):1484–1495.

27. Marks LB, Yu X, Prosnitz RG, et al. The incidence and functional consequences of RT-associated cardiac perfusion defects. *Int J Radiat Oncol Biol Phys.* 2005;63(1):214–223.

28. Prosnitz RG, Hubbs JL, Evans ES, et al. Prospective assessment of radiotherapy-associated cardiac toxicity in breast cancer patients: analysis of data 3 to 6 years after treatment. *Cancer.* 2007;110(8):1840–1850.

29. Zagar TM, Kaidar-Person O, Tang X, et al. Utility of deep inspiration breath hold for left-sided breast radiation therapy in preventing early cardiac perfusion defects: a prospective study. *Int J Radiat Oncol Biol Phys.* 2017;97(5):903–909.

30. Kaidar-Person O, Zagar TM, Oldan JD, et al. Early cardiac perfusion defects after left-sided radiation therapy for breast cancer: is there a volume response? *Breast Cancer Res Treat.* 2017;164(2):253–262.

31. Raj KA, Evans ES, Prosnitz RG, Quaranta BP, Hardenbergh PH, Hollis DR, Light KL, Marks LB. Is there an increased risk of local recurrence under the heart block in patients with left-sided breast cancer? *Cancer J.* 2006;12(4):309–317.

32. Brautigam E, Track C, Seewald DH, Feichtinger J, Spiegl K, Hammer J. Medial tumor localization in breast cancer—an unappreciated risk factor? *Strahlenther Onkol.* 2009;185(10):663–668.

33. Poortmans PM, Collette S, Kirkove C, et al. Internal mammary and medial supraclavicular irradiation in breast cancer. *N Engl J Med*. 2015;373(4):317–327.

34. Whelan TJ, Olivotto IA, Parulekar WR, et al. Regional nodal irradiation in early-stage breast cancer. *N Engl J Med*. 2015;373(4):307–316.

35. Evans ES, Prosnitz RG, Yu X, et al. Impact of patient-specific factors, irradiated left ventricular volume, and treatment set-up errors on the development of myocardial perfusion defects after radiation therapy for left-sided breast cancer. *Int J Radiat Oncol Biol Phys*. 2006;66(4):1125–1134.

36. Latty D, Stuart KE, Wang W, Ahern V. Review of deep inspiration breath-hold techniques for the treatment of breast cancer. *J Med Radiat Sci*. 2015;62(1):74–81.

37. Simonetto C, Eidemuller M, Gaasch A, et al. Does deep inspiration breath-hold prolong life? Individual risk estimates of ischaemic heart disease after breast cancer radiotherapy. *Radiother Oncol*. 2018;131:202–207.

38. Osman SO, Hol S, Poortmans PM, Essers M. Volumetric modulated arc therapy and breath-hold in image-guided locoregional left-sided breast irradiation. *Radiother Oncol*. 2014;112(1):17–22.

39. Darby SC, McGale P, Taylor CW, Peto R. Long-term mortality from heart disease and lung cancer after radiotherapy for early breast cancer: prospective cohort study of about 300,000 women in US SEER cancer registries. *Lancet Oncol*. 2005;6(8):557–565.

40. Thorsen LB, Offersen BV, Dano H, et al. DBCG-IMN: a population-based Cohort Study on the Effect of Internal Mammary Node Irradiation in Early Node-Positive Breast Cancer. *J Clin Oncol*. 2016;34(4):314–320.

41. Lind PA, Marks LB, Hardenbergh PH, et al. Technical factors associated with radiation pneumonitis after local +/− regional radiation therapy for breast cancer. *Int J Radiat Oncol Biol Phys*. 2002, 52(1):137–143.

42. Gagliardi G, Bjohle J, Lax I, Ottolenghi A, Eriksson F, Liedberg A, Lind P, Rutqvist LE. Radiation pneumonitis after breast cancer irradiation: analysis of the complication probability using the relative seriality model. *Int J Radiat Oncol Biol Phys*. 2000;46(2):373–381.

43. Lind PA, Wennberg B, Gagliardi G, Rosfors S, Blom-Goldman U, Lidestahl A, Svane G. ROC curves and evaluation of radiation-induced pulmonary toxicity in breast cancer. *Int J Radiat Oncol Biol Phys*. 2006;64(3):765–770.

44. Choi J, Kim YB, Shin KH, et al. Radiation pneumonitis in association with internal mammary node irradiation in breast cancer patients: an ancillary result from the KROG 08-06 study. *J Breast Cancer*. 2016;19(3):275–282.

45. Kaidar-Person O, Kostich M, Zagar TM, Jones E, Gupta G, Mavroidis P, Das SK, Marks LB. Helical tomotherapy for bilateral breast cancer: clinical experience. *Breast*. 2016;28:79–83.

46. Grantzau T, Overgaard J. Risk of second non-breast cancer among patients treated with and without postoperative radiotherapy for primary breast cancer: a systematic review and meta-analysis of population-based studies including 522,739 patients. *Radiother Oncol*. 2016;121:402–413.

47. Grantzau T, Thomsen MS, Vaeth M, Overgaard J. Risk of second primary lung cancer in women after radiotherapy for breast cancer. *Radiother Oncol*. 2014;111(3):366–373.

CHAPTER 12

Surface Guidance for Stereotactic Radiosurgery and Radiotherapy

Technical Aspects

Laura Cerviño and Grace Gwe-Ya Kim

CONTENTS

12.1 INTRODUCTION: EVOLUTION OF THE STEREOTACTIC RADIOSURGERY TECHNIQUE

Stereotactic radiosurgery (SRS) and stereotactic radiotherapy (SRT) consist of delivering highly conformal high doses of radiotherapy with the goal of ablating a target in the brain in one (SRS) or few (SRT) fractions. It is commonly used for the treatment of one or more brain metastases, achieving an effective local tumor control.[1-4] SRS/SRT is also used for the treatment of other conditions: functional diseases, such as trigeminal neuralgia and Parkinson's disease; vascular lesions, such as arteriovenous malformation; primary benign tumors; and primary malignant tumors, such as glioblastoma multiforme.

The required high doses in SRS/SRT need high conformity of the radiation plan and accuracy of the treatment delivery to spare surrounding critical structures. Conventional SRS treatments required a stereotactic frame that provided a coordinate system of reference for the localization and irradiation of the tumor, an imaging system to locate the target and surrounding critical organs, a localization software used in conjunction with the stereotactic frame, a planning system, and a radiation source. Stereotactic frames consisted of a rigid frame that was bolted to the skull of the patient and fastened to the treatment couch.[5] While these frames provide a great degree of immobilization, they are highly uncomfortable, patients need to wear them for a good part of the day, they can cause infection and stress, and they do not guarantee a complete immobilization, as shown by Ramakrishna et al.[6]

Optically guided frameless SRS approaches were subsequently developed and used with linear accelerators. In this approach, a custom-made thermoplastic mask is used to immobilize the patient, while an optical guidance system is used to track the patient's head motion, using several infrared reflecting fiducials fixed to a bite-block and an infrared camera.[7] The optically guided frameless approach offers increased patient comfort while maintaining the same outcomes as the frame-based approach.[8] However, the use of the bite-block still presents discomfort and can result in patient fatigue and tracking inaccuracies in edentulous patients.

In the last decade or so, optical surface imaging has emerged as an image-guided radiation therapy approach, and has been established as a safe technique to monitor patients during intracranial SRS and SRT. The main advantage of this approach is the use of nonionizing imaging, which can be used as much and as frequently as desired and in real-time throughout the duration of the treatment. While surface imaging has been used for other treatments such as breast, both at free breathing and deep inspiration breath hold, and extremities, among other sites, the use of surface imaging in SRS/SRT has changed the treatment paradigm and has allowed the transition from framed treatments to a more comfortable treatment with open masks, where the patient's head motion is continually monitored using surface imaging to ensure precise positioning is maintained throughout treatment.

In this chapter, we will discuss the system accuracy for surface-guided radiation therapy (SGRT) systems that have been used for SRS/SRT procedures. We will then proceed to discuss the clinical workflow and treatment accuracy considerations.

12.2 SGRT SYSTEM ACCURACY

The American Association of Physicists in Medicine (AAPM) Task Group (TG) 147 report provides recommendations on quality assurance (QA), acceptance, and commissioning of nonradiographic radiotherapy localization and positioning systems, including surface imaging systems.[9] TG-147 reiterates the recommendation from TG-142 to have less than 2 mm combined accuracy of all the systems (linear accelerator, imaging, couch), or 1 mm for SRS and SBRT procedures.[10] Multiple studies have shown the accuracy of SGRT systems and the accuracy of SRS procedures using SGRT.

Cerviño et al.[11] showed submillimeter tracking accuracy using a head phantom with a tracking region of interest (ROI) that includes the eyes, nose, forehead, and temple area and found it comparable to a previous bite-block infrared tracking system. Peng et al.[12] and Wiersma et al.[13] obtained similar results. Wen et al. evaluated the systematic accuracy of SRS in the Edge radiosurgery system (Varian Medical Systems, Palo Alto, CA, USA) using an SGRT system as the surface imaging modality and showed that it could accurately track surface motion of a phantom to within 0.1 mm.[14] The end-to-end localization accuracy of the system was 0.5 ± 0.2 mm with a maximum deviation of 0.9 mm over 90 measurements, and the dosimetric evaluation based on absolute film dosimetry showed a greater than 90% pass rate for all cases using a gamma criteria of 3%/1 mm. They concluded that a frameless image-guided system using SGRT is comparable to robotic or invasive frame-based radiosurgery systems. Mancosu et al. observed that occlusion of one camera pod induced an uncertainty <0.5 mm.[15]

Li et al. also demonstrated that SGRT is capable of detecting 0.1 ± 0.1 mm 1D spatial displacement of a phantom in near real-time and concluded that head-motion monitoring using a near-real-time surface is necessary for frameless SRS.[16] This study also compared two different immobilization systems for use along with SGRT monitoring: a head mold with the Pinpoint bite-block, and a head mold and open mask with an adjustable head board. The open mask with adjustable headboard, compatible with surface-guided SRS, outperformed the Pinpoint system in terms of patient comfort and clinical workflow.

It has been known that SGRT systems have traditionally underperformed with the couch at nonzero angles or with camera occlusion by the gantry. A recent study by Wiant et al. evaluated the accuracy of an improved surface imaging system with an advanced camera optimization calibration technique against a room-mounted orthogonal X-ray imaging system.[17] The study compared shifts detected with both systems of a head phantom at 72 different positions, including couch angles and camera occlusion. Results showed that the loss of accuracy with surface imaging due to camera occlusion was minimal and deemed the technique to be appropriate for SRS.

Some SGRT systems require a commercially provided calibration plate to perform the camera isocenter calibration, either as part of a routine daily QA workflow or when the daily quality check fails prespecified tolerances (1 mm). However, Paxton et al. showed that minor misalignments of the calibration plate with the machine isocenter can lead to erroneously predicted offsets when the couch is rotated,[18] and that additional calibration with a phantom using MV portal imaging reduces these offsets.

12.2.1 Surface Guidance Quality Assurance Specific to SRS/SRT

In addition to the required daily QA for surface imaging that checks the camera calibration, SRS requires more extensive calibration and tests. Specifically, the isocenter coincidence of the surface imaging system, the linear accelerator beam, and other imaging systems need to be verified. For that purpose, surface imaging must be included into an institution's daily image-guided radiation therapy (IGRT) QA program, and a phantom

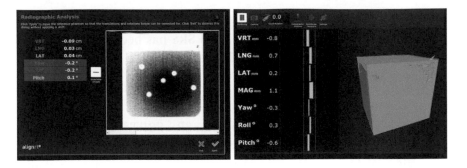

FIGURE 12.1 Radiographic (left) and surface imaging (right) quality assurance test on the same phantom.

that can provide information for all imaging and beam systems should be used for this purpose. As an example, Figure 12.1 shows the comparison of surface imaging and radiographic imaging for the same phantom.

In addition, the Winston-Lutz test should be performed.[19] To include surface imaging in the Winston-Lutz process, a custom phantom for the SRS procedure can be created, like the one shown in Figure 12.2. This phantom consists of a tungsten bead with a simulated face surface attached at a known offset, for surface-based alignment to isocenter. The phantom is first aligned using the surface imaging system, followed by kV and MV imaging to verify the co-location of the tungsten target with the treatment isocenter, in accordance with the conventional Winston-Lutz test.

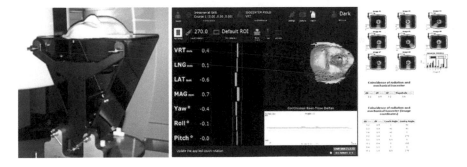

FIGURE 12.2 An example of a surface imaging compatible phantom (left) for performing Winston-Lutz radiation-mechanical-imaging isocenter verification. The phantom is aligned to isocenter using surface imaging (center) and then the radiation and mechanical isocenters are verified using MV imaging (right) according to the conventional Winston-Lutz test.

12.3 CLINICAL WORKFLOW FOR SURFACE-GUIDED SRS/SRT

The clinical workflow for surface-guided SRS has been described in the literature, as in Cerviño et al.[20] and Li et al.[21] A typical workflow is shown in Figure 12.3 and is described below.

12.3.1 Simulation

As per the AAPM Practice Guideline 9.a for SRS-SBRT,[22] the CT slice thickness for SRS should not exceed 1.25 mm and the scan field of view should be optimized for maximum in-plane spatial resolution, while including all necessary anatomy and immobilization hardware in the field of view. The patient should be scanned with the same immobilization as for treatment, which, in most cases, will consist of an open mask. Surface imaging is not required during CT simulation for SRS, and the benefits of using it have not been studied.

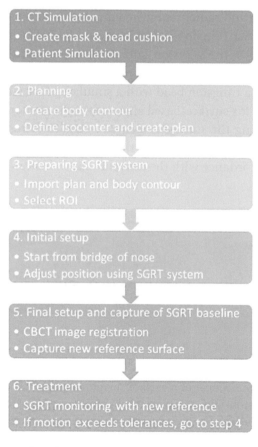

FIGURE 12.3   Typical surface-guided SRS treatment workflow.

12.3.2 Treatment Planning

Treatment planning is performed as usual with other SRS systems. The body contour will later be used for initial patient setup, and, therefore, needs to be as accurate as possible, and might need manual corrections after automatic contouring is used, particularly in the nose area, or if there are artifacts that affect it. The body contour from the radiation therapy plan (from the CT image) and the plan (isocenter and beam/couch angle information) is exported from the treatment planning system and imported into the SGRT system. The accuracy of the plan and body contour import including isocenter position should be verified as part of the pretreatment QA process. An ROI for surface monitoring should be defined on the face area of the imported body contour. The pretreatment QA process is described in more detail later in the chapter.

12.3.3 Patient Setup

Currently, the primary imaging modality for SRS is cone-beam CT (CBCT). The purpose of surface imaging is two-fold. First, surface imaging is used for initial setup. As the patient is positioned on the table, the radiation therapists turn on the surface imaging system and adjust the patient position in 6 degrees of freedom until the displacements shown by the SGRT system are within tolerance (usually below 1 mm for translations and 1° for rotations). Some SGRT vendors offer a three-axis head positioning tool to adjust yaw, roll, and rotation of the head without removing the open-face mask. This stage of the patient's setup is usually done manually, as shown in Figure 12.4, but some SGRT solutions provide a "move couch" feature to automate this process. After initial setup, a pair of orthogonal radiographs may be acquired to check for gross positioning errors and any rotations, followed by a CBCT, which will be registered to the planning CT. The appropriate shifts will be applied, either with a combination of automatic couch movement and manual head rotations, or with the aid of a 6 degrees-of-freedom robotic couch. When adjustment is performed manually, a new CBCT will be acquired, from which new shifts might need to be applied. The use of robotic couches and 2D–3D registration from the orthogonal radiographs makes the setup more streamlined. It is important to continually observe the patient with the SGRT system during imaging to make sure that the patient does not move. If motion is detected, imaging may need to be repeated.

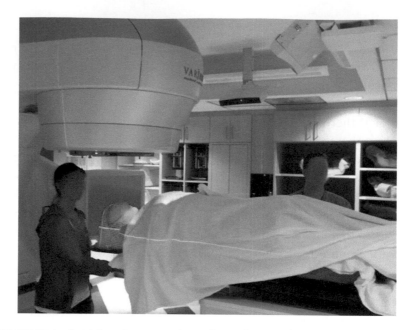

FIGURE 12.4   Initial patient setup is performed manually in the treatment room with surface image guidance.

Immediately after the patient is positioned and verified to be in the treatment position, a new reference surface image is acquired by the SGRT system. Depending on the SGRT system used, a tracking ROI will either need to be redefined or checked, if automatically transferred from the original reference image.

12.3.4 Treatment Delivery

During treatment, the surface image monitoring system remains on. The surface of the patient is tracked in real-time, and displacements with respect to the post-CBCT acquired reference surface are shown, as seen in Figure 12.5. Prespecified tolerances (usually 1 mm for translations and 1° for rotations) are set and used to warn the user or automatically stop the beam in systems that are directly connected to the linear accelerator interface. When tolerances are exceeded and treatment is stopped, two things might happen: (1) the patient returns to within tolerances on their own, (2) the patient position needs to be adjusted. In the latter case, a new CBCT is acquired, followed by shifts if necessary, and a new reference surface is captured to monitor patient position. Treatment can then be resumed.

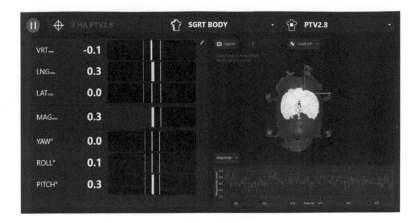

FIGURE 12.5   Surface imaging system display with the surface and the real-time displacements with respect to the reference image.

12.3.5 Risk Assessment

With the increasing complexity of radiotherapy equipment and resources, the AAPM Task Group 100 recommends to perform failure mode and effects analysis (FMEA) in new clinical implementations.[23] Manger et al. assembled a multidisciplinary team to perform an FMEA on the surface-guided SRS/SRT procedure.[24] The team determined 167 failure modes for each step (13 major subprocesses and 91 steps) in the SRS process and estimated risk priority numbers for each step (RPN). Of the 91 steps, 16 were directly related to the use of surface imaging. Then the failure modes were ranked by RPN, and the riskiest failure modes were analyzed via fault tree analysis. Only 1 of the top 25 riskiest failure modes was related to surface imaging (system fails to detect patient movement). Some of the proposed mitigation strategies include having quality assurance of the surface imaging system on treatment day (15 min), creating a daily checklist to ensure that daily quality assurance has passed (5 min), and requiring active monitoring of the surface imaging system by the physicist (no additional time commitment). The RPN for this failure mode was estimated to decrease from 192 to 80 with these mitigation strategies. The authors concluded that the use of surface imaging for monitoring intra-fraction motion during SRS/SRT did not greatly increase the risk of linac-based SRS. A risk analysis for surface-guided SRS and SRT is described in Chapter 14.

12.4 TREATMENT AND ACCURACY CONSIDERATIONS

12.4.1 Treatment Outcomes

Pham et al.[25] and Pan et al.[26] have reported on the positive outcomes of surface guidance as a positioning and monitoring system in SRS of metastatic lesions compared to frame-based and other frameless based treatments. Lau et al.[27] reported on the use of single isocenter SRS treatments using SGRT for patients with multiple metastases. Based on a cohort of 15 patients, the local control at 6 and 12 months was 91.7% and 81.5%, respectively, showing that it may provide similar clinical outcomes compared with conventional radiosurgery. The median treatment time was 7 minutes, highlighting the efficiency of SGRT for delivering accurate SRS treatments to multiple metastases. SGRT has also been used for the SRS treatment of nonmalignant lesions. Paravati et al. showed the use of SGRT as a monitoring modality for the treatment of trigeminal neuralgia in seven patients using 5 mm cones.[28] Outcome results were similar to traditional frame-based systems. Baker and Sullivan published a case report where a 54-year-old woman was treated for trigeminal neuralgia with a conventional IGRT linear accelerator with high-definition multi-leaf collimator and SGRT.[29] The prescription dose for the PTV was 60 Gy to the 80% isodose line. The patient was 80% free of her trigeminal neuralgia after 2 weeks and 100% free at 3 months. A more detailed discussion of surface-guided SRS outcomes is given in Chapter 13.

12.4.2 Mask Selection

Earlier implementations of surface-guided SRS/SRT were based on using a custom head mold and no mask.[11,20] However, the mold making process and the fact that it did not provide enough immobilization for a subset of patients (who were falling asleep and involuntarily moving, for example) led to the use of open-face masks that immobilize the patient while allowing direct surface monitoring of the patient's face. When selecting the mask, the user needs to consider the ROI to be tracked. Since the ROI in SRS treatments consists of the face, temporal region, eyes, and nose, an open mask such as the one in Figure 12.6 should be used. Kügele et al. evaluated the system accuracy using eight different types of commercial open masks, concluding that all of them were capable of providing the same accuracy.[30]

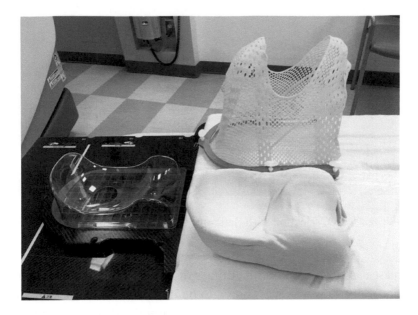

FIGURE 12.6   Example of an open-face mask with custom and plastic headrests commonly used in surface-guided SRS.

12.4.3 ROI Selection

The use of current surface imaging systems for SRS procedures is based on the hypothesis that the head (and brain) is a solid body where motion can be measured from the outside. The selection of the ROI is then crucial to follow that hypothesis as closely as possible. The ROI should not contain any rigid non-body parts such as the mask nor any movable body parts such as the lips or jaw. Consideration should also be taken to remove anatomy from the ROI that exhibits deformation that could void the above hypothesis and to leave a small gap between the mask and the ROI edges. The consensus among multiple clinics is to use an ROI that covers the forehead, eyes, temporal region, and nose, as shown in Figure 12.7. In some patients, eye blinking might affect the detected motion. In those cases, the clinician can choose to edit the ROI to remove the eyes as well with minimal change in tracking performance.

Besides considerations of anatomical features, the ROI selection can affect accuracy and time response. The surface in the imaging system is made up of multiple faces and vertex points. Wiersma et al. showed that the processing frequency (fps) of the 3D surface system linearly

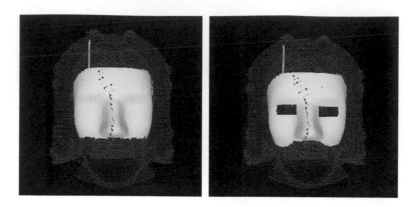

FIGURE 12.7 ROI selection with (left) and without (right) eyes.

decreases as a function of the number of ROI vertex points.[13] The tracking accuracy, however, was shown to be independent of the ROI size as long as the ROI was sufficiently large and contained enough features for registration.

12.4.4 Considerations for Treating Single vs. Multiple Targets

Yock et al. characterized the sensitivity of treatment plans with a single metastasis per isocenter and multiple metastases per isocenter to uncorrected patient motion,[31] and retrospectively estimated the plan-specific action limits for surface image-guided SRS intra-fraction monitoring. They evaluated the effect of a combination of translations and rotations (0–3 mm/0–3°) on treatment quality, using the proportion of the gross tumor volume that remained within the 100% prescription isodose line as a metric. The retrospective study used data from 29 patients (72 brain metastases in total), with 25 targets treated individually and 47 metastases treated in a plan simultaneously with at least one other metastasis. They modeled targets as spheres. In their results, they showed that both single-metastasis per isocenter and multiple metastases per isocenter plans exhibited compromised target coverage when translations and rotations were present. This effect was considerably larger in multiple metastases per isocenter plans (2.3% and 39.8%, respectively). They concluded that the use of multi-isocenter treatment plans should influence motion management, and that patient-specific action limits could be derived from a graphical representation of the effect of translations and rotations on any particular plan.

KEY POINTS

- Surface imaging has proven to be an adequate motion management tool in SRS/SRT, where the accuracy of currently available SGRT systems satisfies SRS-specific clinical guidelines.

- The use of surface imaging for SRS treatments requires additional QA tests and the integration of surface imaging into the daily IGRT QA program.

- An open-face mask approach is preferred for SRS/SRT treatments, as it provides sufficient immobilization while allowing good surface tracking for the treatment, with the added benefit of enhanced patient tolerability and comfort.

- A typical workflow for using surface guidance in SRS has been shown, including specific steps related to the implementation of surface imaging in the clinic.

REFERENCES

1. Andrews DW, Scott CB, Sperduto PW, et al. Whole brain radiation therapy with or without stereotactic radiosurgery boost for patients with one to three brain metastases: phase III results of the RTOG 9508 randomised trial. *Lancet*. 2004;363:8.
2. Kondziolka D, Patel A, Lunsford LD, Kassam A, Flickinger JC. Stereotactic radiosurgery plus whole brain radiotherapy versus radiotherapy alone for patients with multiple brain metastases. *Int J Radiat Oncol Biol Phys*. 1999;45(2):427–434.
3. Sneed PK, Suh JH, Goetsch SJ, et al. A multi-institutional review of radiosurgery alone vs. radiosurgery with whole brain radiotherapy as the initial management of brain metastases. *Int J Radiat Oncol Biol Phys*. 2002;53(3):519–526.
4. Aoyama H, Shirato H, Tago M, et al. Stereotactic radiosurgery plus whole-brain radiation therapy vs stereotactic radiosurgery alone for treatment of brain metastases: a randomized controlled trial. *JAMA*. 2006;295(21):2483.
5. Lightstone AW, Benedict SH, Bova FJ, Solberg TD, Stern RL. Intracranial stereotactic positioning systems: report of the American Association of Physicists in Medicine Radiation Therapy Committee Task Group No. 68: AAPM Task Group 68 Report. *Med Phys*. 2005;32(7Part1):2380–2398.
6. Ramakrishna N, Rosca F, Friesen S, Tezcanli E, Zygmanszki P, Hacker F. A clinical comparison of patient setup and intra-fraction motion using frame-based radiosurgery versus a frameless image-guided radiosurgery system for intracranial lesions. *Radiother Oncol*. 2010;95(1):109–115.

7. Bova FJ, Meeks SL, Friedman WA, Buatti JM. Optic-guided stereotactic radiotherapy. *Med Dosim.* 1998;23(3):221–228.
8. Nath SK, Lawson JD, Wang J-Z, et al. Optically-guided frameless linac-based radiosurgery for brain metastases: clinical experience. *J Neuro Oncol.* 2010;97(1):67–72.
9. Willoughby T, Lehmann J, Bencomo JA, et al. Quality assurance for non-radiographic radiotherapy localization and positioning systems: report of Task Group 147. *Med Phys.* 2012;39(4):1728–1747.
10. Klein EE, Hanley J, Bayouth J, et al. Task Group 142 report: quality assurance of medical accelerators. *Med Phys.* 2009;36(9Part1):4197–4212.
11. Cerviño LI, Pawlicki T, Lawson JD, Jiang SB. Frame-less and mask-less cranial stereotactic radiosurgery: a feasibility study. *Phys Med Biol.* 2010;55(7):1863.
12. Peng JL, Kahler D, Li JG, et al. Characterization of a real-time surface image-guided stereotactic positioning system. *Med Phys.* 2010;37(10):5421–5433.
13. Wiersma RD, Tomarken SL, Grelewicz Z, Belcher AH, Kang H. Spatial and temporal performance of 3D optical surface imaging for real-time head position tracking. *Med Phys.* 2013;40(11):111712.
14. Wen N, Snyder KC, Scheib SG, et al. Technical note: evaluation of the systematic accuracy of a frameless, multiple image modality guided, linear accelerator based stereotactic radiosurgery system. *Med Phys.* 2016;43(5):2527.
15. Mancosu P, Fogliata A, Stravato A, Tomatis S, Cozzi L, Scorsetti M. Accuracy evaluation of the optical surface monitoring system on EDGE linear accelerator in a phantom study. *Med Dosim.* 2016;41(2):173–179.
16. Li G, Ballangrud A, Chan M, et al. Clinical experience with two frameless stereotactic radiosurgery (fSRS) systems using optical surface imaging for motion monitoring. *J Appl Clin Med Phys.* 2015;16(4):149–162.
17. Wiant D, Liu H, Hayes TL, Shang Q, Mutic S, Sintay B. Direct comparison between surface imaging and orthogonal radiographic imaging for SRS localization in phantom. *J Appl Clin Med Phys.* 2019;20(1):137–144.
18. Paxton AB, Manger RP, Pawlicki T, Kim G-Y. Evaluation of a surface imaging system's isocenter calibration methods. *J Appl Clin Med Phys.* 2017;18(2):85–91.
19. Lutz W, Winston KR, Maleki N. A system for stereotactic radiosurgery with a linear accelerator. *Int J Radiat Oncol Biol Phys.* 1988;14(2):373–381.
20. Cerviño LI, Detorie N, Taylor M, et al. Initial clinical experience with a frameless and maskless stereotactic radiosurgery treatment. *Pract Radiat Oncol.* 2012;2(1):54–62.
21. Li G, Ballangrud A, Kuo LC, et al. Motion monitoring for cranial frameless stereotactic radiosurgery using video-based three-dimensional optical surface imaging. *Med Phys.* 2011;38(7):3981–3994.
22. Halvorsen PH, Cirino E, Das IJ, et al. AAPM-RSS medical physics practice guideline 9.a. for SRS-SBRT. *J Appl Clin Med Phys.* 2017;18(5):10–21.
23. Huq MS, Fraass BA, Dunscombe PB, et al. The report of Task Group 100 of the AAPM: application of risk analysis methods to radiation therapy quality management: TG 100 report. *Med Phys.* 2016;43(7):4209–4262.

24. Manger RP, Paxton AB, Pawlicki T, Kim G-Y. Failure mode and effects analysis and fault tree analysis of surface image guided cranial radiosurgery. *Med Phys*. 2015;42(5):2449–2461.
25. Pham N-LL, Reddy PV, Murphy JD, et al. Frameless, real-time, surface imaging-guided radiosurgery: update on clinical outcomes for brain metastases. *Transl Cancer Res*. 2014;3(4):351–357.
26. Pan H, Cerviño LI, Pawlicki T, et al. Frameless, real-time, surface imaging-guided radiosurgery: clinical outcomes for brain metastases. *Neurosurgery*. 2012;71(4):844–851.
27. Lau SKM, Zakeri K, Zhao X, et al. Single-isocenter frameless volumetric modulated arc radiosurgery for multiple intracranial metastases. *Neurosurgery*. 2015;77(2):233–240.
28. Paravati AJ, Manger R, Nguyen JD, Olivares S, Kim G-Y, Murphy KT. Initial clinical experience with surface image guided (SIG) radiosurgery for trigeminal neuralgia. *Transl Cancer Res*. 2014;3(4):333–337.
29. Baker BC, Sullivan TA. Trigeminal rhizotomy performed with modern image-guided linac: case report. *Cureus*. 2013;5(9):e139.
30. Kügele M, Konradsson E, Nilsing M, Ceberg S. EP-1958 Eight different open face masks compatibility with surface guided radiotherapy. *Radiother Oncol*. 2019;133:S1068.
31. Yock AD, Pawlicki T, Kim G-Y. Prospective treatment plan-specific action limits for real-time intrafractional monitoring in surface image guided radiosurgery. *Med Phys*. 2016;43(7):4342.

CHAPTER 13

Clinical Outcomes of Intracranial Surface-Guided Stereotactic Radiosurgery

Douglas A. Rahn III

CONTENTS

13.1 INTRODUCTION

Stereotactic radiosurgery (SRS) has been used for decades in the treatment of benign and malignant intracranial tumors. Technological advancements have moved from invasive frame-based systems fixed to the skull, to frameless but rigid thermoplastic mask systems using repeated X-rays or bite-block systems for localization, to more recent use of surface imaging systems for initial setup and intra-fractional motion monitoring. This chapter reviews the rationale for SRS with surface-guided radiation therapy (SGRT) and published outcomes from the first 10 years of its clinical use in the treatment of various benign and malignant diagnoses.

13.2 RATIONALE FOR SURFACE-GUIDED RADIOSURGERY

Pioneering technology for intracranial SRS originally used a stereotactic head frame rigidly pinned to the skull as a means of patient immobilization. The Lars Leksell Gamma Knife system (Elekta Ltd., Stockholm, Sweden) was the original equipment to use this approach.[1] There are some disadvantages and risks associated with head frames. There is a risk of patients developing headache or pain. There is a risk of skin or soft tissue injury including a small risk of skin infection from head frames.[2] Frame-based treatments typically involve same-day simulation and treatment, therefore the patients need to remain in the head frame for several hours. Some patients may not be able to tolerate the frame, such as pediatric patients or those with severe anxiety. There is also a cost of time and staff resources when coordinating appointments for frame placement in the neurosurgery clinic, the magnetic resonance imaging (MRI) session, and computed tomography (CT) simulation and treatment in the radiation therapy clinic on the same day.

Taken together these discomforts and risks represent a nontrivial disadvantage to rigid immobilization. The procedural burden of using a head frame can also be a disincentive to treat larger brain metastases with three- or five-fraction prescriptions, where such prescriptions may allow for an improved therapeutic window of treatment. Proponents of rigid head frames would mention that the intra-fractional motion may be better restricted with a head frame and that no additional daily imaging exposures from cone-beam computed tomography (CBCT) scans or repeated planar X-rays are needed.

Due to the limitations of frame-based SRS systems stated above, frameless stereotactic systems were developed. The ExacTrac frameless SRS system (BrainLab AG, Munich, Germany) immobilizes the patient in a thermoplastic mask and uses repeated X-ray imaging to localize the patient. The ExacTrac system consists of two oblique kV X-ray units that are capable of imaging the patient at non-coplanar couch positions. The disadvantages to this technology are that it does not provide continuous real-time monitoring and also that patient positioning by radiation therapists does not have in-room feedback, and hence may require a trial-and-error approach of checking setup images and refining the patient's setup position.

Bite-block systems with infrared reflectors for intra-fractional monitoring have been used in the past,[3] although there are some disadvantages to this approach. Such systems require that the patient has adequate dentition, require a rigid mask and so may be difficult to tolerate by patients. A bite-block system was commercially available (Frameless Array, Varian Medical Systems, Palo Alto, California), although it is no longer supported. Other bite-block devices are available to assist with patient immobilization, but without intra-fractional monitoring capabilities provided by the infrared reflectors.

Other SRS systems discussed above are contrasted with SGRT, as summarized in Table 13.1.

Linear accelerator-based SRS carries an added challenge of requiring extremely precise patient positioning in order to match the treatment setup position to the original CT simulation position, in contrast to frame-based SRS. When positioning a patient for radiosurgery using a linear accelerator-based system, the patient may require repositioning to match the CT simulation images. Lateral, longitudinal, and vertical offsets can be corrected by a standard three-axis linear accelerator patient support. If the patient's setup requires rotation, pitch, or yaw corrections to align with the simulation images, then a manual repositioning is required, unless a secondary technology such as a 6 degrees-of-freedom (6DOF) adjustable couch, or manual head adjuster, is available. In either case, there is an added cost of additional time on treatment or an additional equipment purchase. However,

TABLE 13.1   Comparison of Radiosurgery Systems

Frame-Based	Surface-Guided	Repeated X-Ray Based Monitoring
• May cause headache or pain	• Painless	• Painless
• May cause skin or soft tissue injury	• No risk of skin or soft tissue injury	• No risk of skin or soft tissue injury
• Duration of several hours in the head frame and immobilized	• Treatment appointments as short as 15 minutes	• Short treatment appointment times
• Possible poor tolerance of the frame among pediatric patients or those with severe anxiety	• Improved procedure tolerance with open-face thermoplastic mask	• May or may not have open-face thermoplastic mask
• Cost of time and staff resources when coordinating appointments for frame placement, the magnetic resonance imaging, and the treatment delivery all on the same day	• No need to coordinate treatment appointments in other Departments on the day of treatment	• No need to coordinate treatment appointments in other Departments on the day of treatment
• Does not require monitoring of patient position during setup or treatment delivery	• Provides continuous real-time monitoring of patient position during setup and treatment delivery	• Does not provide continuous real-time monitoring during setup or treatment delivery

SGRT systems provide real-time feedback to radiation therapists in the treatment vault positioning a patient prior to imaging, which can reduce overall setup time for SRS.

13.3 PERCEIVED BARRIERS TO USING SGRT FOR INTRACRANIAL SRS

Head frames and other frameless systems that utilize repeated planar X-ray imaging have been in use for decades and have established reliability, and it is fair to ask why these methods should be abandoned in favor of SGRT. Also, many clinicians are skeptical that the surface anatomy of the face can accurately correlate with the location of an intracranial tumor with submillimeter accuracy. Obviously, it is not a technological leap that should be taken on faith, without rigorous testing and validation.

The rationale for SGRT described in the preceding section has proven to be enticing enough to warrant design and implementation of the technology. Certainly, there has been exhaustive physics research testing and validation of the accuracy of SGRT for radiosurgery, which are described in other chapters of this book. Despite this justification, there remain skeptical voices questioning if the technology can yield clinical outcomes akin to frame-based systems. Fortunately, there is a growing body of clinical outcomes research supporting the use of SGRT for SRS, which will be described later in this chapter.

13.4 CLINICAL PROCEDURE FOR USING SGRT FOR INTRACRANIAL SRS

The feasibility of using surface guided radiotherapy for radiosurgery was first described in 2010[4,5] using the AlignRT SGRT system (Vision RT, London, UK), followed by a detailed description of the clinical procedure and initial experience in 2012.[6] The procedure has undergone some modifications in the intervening time, but the principles have remained consistent since SGRT with open-face thermoplastic masks has become the standard methodology for radiosurgery at many institutions.

Initially, patients undergo a contrast-enhanced MRI scan prior to the treatment date. The scan should be ideally less than 7 days prior to treatment but not more than 14 days prior to the treatment. A CT simulation is performed with a customized head cushion such as Accuform (CIVCO Medical Solutions, Kalona, IA) with a thermoplastic open-face mask covering the forehead and chin, but exposing the central and lateral face. A noncontrast CT simulation scan is then performed. The CT images and the MRI images are fused by matching the bony anatomy of the skull, with the accuracy being checked by the physicist and physician. The treating physician contours the target volume(s) and the organs-at-risk (OARs). Typically, a 1 mm planning target volume (PTV) margin is added to the gross tumor volume (GTV), in contrast to frame-based SRS systems where there is typically no added margin to the GTV. The body surface is automatically contoured, and this structure is exported to the SGRT system for use as the reference surface for initial patient setup. The prescription and treatment plan is approved prior to the day of treatment, and quality assurance is performed by the physicist.

On the day of treatment, the patient is positioned on the treatment couch and immobilized with the head rest and mask. The SGRT system (e.g., AlignRT with a region-of-interest set as the exposed central face of the patient) is then used by the therapy staff in real-time inside the treatment room to finely adjust the position and rotation of the patient in 6DOF. On-board kV imaging and CBCT imaging are then acquired to ensure correct patient positioning with respect to the treatment isocenter. After any necessary shifts are made, the SGRT system acquires a new reference surface of the patient, which is used for subsequent intra-fractional motion management for that fraction. Any deviation from the set tolerance (typically, 1 mm translation or 1° rotation) requires the treatment to be stopped, either manually or automatically if the SGRT system is interfaced with the linear accelerator. Once the patient has been repositioned if necessary, radiographic imaging is used to verify.

13.5 COMPARISON OF SURFACE GUIDED RADIOTHERAPY IMMOBILIZATION EQUIPMENT

Li et al.[7] used the AlignRT system to compare the initial head alignment and intra-fractional motion of two immobilization systems: the Pinpoint system (Aktina Medical, Congers, New York) using a bite-block device and the Freedom system (CDCR Systems, Calgary, Canada) using an open-face mask. They tested the effects of patient movement on the accuracy of the systems. Both systems were able to accurately position patients and monitor intra-fractional motion. They found that the Freedom system outperformed the Pinpoint system in patient comfort and clinical workflow. The authors recommend that the head be in a neutral position at the time of simulation. Results from this study support the use of open-face mask systems for SGRT-SRS.

13.6 APPROPRIATE PATIENTS

Virtually any patient that is a candidate for SRS with a head frame or frameless SRS with repeated X-ray imaging would also be a candidate for SGRT-SRS. Certain restrictions are universal for SRS candidacy, such as an inability to lie flat due to respiratory symptoms or an inability to remain still due to claustrophobia or other psychological or neurologic conditions (although this may be overcome with anesthesia in appropriate cases). Additionally, the open-face masks used for SRS with SGRT can sometimes help alleviate claustrophobia.

Regarding diagnoses appropriate for radiosurgery with SGRT, one could argue that the indications should not matter: if the technology is appropriate for intracranial SRS, then it is appropriate for all indications for intracranial SRS. While this may be a defensible position, it is reassuring to see the wide array of diagnoses that have been reported to be successfully treated with SGRT for SRS. Many of these diagnoses carry nuances of patient-specific factors, treatment planning or predicted outcome, which warrant discussion individually. This will be the subject of later sections in this chapter.

13.7 FAILURE MODE AND EFFECTS ANALYSIS

In 2015, Manger et al. conducted a study of failure mode and effects analysis (FMEA) for surface-guided cranial radiosurgery.[8] They created a process map of the clinical procedure and a fault tree analysis (FTA). Risk priority numbers (RPNs) were estimated for the failure modes and were subsequently ranked by priority. Additionally, the fault tree analysis was used to ascertain the root factors of the highest risk failure modes. Collectively, this information was used to guide safety protocols for the clinical procedure. RPNs were reestimated after mitigation strategies were implemented.

The potentially highest risk step in the SRS process that was specific to surface-guidance was "Monitor SIG[SGRT]-indicated offsets to ensure patient position is within tolerance." The second highest was "Ensure surface imaging cameras are within tolerance." Of the 91 total steps in the SRS process, there were 15 steps specific to SGRT.

They concluded that surface guidance did not increase the risk of the SRS process; and in some cases it reduced the risk associated with the SRS process. A more in-depth discussion of risk analysis with SRS and SGRT is presented in the following chapter.

13.8 CLINICAL OUTCOMES FOR SPECIFIC DIAGNOSES

The following sections review published reports of institutional series using surface-guided radiotherapy for treatment of various benign and malignant diagnoses. The majority of these studies use the methodology outlined in Section 13.4. A summary of these studies is presented in Table 13.2.

13.9 BRAIN METASTASES

Pan et al.[9] described an institutional series of 44 patients with brain metastases (115 total lesions) treated with surface-guided radiosurgery from 2009 to 2011. Patients had a median of two metastases treated with a median dose of 20 Gy in a single fraction. In patients with multiple lesions, a single plan was used (i.e., a single isocenter). The median treatment time from beam on to beam off was 15 minutes. The patient position was monitored and the beam was held if motion exceeded the threshold tolerance. The authors noted that patients usually reverted quickly to a position within tolerance without external intervention. A posttreatment CBCT scan was taken for some patients, but this process was omitted when there was no observed evidence of systemic shift or error.

TABLE 13.2   Reported Clinical Outcomes of Intracranial Radiosurgery Using Surface-Guided Radiation Therapy (SGRT)

Reference	Diagnosis	Patients (n=)	Lesions (n=)	Result	Acute Toxicity	Late Toxicity
Pan et al.[9]	Brain metastases	44	115	76% 12 mo local control	18% (5% grade 3)	5% (2.5% grade 3)
Pham et al.[10]	Brain metastases	163	490	79% 12 mo local control	not reported	not reported
Lau et al.[11]	Brain metastases	15	62	81.5% 12 mo local control	6.7% (1 patient with seizure), no grade 3	none
Lau et al.[12]	Benign skull base tumors	48	48	98% 5 yr local control	no grade 3	no grade 3
Paravati et al.[13]	Trigeminal neuralgia	7	7	100% partial response, 57% complete response	none	one patient with self-limited dysesthesia of face
Wong et al.[14]	Pediatric germinoma	1	1	normalization of tumor lab markers	fatigue	none

Of the 88 lesions evaluable, the 6- and 12-month local control was 90% and 76%, respectively. Grade 3 or greater toxicity occurred in 9% of patients. The authors concluded that surface-guided radiosurgery compares favorably with similar reports of frame-based SRS, with the added benefit of improved patient comfort and shorter treatment time. Pham et al.[10] published an updated larger series of 163 patients (490 lesions) with intact brain metastases and post-operative cavities treated with radiosurgery using SGRT with the AlignRT system. An open-face mask on the forehead and jaw was used in conjunction with a customized foam headrest. During treatment, if movement exceeded the tolerance of 1–2 mm or 1 degree of rotation, then the beam was held automatically or manually, depending on the equipment available. The median time from beam on to beam off was 15 minutes. The actuarial 6- and 12-month local control rates were 90% and 79%, respectively. The authors concluded that intracranial radiosurgery with SGRT can produce outcomes comparable to those with conventional frame-based and frameless SRS techniques while providing greater patient comfort with an open-face mask and fast treatment time.

Lau et al.[11] described a similar series of patients with brain metastases, who were treated exclusively with volumetric modulated arc therapy (VMAT) and single-isocenter treatment plans using surface-guided radiosurgery. They reported a series of 15 patients with 62 total metastases treated. Local control at 6 months was 91.7%. The advantage to using VMAT in this setting was to reduce treatment time, which in turn improves patient comfort and reduces the chance of intra-fractional motion. The authors found a median time of 7.2 minutes from beam on to beam off. One patient (6.7%) reported a seizure after treatment, but there was no Grade 3 or greater acute or late toxicity.

13.10 BENIGN SKULL BASE TUMORS

Skull base tumors can be challenging to treat with radiosurgery given the proximity to critical structures such as the brainstem, cranial nerves, and cochleae. These lesions may also be geometrically more irregular than brain metastases (which are often spherical), which can make treatment planning more challenging. In order to justify treatment of these benign lesions, it is important for radiosurgery to maintain a very low risk of toxicity.

Lau et al.[12] conducted a retrospective review of 48 consecutive patients with benign skull base tumors who were treated between 2009 and 2011 with SRS using SGRT. The series included meningioma ($n = 22$), vestibular schwannoma ($n = 20$), and pituitary adenoma ($n = 6$). The median follow-up was 65 months. Doses were 12–13 Gy in a single fraction, 8 Gy × 3 fractions, and 5 Gy × 5 fractions. The tumor control rate at 5 years was 98%, with no reported grade 3 or higher toxicity. The authors concluded that these results are highly comparable to frame-based SRS reports, and that the results support the use of SGRT for SRS of skull base lesions.

13.11 TRIGEMINAL NEURALGIA

SGRT-SRS has also been used in cases of trigeminal neuralgia. Given the extremely high doses used in this scenario (80–90 Gy) and the proximity to the brainstem, this procedure requires the utmost precision.

Paravati et al.[13] reported their initial experience with SGRT for trigeminal neuralgia. Seven patients were treated to a dose range of 80–90 Gy in a single fraction normalized to the 80% isodose line. The brainstem maximum point dose ranged from 9 to 12 Gy. Plans were delivered using 13 noncoplanar arcs with a 5 mm cone. With a median follow-up of 31.4 months, all patients reported pain improvement, and four of the seven reported complete resolution of pain. There were two cases of pain recurrence and one case of intermittent facial numbness, which resolved after 9 months. The authors concluded that the results compared favorably to other frame-based series of SRS for trigeminal neuralgia, and that this application of SGRT is a significant leap forward in patient comfort and treatment time.

13.12 PEDIATRICS

Wong et al.[14] described a case study of a 17-year-old male with a pineal germ cell tumor treated with radiosurgery using SGRT. The patient was initially treated with induction chemotherapy followed by craniospinal radiotherapy (36 Gy) with an intensity modulated radiation therapy (IMRT) boost to the pineal gland (54 Gy). The patient presented with a local recurrence nearly 2.5 years later and started salvage chemotherapy. MRI showed a 1.7 cm pineal tumor, which was treated with fractionated radiosurgery to a dose of 25 Gy in 5 fractions. Surface imaging was used for intra-fractional motion monitoring. The patient tolerated the treatment well and tumor marker laboratory levels returned to normal.

The authors highlighted potential advantages of this surface-guided frameless radiosurgery approach, including increased patient comfort, ability to fractionate treatment, greater time for the multidisciplinary team review of imaging, contours, and dosimetry, and shorter daily treatment appointments.

SGRT using open-face mask immobilization for SRS can be quite tolerable for pediatric patients, in the clinical experience at UC San Diego and other institutions. In some cases it can allow anesthesia to be reduced or eliminated. Pediatric applications of SGRT are described in more detail in Chapter 19.

13.13 FUTURE DIRECTIONS

SGRT technology has become more widely adopted in recent years, and there are now multiple commercially available systems. Clinicians and developers of technology have the opportunity to work together to improve the clinical protocols for SGRT in order to achieve better outcomes for patients. Some ideas for potential avenues of exploration are presented below.

13.14 PLAN-SPECIFIC ACTION LIMITS FOR INTRA-FRACTIONAL MONITORING

Yock et al.[15] conducted a study to evaluate the sensitivity of individual radiosurgery plans to intra-fractional motion. They noted that action limit values are usually constant across patients even though motion affects target coverage differently in individual radiosurgery plans. They studied these effects for 29 patients with 72 total brain metastases. Of these, 25 were single-metastasis per isocenter (single-met-per-iso) plans and 47 were multiple-metastases per isocenter (multimet-per-iso) plans. They calculated the proportion of the GTV remaining within the 100% isodose line under various translations and rotations (0.0–3.0 mm, 0.0–3.0 degrees). They found that the multimet-per-iso plans had more target compromise than single-met-per-iso plans (39.8% vs. 2.3%), using a test scenario of translational and rotation error. This effect was primarily due to the greater sensitivity to rotational errors observed in multimet-per-iso plans than in single-met-per-iso plans. In addition to highlighting the value of intra-fraction motion monitoring for all SRS patients, multimet-per-iso plans may benefit from more stringent action limits for interrupting treatment due to motion detected by the surface imaging system.

The methods in this study can be used to create plan-specific action limits to maintain 100% target coverage in SRS plans treated with SGRT. This concept has not yet gained widespread clinical practice, but it raises a very valid clinical question: Why should every SRS plan have the same tolerance levels for translational and rotational errors? Tailoring these parameters to the individual treatment plan would allow for more accurate treatment.

13.15 INTEGRATION WITH TREATMENT COUCHES ALLOWING FOR AUTOMATED CORRECTION WITH 6 DEGREES OF FREEDOM

Current SGRT-SRS treatment procedures without 6DOF couches (as described, for example, earlier in this chapter) require therapists to manually adjust the patient's position until it matches the simulation position. In-room monitors give therapists feedback and help the process, including offering real-time 6DOF shift information from the SGRT system, but there remains some cost of time and potential for small inaccuracies. There can also be intra-fractional patient movement requiring the beam to be stopped and the patient to subsequently be repositioned.

There are commercially available radiotherapy treatment couches that are able to move with 6DOF. Some SGRT solutions integrate with the treatment couch to enable automatic correction of the patient's position, in up to 6DOF, both before treatment and intra-fraction when the beam is not being delivered. Additional automation, possibly during treatment, could yield faster and more accurate treatment.

13.16 USE OF SGRT IN FRACTIONATED INTRACRANIAL RADIOTHERAPY

As procedures for SGRT have become more efficient and commonplace in the clinic, it has become practical to use the technology for conventionally fractionated intracranial radiotherapy. Currently, a typical expansion from clinical target volume (CTV) to PTV in, for example, an IMRT plan for glioblastoma would be 3 mm. A reduction to 1 mm may at first sound minor, but given the large treatment volumes in this scenario the reduction in PTV volume and, in turn, potential toxicity could in fact be clinically meaningful.

KEY POINTS

- Surface-guided radiation therapy (SGRT) has been successfully implemented for radiosurgery of benign and malignant intracranial tumors over the past 10 years.

- Results from clinical outcomes studies show efficacy and toxicity of radiosurgery with SGRT to be comparable to frame-based radiosurgery systems.

- SGRT systems for radiosurgery allow for faster treatment times and greater patient comfort than other radiosurgery systems.

- There are no clinical indications for intracranial radiosurgery that could not be treated with SGRT.

REFERENCES

1. Leksell L. The stereotactic method and radiosurgery of the brain. *Acta Chir Scand.* 1951;102:316–319.
2. Safaee M, Burke J, McDermott MW. Techniques for the application of stereotactic head frames based on a 25-year experience. *Cureus.* 2016;8(3):e543.
3. Lawson JD, Wang JZ, Nath SK, Rice R, Pawlicki T, Mundt AJ, Murphy K. Intracranial application of IMRT based radiosurgery to treat multiple or large irregular lesions and verification of infra-red frameless localization system. *J Neuro Oncol.* 2010;97(1):59–66.
4. Cervino LI, Pawlicki T, Lawson JD, Jiang SB. 2010. Frame-less and mask-less cranial stereotactic radiosurgery: a feasibility study. *Phys Med Biol.* 2010;55(7):1863.
5. Peng JL, Kahler D, Li JG, Samant S, Yan G, Amdur R, Liu C. Characterization of a real-time surface image-guided stereotactic positioning system. *Med Phys.* 2010;37(10):5421–5433.

6. Cerviño LI, Detorie N, Taylor M, Lawson JD, Harry T, Murphy KT, Mundt AJ, Jiang SB, Pawlicki TA. Initial clinical experience with a frameless and maskless stereotactic radiosurgery treatment. *Pract Radiat Oncol.* 2012;2(1):54–62.

7. Li G, Ballangrud A, Chan M, et al. Clinical experience with two frameless stereotactic radiosurgery (fSRS) systems using optical surface imaging for motion monitoring. *J Appl Clin Med Phys.* 2015;16(4):149–162.

8. Manger RP, Paxton AB, Pawlicki T, Kim GY. Failure mode and effects analysis and fault tree analysis of surface image guided cranial radiosurgery. *Med Phys.* 2015;42(5):2449–2461.

9. Pan H, Cerviño LI, Pawlicki T, et al. Frameless, real-time, surface imaging-guided radiosurgery: clinical outcomes for brain metastases. *Neurosurgery.* 2012;71(4):844–851.

10. Pham NL, Reddy PV, Murphy JD, et al. Frameless, real-time, surface imaging-guided radiosurgery: Update on clinical outcomes for brain metastases. *Transl Cancer Res.* 2014;3(4):351–357.

11. Lau SK, Zakeri K, Zhao X, et al. Single-isocenter frameless volumetric modulated arc radiosurgery for multiple intracranial metastases. *Neurosurgery.* 2015;77(2):233–240.

12. Lau SK, Patel K, Kim T, et al. Clinical efficacy and safety of surface imaging guided radiosurgery (SIG-RS) in the treatment of benign skull base tumors. *J Neurooncol.* 2017;132(2):307–312.

13. Paravati AJ, Manger R, Nguyen JD, Olivares S, Kim GY, Murphy KT. Initial clinical experience with surface image guided (SIG) radiosurgery for trigeminal neuralgia. *Transl Cancer Res.* 2014;3(4):333–337.

14. Wong K, Opimo AB, Olch AJ, et al. Re-irradiation of recurrent pineal germ cell tumors with radiosurgery: Report of two cases and review of literature. *Cureus.* 8(4):e585.

15. Yock AD, Pawlicki T, Kim GY. Prospective treatment plan-specific action limits for real-time intrafractional monitoring in surface image guided radiosurgery. *Med Phys.* 2016;43(7):4342.

CHAPTER 14

Risk Analysis
for SGRT-Based SRS/SRT

Ryan Manger

CONTENTS

14.1 INTRODUCTION

As has been discussed in previous chapters, surface imaging may be used in the stereotactic radiosurgery (SRS) or stereotactic radiotherapy (SRT) setting instead of a frame-based approach. SRS/SRT is a high-risk procedure that delivers large radiation doses near critical structures in few (1–5) fractions, hence the risk of implementing a frameless, surface image guided, radiosurgery (SGRT-SRS) procedure must be considered. There are several methods available for conducting a risk analysis. The American Association of Physicists in Medicine (AAPM) Task Group 100 recommends using failure modes and effects analysis (FMEA) in conjunction with a fault tree analysis (FTA) to conduct the risk analysis of a process.[1] This chapter will describe an FMEA and FTA of the clinical implementation of the SGRT-SRS process.[2]

14.2 RISK ASSESSMENT WITH FMEA AND FTA

FMEA is a systematic method of identifying component and process problems before they occur.[3] While medical physicists have always analyzed processes for potential failures, the FMEA process standardizes the approach and establishes a common language that can be used to communicate between members of the department and hospital administration. In general, these are the following steps in an FMEA:

1. Create a process map that describes the steps involved in the proposed treatment process.

2. For each step in the proposed treatment process, ask "What could go wrong?" The result of this is a series of failure modes. There could be multiple failure modes for each process step.

3. For each failure mode, ask "How could this have gone wrong?" The result of this is a number of causes for each failure mode. There could be multiple failure pathways for each failure mode.

4. Determine the severity (S), probability of occurrence (O), and likelihood of detection (D) values for each failure mode and pathway and calculate the risk priority number (RPN) for each failure mode/cause combination.

5. Use the risk priority number to rank the failure modes. Review the top failure modes as part of the risk assessment.

These causes and effects were used when determining the risk priority numbers (RPNs) of each failure mode. The RPN is an arbitrary metric of risk that is computed as the product of the occurrence (O), severity (S), and detectability (D). In this FMEA, the O, S, and D values range from 1 to 10. Hence, RPN ranges from 1 to 1000, where a greater RPN signifies a riskier failure mode. Specific details about the severity, detectability, and occurrence scoring are provided later in this chapter.

Sometimes, FTA can be used in conjunction with FMEA to create a robust or more comprehensive risk assessment. FTA is a deductive analysis approach for resolving an undesired event into its causes and is a backward looking analysis,[3] looking backward at the causes of a given event (deductive). FTA is not an FMEA, which assesses different effects of single basic failures and pathways (inductive). The result of an FTA is a Fault Tree, which is the logical model of the relationship of a failure to more basic events. The top event of the fault tree is the undesired event. The middle events are intermediate events. The bottom of the fault tree is the causal basic events or primary events. The logical relationships of the events are shown by logical symbols, called gates. Therefore, an FTA explicitly shows all the different relationships that are necessary to result in the top event (i.e., failure). Human errors are classified into two basic types: errors of omission and errors of commission. An error of omission is not doing a correct action whereas an error of commission is doing an incorrect action. Human errors are modeled as basic events in a Fault Tree, similarly to component failures.

14.3 SGRT-SRS PROCESS

The SGRT-SRS FMEA was conducted by a multidisciplinary team of one physician, four physicists, one dosimetrist, and two therapists. The process map was composed of 13 subprocesses: (1) consultation and treatment strategy, (2) CT simulation, (3) image registration and data transfer, (4) preplanning, (5) planning, (6) plan review, (7) postplanning, (8) patient-specific quality assurance, (9) chart review, (10) pretreatment, (11) patient setup, (12) treatment delivery, and (13) posttreatment. Each of these subprocesses consisted of several steps. The SGRT-SRS process considered in the FMEA was composed of 91 steps and is listed in Figures 14.1 and 14.2 (figures from Manger 2015). Each step is labeled with the staff member performing the task, and the asterisk designates steps that are specific to SGRT. Of the 91 steps, 15 were SGRT specific and were mostly in the pretreatment, patient setup, and treatment delivery subprocesses.

In this SGRT-SRS process, the SGRT system is used to reduce the translational and rotational discrepancies during initial patient positioning and to monitor the patient during treatment. For initial patient setup, the SGRT system is used as a reference for aligning the patient to the planning CT body contour. Radiographic imaging is then used to verify patient position and provide further translational and/or rotational adjustments if necessary. If shifts are indicated on cone-beam CT (CBCT), they are carried out because the CBCT to planning CT match is considered the gold standard for patient setup. The SGRT reference surface is acquired to match the patient position after any CBCT shifts. SGRT is then used to monitor the patient for intra-fraction motion throughout the entire fraction of treatment.

SUB - PROCESSES

Consult & treatment strategy	CT Simulation	Image registration and transfer	Preplanning	Planning	Plan Review	Postplanning
1. Patient file created in medical record system (N)	7. Dispo form consulted for sim guidance (T)	20. Sim CT imported to treatment planning system (D)	25. Populate structure set with relevant structures (D)	34. Choose beam arrangement based on tumor location/size (D)	39. Review OAR statistics (DVH, etc.) (M)	43. Insert setup fields and generate DRRs (D)
2. Patient consultation (M)	8. Patient identified (T)	21. High resolution MR images imported to TPS (D)	26. Contour the body (D)	35. Set calculation grid size to be finer than default (SRS) (D)	40. Review PTV max, min, mean, coverage, CI, GI, HI (M)	44. Provide setup notes (shifts, PTV margin, etc.) (D)
3. Treatment strategy determined (M)	9. Scanning table prepared with setup devices (T)	22. Planning CT & MR evaluated (slice size, scan range) (D)	27. Set origin to nasion (D)	36. Set optimization objectives and NTO (D)	41. Make adjustments or submit plan for physics review (D)	45. Create a plan report document in patient's record (D)
4. Consult noted entered in record (M)	10. Patient positioned on couch with sagittal laser at midline (T)	23. Images labeled with acquisition date and technique. (D)	28. Create and properly label new course/plan (D)	37. Optimize using appropriate optimizer (VMAT vs static) (D)	42. Review and approve or request revisions (P).	46. Create a plan report document in patient's record (D)
5. Patient consent acquired (M)	11. Accuform cushion molded to neck and headrest (T)	24. MR registered to planning CT using rigid registration. (D)	29. If previously treated, register CT with previous tx CT (D)	38. Normalize (PTV $V_{100} = 98$–100%) (D)		47. Planning approve & select contours to be overlaid on DRRs (D)
6. MD dispo form created (scanning protocol, fusion requests, etc.) (M)	12.* Open-faced mask molded to patient's face and connected to frame. (T)		30. Provide Rx, PTV & PRV margins, and constraints (M)			48. Send plan to linac (D)
	13. SRS protocol selected (1.25 mm slice size) (T)		31. Contour critical structures and PRVs (M)			49.* Export body contour and plan for SIG system (D)
	14. Scouts used to determine scan range (T)		32. Create control structures (D)			50. Submit plan for patient-specific QA (D)
	15. CT scan acquired (T)		33. Insert Rx and target in the plan (D)			51. Submit plan for physics chart check (D)
	16. Tx start date and location scheduled (T)					
	17. Simulation notes documented (T)					
	18. CT image pushed (T)					
	19. Mask and head cushion labeled and delivered to linac (T)					

*: SIG-RS specific step
M: physician, P: physicist, D: dosimetrist, T: therapist, N: Nurse

FIGURE 14.1 The sub-processes and steps of SGRT-SRS (part 1). (Adapted from Manger, R.P. et al., *Med. Phys.*, 42, 2449–2461, 2015.)

SUB - PROCESSES

Patient-specific quality assurance	Chart review	Pretreatment	Patient Setup	Treatment delivery	Posttreatment
52. Create a verification plan. (D)	55. Physics plan review (P)	59. Ensure SRS QA has been completed (Winston-Lutz, etc.) (P)	67. Confirm patient identity (T)	80.* Capture a new SIG reference surface (P)	88. Confirm future visits with patient if applicable (SRT) (T)
53. Deliver verification plan and determine if it is within tolerance. (P)	56. Input imaging order and schedule imaging (kV/kV and CBCT) (P)	60. Ensure daily integrated IGRT QA has been performed (P)	68. Position patient on table with iso at the nasion (T)	81.* Edit ROI on new reference surface (P)	89. Escort patient to on treatment visit (OTV) if necessary (T, M)
54. Update plan document with QA results. (P)	57. Approve plan for treatment (P)	61.* Ensure surface imaging cameras are within tolerance (P)	69.* Begin surface monitoring using body contour as reference (P)	82.* Begin monitoring using new reference surface (P)	90. Dispose of custom immobilization devices following last tx (T)
	58. Peer review the case at chart rounds. (M,P)	62. Perform pre-treatment therapist chart review (T)	70.* Manually adjust head to reduce rotational and translational offsets (T)	83. Deliver treatment arcs/fields (T)	91. Perform physics final chart check (P)
		63.* Create patient file in surface imaging system (resolution, skin tone) (P)	71. Place mask over patient's head (T)	84.* Monitor SIG indicated offsets to ensure pt position is within tolerance (P,T)	
		64.* Import body contour and plan to SIG system (P)	72.* Fine tune head position with head adjuster until offsets are within 0.5 mm & 0.5° (T)	85.* Beam off if SIG offsets exceed tolerance (~1 mm) (P,T)	
		65.* Select the ROI to be used for SIG setup (P)	73. Take orthogonal setup images (T)	86.* Rotate reference surface for arcs/fields with couch rotations (P)	
		66. Prepare treatment table for patient (T)	74. Make rotational adjustments if necessary (T)	87. After all fields have been delivered, mark the treatment as complete (T)	
			75. Register orthogonal setup images with DRRs (T)		
			76. Take CBCT using 1 mm slices (T)		
			77. Register CBCT to sim CT (T)		
			78. Review registration (M, P)		
			79. Apply CBCT couch shifts (T)		

*: SIG-RS specific step
M: physician, P: physicist, D: dosimetrist, T: therapist, N: Nurse

FIGURE 14.2 The sub-processes and steps of SGRT-SRS (part 2). (Adapted from Manger, R.P. et al., *Med. Phys.*, 42, 2449–2461, 2015.)

14.4 FAILURE MODES AND RISK PRIORITY NUMBERS

Each step in the SGRT-SRS process had one or more failure modes that were contributed by the multidisciplinary team. In total, 167 failure modes were determined. Each failure mode was associated with a list of potential causes and effects. An example of this is provided in Table 14.1.

The team was instructed to use a standard system for setting the O, S, and D of a failure mode (Tables 14.2 through 14.4); this ensured that the data were normalized and scaled appropriately.

After computing the RPN for each failure mode, the failure modes were ranked by RPN to see which steps carried the greatest risk. Of the top

TABLE 14.1   Failure Modes for the 84th Step in the SGRT-SRS Process

Step	Potential Failure Modes	Potential Cause of Failure	Potential Effects of Failure
84. Monitor SGRT indicated displacements to ensure patient position is within tolerance	Not performed	Inattention	Geometric miss
	SGRT system fails to detect patient movement	SGRT system failure	Geometric miss
	SGRT system indicates movement, yet patient did not move	SGRT system failure	Make unnecessary shifts
	Not all metrics were being monitored	Inattention	Geometric miss

Source: Manger, R.P. et al., *Med. Phys.*, 42, 2449–2461, 2015.

TABLE 14.2   Occurrence Value System

Value	Qualitative	Frequency
1	Failure unlikely	1/10,000
2		2/10,000
3	Relatively few failures	5/10,000
4		1/10,000
5		<0.2%
6	Occasional failures	<0.5%
7		<1%
8	Repeated failures	<2%
9		<5%
10	Failure inevitable	>5%

Source: Manger, R.P. et al., *Med. Phys.*, 42, 2449–2461, 2015.

TABLE 14.3 Severity Value System

Value	Qualitative	Categorization
1	No effect	
2	Inconvenience	Inconvenience
3		
4	Minor dosimetric error	Suboptimal plan or treatment
5	Limited toxicity or under-dose	Wrong dose, dose distribution,
6		location, or volume
7	Potentially serious toxicity or under-dose	
8		
9	Possible very serious toxicity	Very wrong dose, dose
10	Catastrophic	distribution, location, or volume

TABLE 14.4 Detectability Value System

Value	Estimated Probability of Going Undetected (%)
1	0.01
2	0.2
3	0.5
4	1.0
5	2.0
6	5.0
7	10
8	15
9	20
10	>20

25 RPN-ranked failure modes, only one was related to SGRT, and it was ranked 8th overall. The top five RPN-ranked, SGRT-specific failure modes are listed in Table 14.5. The first column lists the overall RPN rank of the failure mode, and the second column lists the step where the failure mode may occur.

Of the top five RPN-ranked failure modes, four may occur during step 84, "Monitor the SGRT indicated displacements to ensure patient position is within tolerance." This is a critical step in the SGRT-SRS process that should be emphasized when implementing SGRT-SRS in the clinic because it has many high-risk failure modes. During treatment, the surface monitoring system displays a series of real-time displacements (or deltas) that indicate how much the patient has moved. The system provides four translational (cranial-caudal, left-right, anterior-posterior, and the vector

TABLE 14.5   Top Five SGRT-Specific Failure Modes Ranked by RPN

Rank	Step	Potential Failure Modes	Potential Cause of Failure	Potential Effects of Failure	O	S	D	RPN
8	84. Monitor SGRT indicated displacements to ensure patient position is within tolerance	SGRT system fails to detect patient movement	SGRT system failure	Geometric miss	3	8	8	192
26	84. Monitor SGRT indicated displacements to ensure patient position is within tolerance	Not done	Inattention	Geometric miss	4	8	4	112
28	61. Ensure surface imaging cameras are within tolerance	Not checked	Inattention	System may be out of tolerance	6	4	4	96
28	84. Monitor SGRT indicated displacements to ensure patient position is within tolerance	Not all metrics were monitored	Mental lapse	Patient position out of tolerance on the unmonitored axis	4	6	4	96
32	84. Monitor SGRT indicated displacements to ensure patient position is within tolerance	SGRT system indicates movement, yet patient did not move	SGRT system isocenter drift	Prolong treatment to investigate movement	10	3	3	90

Source: Manger, R.P. et al., *Med. Phys.*, 42, 2449–2461, 2015.

sum of the three) and three rotational (pitch, yaw, and roll) displacements. During step 84, the physicist and therapists are expected to monitor these displacements to ensure the patient has not moved outside of the institutional tolerance (1 mm and 1° for our institution).

The most frequent failure mode experienced during this step is, "SGRT system indicates movement, yet patient did not move." This typically occurs when rotating the patient support device to deliver a non-coplanar treatment field or arc. An effective risk mitigation strategy for this failure mode is to rotate the patient support device back to 0° (coplanar) and assess the displacements. If the indicated displacements are still out of tolerance, a radiographic image is acquired (CBCT or orthogonal kV pair), shifting if necessary, then reacquiring a new reference surface and proceeding with treatment.

Two of the failure modes related to step 84 (ranks 26 and 28) are human failures that are caused by not monitoring one or more of the indicated displacements. These failure modes have a relatively high RPN because they can lead to a geometric miss, which is a severe effect.

The final failure mode related to step 84 was the highest RPN-ranked failure mode, "SGRT system fails to detect patient movement." To further examine this failure mode, an FTA was conducted.

14.5 FAULT TREE ANALYSIS

FTA is used to determine the underlying causes of a failure mode. The FTA for the highest-ranking SGRT-specific failure mode is presented in Figure 14.3. The fault tree begins with a failure mode (the first box on the left side of Figure 14.3). A top-down approach is then

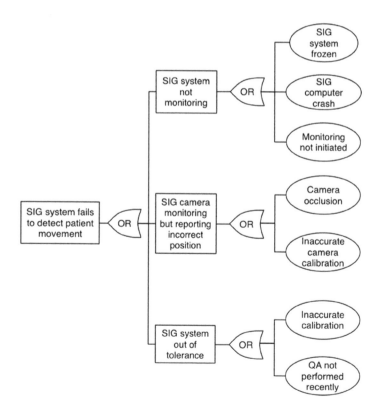

FIGURE 14.3 FTA for the top RPN-ranked, SGRT-specific failure mode. (Adapted from Manger, R.P. et al., *Med. Phys.*, 42, 2449–2461, 2015.)

taken to deduce the possible causes that may lead to the failure mode. The causes at a given level in the fault tree are linked through AND or OR logic gates depending on if a specific combination of failures must occur to lead to the preceding failure mode. The causes at this second level of the diagram are then further analyzed to determine if they have other underlying causes. This process is continued until no more fundamental causes can be determined. The fundamental causes are then used to guide mitigation efforts. For instance, in the case of Figure 14.3, two of the fundamental causes are "inaccurate calibration." Naturally, the risk mitigation efforts should include some means of ensuring the cameras are calibrated correctly.

14.6 RISK MITIGATION

The purpose of risk mitigation is to improve the detectability, decrease the occurrence, and reduce the severity of the failure modes. Since only one of the SGRT-specific failure modes carried a relatively high amount of risk, the risk mitigation efforts undertaken at clinics utilizing SGRT for radiosurgery should be primarily focused on combating the failure mode—"SGRT system fails to detect patient movement." The fundamental causes of this failure mode determined using the FTA shown in Figure 14.3 were: system frozen, computer crash, monitoring not initiated, camera occlusion, inaccurate camera calibration, and QA not performed recently. Considering these fundamental causes, the recommended mitigation strategies to reduce the risk of the SGRT system failing to detect patient movement are: (1) daily quality assurance of the positional accuracy of the SGRT camera via a megavoltage (MV)-SGRT coincidence test using a phantom with a known geometry (such a test is described in Chapter 12); (2) ensure that one of the therapists is monitoring the patient position using the closed-circuit television to detect gross movements in the event of a computer crash, software freeze, or surface monitoring not being initiated; and (3) ensure a physicist is present during treatment to monitor the surface imaging system.

It should be noted that SGRT in itself can be seen as a tool for risk mitigation when applied to radiosurgery. It can be used to verify that the proper radiographic shifts have been applied, to verify the proper couch rotations have been executed, and to reduce the risk of treating the wrong patient (since the region of interest (ROI) relies on face shape).

KEY POINTS

- SRS/SRTs are high-risk treatments due to the high radiation doses delivered over few fractions. The treatment process implemented in a clinic, including the technology utilized, warrants a risk analysis.

- FMEA is a risk analysis method endorsed by AAPM TG-100 for guiding the quality improvement and quality assurance of a clinical process.

- The use of SGRT does not greatly increase the risk of the linac-based SRS process.

- SGRT helps reduce the risk of the SRS process in some cases.

- The SGRT-specific failure mode carrying the greatest risk is the failure of the SGRT system to detect patient movement. This can be mitigated by ensuring daily quality assurance of the positional accuracy of the SGRT camera, ensuring that the therapists are also monitoring patient position using the closed-circuit television, and ensuring that a physicist is present during treatment to monitor the surface imaging system.

REFERENCES

1. Huq MS, Fraass BA, Dunscombe PB, et al. The report of Task Group 100 of the AAPM: application of risk analysis methods to radiation therapy quality management. *Med Phys*. 2016;43(7):4209–4262. doi:10.1118/1.4947547.
2. Manger RP, Paxton AB, Pawlicki T, Kim G-Y. Failure mode and effects analysis and fault tree analysis of surface image guided cranial radiosurgery. *Med Phys*. 2015;42(5):2449–2461. doi:10.1118/1.4918319.
3. Tague NL. *The Quality Toolbox*. 2nd ed. Milwaukee, WI: ASQ Quality Press; 2005.

CHAPTER 15

SBRT and Respiratory Motion Management Strategies with Surface Guidance

Michael J. Tallhamer

CONTENTS

15.1 INTRODUCTION

In the past decade technological developments in image guidance, motion management, target delineation, and highly conformal delivery techniques have accelerated the use of extracranial stereotactic treatments, commonly referred to stereotactic body radiation therapy (SBRT), to become part of many clinical practices. SBRT techniques have now been successfully applied to a wide variety of treatment sites, including lung,[1,2] liver,[3] pancreas,[4] and other sites.[5–9]

Because of the high doses delivered and the short fractionations associated with SBRT treatments, it is extremely important that the CT simulation geometry used during planning be replicated during the treatment as close as possible. The use of immobilization devices[10,11] and the clinical implementation of motion management techniques have been viewed as crucial elements in moving toward this goal. Additionally, the continued advancement of on-board image-guided radiation therapy (IGRT) technologies like orthogonal kV, MV, and cone-beam CT (CBCT) has greatly reduced the required complexity of immobilization devices used for SBRT while increasing the overall accuracy of the patient setup.[12]

Respiratory management strategies for SBRT treatments are put in place due to several complicating factors not commonly encountered in the intracranial applications of stereotactic radiosurgery/stereotactic radiotherapy (SRS/SRT). Similarly, respiratory management for standard fractionation treatments are not typically used due to the random nature of patient setup over a protracted treatment course and the tendency to average out over the course of treatment. Some of the complicating factors associated with SBRT treatments are:

1. The motion of internal targets

2. The motion of nearby organs-at-risk (OAR) and other normal tissues

3. The difficulty in determining a rigid stereotactic frame of reference like that of the stereotactic frame used in the intracranial SRS techniques.

The first and second challenges to SBRT plan development listed above relate to the motion of internal targets, organs, and other normal tissues and can be due to several factors including but not limited to: standard cardiac activity, organ filling/emptying, gastrointestinal activity, and respiratory motion.[13] Of these sources of motion, the most consequential to SBRT planning and delivery is most often that of respiration because of its tendency to produce large positional displacements for whole or partial organs systems, other normal tissues and potential targets throughout the thorax and abdomen. SGRT systems can be used in various ways to address these motion management concerns during both CT simulation and treatment delivery reducing the impact they can have on the quality and accuracy of the overall treatment.

During CT simulation, the role of SGRT systems in addressing these challenges depends on the desired mode of control for the respiratory motion. If the clinician desires to limit the respiratory motion through passive control of the patient's respiration, the SGRT system can be used in a training or coaching capacity to directly monitor the patient's respiratory cycle and provide visual feedback to the radiation oncology team. This allows clinicians to both direct and provide feedback on the patient's adherence to an institution's specific breath control protocols, like deep inspiration breath hold (DIBH), prior to initiating a CT simulation scan. Likewise, if the clinician desires more active or direct control of the patient's respiratory motion, while also gaining an immobilization advantage, they may choose to use a uniform abdominal compression technique with a thermoplastic body mask or compression belt under expiration breath hold (EBH). Again, this would place the SGRT system in a training or coaching capacity where a clinician can evaluate the compliance of the patient to a selected breath control protocol prior to initiating the CT simulation scan. Finally, clinicians may opt to use the SGRT system as a means of monitoring and recording the respiratory phase information during a four-dimensional (4D)-CT scan. The recorded phase information can then be exported in an open format to the CT vendor's hardware or similar third-party solution. The recorded phase-based information can then be used to bin the 4D data into a series of phase delineated three-dimensional (3D) datasets for planning.

SGRT systems uniquely address the third challenge to SBRT treatment delivery, namely, the difficulty in determining a rigid stereotactic frame of reference. This is fundamentally different than immobilization-centric approaches. SGRT uses the external surface of the patient obtained directly from the digital imaging and communications in medicine (DICOM) planning dataset as a primary coordinate reference frame, eliminating the need for a secondary frame of reference and rigid fixation between it and the patient. As will be discussed in later sections, the SGRT setup uses the real-time 6 degrees-of-freedom (6DOF) information provided by the system to directly localize the patient to isocenter using the DICOM surface information from the planning system. This not only provides a direct reference between the patient surface and the isocenter coordinate but also allows the radiation therapy team to properly correct the patient's posture prior to the IGRT localization. Small adjustments from initial SGRT setup can then be made using an appropriate IGRT localization technique in order to localize to the internal target position. After standard IGRT localization, the SGRT system can seamlessly switch roles to become the primary or secondary motion management device referencing back to the primary coordinate frame of the patient rather than relying on the body frame to rigidly immobilize the patient throughout treatment. This makes SGRT systems for SBRT typically quicker and more comfortable for the patient, having eliminated the need for the body frame, making intra-fraction motion less probable without the need to physically limit the motion through rigid patient immobilization.

SGRT technologies can be implemented in all phases of SBRT treatment development including CT simulation, initial treatment setup, IGRT, and treatment as real-time motion management. In the following sections, the various roles of SGRT systems in addressing the challenges associated with SBRT treatments will be discussed.

15.2 CT SIMULATION AND TARGET DELINEATION

The consistency and reproducibility of the setup created during the CT simulation is crucial to the success of any treatment technique. SBRT treatments start with a well-defined set of parameters or instructions for CT simulation. The quality of data obtained during the CT simulation process will set the groundwork for the successful development of any SBRT treatment. Because of the ablative nature of SBRT treatments, well-defined targets are considered ideal. Unlike traditional radiation therapy techniques, where a uniform dose is prescribed to the planning target volume (PTV), the dose distribution for SBRT PTVs can contain hot spots in excess of 125%–135% of the prescribed dose. Therefore, the size of the PTV margin should be determined carefully as that margin may contain sensitive normal tissues. If this is the case, those normal tissue volumes are susceptible to being exposed to these higher than normal hot spots. It should also be noted that organs and other normal tissues may move in and out of the delineated PTV with respiration or other physiological motion. It is therefore recommended that normal tissues as well as target volumes be evaluated for motion-induced exposure to the high-dose region. Likewise, the volume of normal tissue outside of the PTV receiving high dose should be minimized in order to lower the risk of toxicity related to these types of treatment.

The need to keep the amount of normal tissue inside of the PTV to a minimum requires precise delineation of not only treatment targets but also the surrounding normal tissues. One significant issue to be addressed during the CT simulation process is the introduction of undesirable artifacts into the resulting planning images due to motion originating from several physiological processes. These motion artifacts can make visualizing the potential targets, normal tissues, and other surrounding organs difficult and therefore hard to precisely delineate during the contouring process. The need to also characterize and understand the time-dependent positions of these tissues is also a confounding variable when it comes to properly delineating and constructing the structures for treatment. Therefore, there have been a number of techniques and technologies developed which allow physicists to characterize, reduce, or in some cases completely remove these motion artifacts in the resulting planning CT scans.

While respiration is often a primary and significant source of motion in SBRT treatments, its periodic nature makes it reasonably predictable. This periodic nature has resulted in many efforts to exploit the physiology of respiration in order to eliminate or reduce its impact on the quality of the CT images. This is either done through direct measurement of the respiratory cycle or through cruder attempts to physically limit respiratory induced motion. Motion management techniques, including abdominal compression, breath-hold, and respiratory gating, can be essential in the simulation process for these highly conformal and hypofractionated treatments. The American Association of Physicists in Medicine (AAPM) has published Task Group (TG) 76 which describes several detailed strategies for imaging moving targets including breath-hold techniques, slow CT, 4D-CT acquisition, and others.[13] There now exist a number of commercially available technologies for use during the CT simulation process which either track tumor movement directly or indirectly through a surrogate of the respiratory cycle. Tracking with these technologies is typically achieved through use of some apparatus placed on or in the patient during the CT simulation. These apparatuses can take many forms based on the technological approach used for monitoring the respiratory cycle such as a spirometry, external reflective infrared marker blocks, belt-type strain gauge devices, and many others.

Optical imaging systems are foundational in SGRT and are unique in both application and implementation when compared to the other aforementioned motion management technologies. These optical systems can be used in a variety of ways during the CT simulation process to facilitate many of the techniques discussed in the AAPM TG-76 report. While other systems exist for surface-guided 4D-CT acquisition such as the Sentinel system (C-RAD AB, Uppsala, Sweden), this chapter focuses on the GateCT system (Vision RT, Ltd., London, UK). A brief overview of the GateCT system, which will be referred to throughout the chapter, will serve to illustrate the general approach surface guidance systems take when dealing with respiratory motion tracking. The GateCT system is a stand-alone optical surface tracking system that facilitates the real-time tracking of points and/or regions on the 3D-reconstructed patient surface during CT simulation. The consequences of the system design, implementation, and assumptions made during the calibration process will be discussed along with quality assurance (QA) considerations for determining proper clinical operation.

15.3 SYSTEM OVERVIEW

15.3.1 GateCT® Basic Features

- The system uses a single centrally located stereoscopic camera system, typically calibrated to the center of the CT scanner's central imaging plane.

- The system uses an integrated gating controller (IGC) or external X-ray detector to interface with most vendor's CT scanner in order to synchronize the data to the completed 4D-CT scan.

- The system is used to characterize and record a patient's respiratory motion by monitoring a point and/or region, sometime referred to as a patch, on the 3D reconstructed surface of the patient as they move through the CT scanner. The respiratory-synchronized information obtained by tracking the patient's surface during a 4D-CT scan is then exported to the CT vendor's proprietary platform in the common and open *.vxp format and used in the 4D-CT reconstruction process.

- The GateCT system has a "contactless" or observer-based approach to the common 4D-CT tracking and acquisition problem. This means that there is no apparatus or surrogate that "rides" the couch and/or patient through the CT during CT data acquisition.

Different SGRT vendors' CT-based products may or may not support some of these features but the general strategies of using SGRT for respiratory motion management should apply to all systems.

A contactless optical tracking approach has some advantages over the traditional approaches to acquiring the correlated respiratory information needed during 4D-CT simulation. Namely:

- The respiratory surrogate is the actual 3D-reconstructed surface of the patient.

- There is no need to modify an apparatus or the patient setup to accommodate data acquisition requirements at the time of CT simulation only to have to then address these compromises at the time of planning and/or treatment delivery.

- The tracking point or region placement is completed digitally from the console and is not limited by patient body morphology or by other complicating factors related to a surrogate device's geometry in relation to the patient position on the CT couch. This eliminates the need to repeatedly enter the room to modify physical devices used in the tracking process.

While having its benefits, the observer-based approach also requires a clinician to think differently when calibrating and operating the system when compared to the other previously mentioned 4D-CT techniques. The SGRT-based approach relies on a disconnected geometry where there is no physical connection to the moving patient frame of reference. This disconnect between the static room-based coordinate system and the moving patient coordinate system can introduce new QA steps required to ensure that the camera geometry, as calibrated, and the reported speed of the moving frame of reference within the optical field of view are well understood.

In order to ensure system accuracy, the SGRT camera must be correctly calibrated to the CT scanner's geometry by obtaining a series of images of a calibration plate. By processing these images, the software is then able to construct the camera's internal coordinate system. As a safety measure, if the calibration routine is not carried out successfully, it should not allow the user to retrieve patient data during a clinical session. These SGRT-specific QA steps should be incorporated into the institution's routine QA processes for CT simulation. Recommendations for CT simulation QA can be found in the report of AAPM Task Group 66.[14]

15.4 4D-CT ACQUISITION

Retrospectively correlated CT scans are considered to be the closest thing to a "true" 4D-CT scan as the entire respiratory cycle is typically captured during the CT simulation. For that reason, these scans have become increasingly popular for SBRT where respiratory motion is of primary concern. In these scans, a CT scan volume representing the anatomy over the entire free-breathing respiratory cycle is acquired so that internal motion can be assessed and accounted for during treatment planning. During a retrospectively correlated CT scan, a patient is scanned while under their natural breathing pattern without any imposed breathing requirements. For lung SBRT patients, this may be preferable as many of these patients may already have compromised pulmonary function as a result of their disease, making many common breath control protocols difficult. Retrospectively correlated CT scans are completed in a helical fashion with a very low pitch or an axial fashion with a cine duration long enough to capture at least one breathing period. This allows for multiple images to be acquired through the patient volume with the intent of capturing all phases of respiration. The imaging data is collected concurrently with an external respiratory signal in order to label and bin the data after the acquisition is complete. When working with an SGRT system, the respiratory signal is collected by directly tracking a location on the 3D reconstructed surface of the patient rather than relying on surrogate markers or mechanical sensors that must be placed on the patient. After acquisition, the respiratory trace information is exported to the CT vendor's platform and used to label the reconstructed images with the respiratory phase in which they were acquired. In some cases, the raw CT dataset and the respiratory trace information may be exported to a third-party application where this process is completed. The reconstructed images are not necessarily captured at the same phase of the respiratory cycle. Therefore, the phase tag used to label the image is representative of the phase window, say the 0% to 10% phase or the 50% to 60% phase of a 10 phase image set, in which it was acquired rather than the exact phase value. This label is used in the binning process where images of the same phase bin label are collected into a series of discrete 3D datasets.

Optically based systems are often favored when implementing a 4D-CT program primarily due to their noninvasive approach. The promise of quick setups and relatively simple implementation has made them very common in the acquisition of 4D-CT data for SBRT. SGRT systems like

GateCT facilitate the acquisition of the respiratory trace information during the 4D-CT simulation process through monitoring of a 3D reconstruction of the patient surface using a projected pseudorandom speckle pattern of visible light. This pattern is projected onto the patient surface where a stereoscopic camera pod uses the pattern to both reconstruct the patient 3D geometry and to track oscillating changes in amplitude at various locations on the surface as identified by the operator. This allows the operator to select any point on the visible surface of the patient where they feel the respiratory trace information is most stable while also selecting a monitoring point, which is used to monitor for gross patient movement during the 4D-CT scan. The SGRT approach is not limited to areas where a device or apparatus may be stably affixed to the patient's anatomy for tracking as is the case with other optical systems. If the surface is visible throughout the CT scan range, the system should successfully track the respiratory information using the tracking point selected. However, if the surface is obstructed at any point during the scan by sheets, clothing, or immobilization devices, the obstructed area will not be available for tracking purposes. This system has the advantage of improving patient setup by avoiding the common need to compromise the physician-ordered immobilization devices in order to accommodate yet another vendor apparatus required to acquire high-quality respiratory information only to have to address that compromise in immobilization during the treatment delivery phase. The device-less approach adopted by the SGRT vendors often allows for higher quality respiratory trace information to be acquired throughout the 4D-CT scan process. These systems allow tracking of locations along the visible surface of the patient that, while stable, are often prohibitive to other technologies due to the immobilization requirements of the treatment, the patient's body morphology, and/or the tracking device geometry.

A procedure for collecting the respiratory data during a retrospectively correlated 4D-CT scan is presented below. The reader can use this generic procedure together with their SGRT vendor's user manual to develop an appropriate 4D-CT protocol for use with their particular equipment and other department resources.

1. At the time of simulation, the patient should be brought in and the 4D-CT simulation process is articulated so that they understand the process and are prepared to respond to any instructions during the scan. Since every clinic has variations on the 4D-CT process, it

is important to articulate expectations to the patient so as to avoid confusion, such as if and when breath control instructions will be given or if the patient should respond to certain commands during simulation or simply perform them without acknowledgment.

2. The patient should be immobilized in a way that ensures that a sufficient area of the patient surface will be exposed for surface camera visualization.

   a. SBRT immobilization can include breath limiting devices such as a compression belt, paddle, or body thermoplastic mask. These may obstruct portions of the patient surface so attention should be paid to the amount of surface visible to the camera. Again, a test surface can be acquired at any time to validate the amount of visible surface prior to the actual 4D-CT scan.

   b. Once immobilized, but prior to starting the simulation process, a surface can be acquired to verify proper visibility of the patient as positioned in the CT scan geometry.

      i. A preliminary tracking surface can also be acquired to use as a coaching surface to gauge patient adherence to instructions prior to initiating the actual 4D-CT scan.

      ii. This step can become important if there are concerns about the patient setup or patient compliance and can help the operator avoid potential loss of data during the 4D-CT scan, or unnecessary irradiation from an incomplete or repeat scan.

3. The operator should select a proper retrospectively correlated 4D-CT scan protocol for the patient being simulated.

4. The parameters of the scan should be verified as appropriate for the patient and the final CT couch speed.

5. Prior to initiating the 4D-CT scan, a reference surface is captured at the starting position for the 4D-CT scan.

   a. It is important that a user understand the SGRT approach as described in the system's manual. The reference surface must be captured at the starting point for the 4D-CT scan. This is the reference location from which the system will track the patient during the scan itself. If a reference is captured at one location and then the patient is moved to another location prior to starting the 4D-CT, the tracking points will not remain in the proper locations.

6. The operator can select a respiratory tracking location anywhere on the visible surface of the patient using the mouse and keyboard at the console. A secondary monitoring point can be selected on a stable portion of the patient's anatomy to watch for any gross positional changes in the patient's position during the 4D-CT.

7. SGRT tracking is started and the system is put in a waiting state.

8. The 4D-CT scan is initiated and images are acquired. The X-ray on signal is transmitted from the CT scanner via a junction box to the SGRT 4D-CT acquisition module for retrospective binning analysis.

9. After the scan is completed, the respiratory trace information should be reviewed and edited prior to export to the CT vendor's hardware or third-party application for analysis.

The outlined simulation process above can be completed in approximately 20 minutes. This does not account for 4D-CT binning and reconstruction time as it varies widely across CT vendor platforms. The reader is encouraged to map out the 4D-CT process at their respective institutions or review existing 4D-CT procedures within their various departments prior to implementation of an SGRT-based 4D-CT program. Quality assurance concerns like those listed in this work should also be addressed prior to clinical implementation to make sure the technology is properly understood within the framework of a specific process. That being said, SGRT technologies can integrate extremely well with the most common 4D-CT protocols.

15.5 CT SIMULATION WITH BREATH-HOLD TECHNIQUES

Even with the advancement of IGRT technologies like CBCT and kV on-board imaging, accurate target localization and reliable prediction of respiratory motion remains a challenge when treating tumors that move with respiration. Voluntary breath-hold techniques are another method for controlling the respiratory motion seen during SBRT. Breath-hold techniques include DIBH which was discussed in Chapter 10 and exhalation breath hold (EBH). Whereas in breast applications, the goal of DIBH is to move the chest wall and breast away from the heart, in SBRT applications, the goal of DIBH and EBH is to minimize respiratory motion of the treatment target. DIBH tends to be more easily tolerated by patients and can be held for longer than EBH. However, the inhalation level of a DIBH and subsequent position of the diaphragm can be more variable. For targets close to the diaphragm (such as lower lung) or within the liver, EBH may be a more effective technique for minimizing target motion. To utilize breath-holds during radiation delivery, a patient must have relatively good respiratory function and the ability to comply with any coaching instructions that may be part of an institution's breath-hold protocols. A breath-hold technique can provide additional reductions in the amount of normal tissue contained within the PTV for SBRT treatments. By restricting radiation delivery to a fixed position within the inspiration phase of respiratory cycle, breath-holds can provide some advantages over other 4D approaches to the management of motion-related concerns for SBRT patients. The broader use of breath-hold techniques for SBRT has been further facilitated with the recent arrival of high dose-rate capabilities through technologies such as the flattening filter free (FFF) modes on modern linacs.

During CT simulation for SBRT treatments where the breath-hold is intended as the primary form of respiratory motion management, the reproducibility and sustainability of the breath-hold position should be determined prior to acquiring the treatment planning CT scan. Methods for achieving the proper breath-hold position depend on variety of site-specific factors such as the specific respiratory monitoring system being used, the immobilization system requested, treatment site, and the intended treatment delivery technique (e.g., VMAT versus static field delivery).

A patient should be consulted prior to the CT simulation for SBRT with a breath-hold. During that consultation, the breath-hold technique along with its motion management goals and benefits should be explained and discussed with the patient. If the patient appears to understand the breath-hold technique and is a good candidate, they can be given a series of breathing exercises to help them prepare for the CT simulation. Following initial consultation, when the patient returns for CT simulation, a verbal walk-through of the breath-hold simulation process should be discussed to gauge the patient's comfort level and the general plausibility of their successful participation in the process. Contraindications such as sickness, nasal congestion, and seasonal allergies on the day of the simulation should be brought to the physician's attention prior to initiating the CT simulation. It should be noted with the patient that the breath-hold technique is one of many respiratory motion management techniques used in radiation oncology. The goal is to reduce any anxiety the patient may experience if they are unable to achieve a reproducible and sustainable breath-hold position during coaching and the patient should go into the simulation process as relaxed as possible. Avoiding the impression of a pass/fail environment is always desirable. If the patient feels like they are failing due to an inability to hold their breath for the desired amount of time, they may become frustrated or upset and any further efforts to work around the issue tend to cause more stress. Avoiding this scenario is key to a good setup and successful breath-hold session for SBRT simulation and later treatment.

Following the initial instructions and basic walk through of the breath-hold simulation process with the patient, the radiation therapy team should then set up the patient on the CT couch within the intended immobilization system. A sufficient area of the patient's surface, in or around the treatment site, should be exposed so that the SGRT cameras are able to perform surface tracking and registration. This will allow clinicians to monitor breathing the scan as well as during a pre-simulation coaching session where the reproducibility and sustainability of the breath-hold position will be determined. If the immobilization system ordered by the physician interferes with the desired breath-hold monitoring setup, the physicist and/or the physician should be consulted to rectify the situation or determine a proper compromise while maintaining the intended goals of the immobilization requested.

After being properly immobilized and positioned on the CT simulator, the patient should be brought to a monitoring position that is within the field of view of the SGRT camera(s). This position should be determined and documented during the system's commissioning and used as a standard position for all coaching sessions to avoid delays in the CT simulation process. Once any coaching is finished and a dry run is completed, the SGRT system can be used to monitor the patient by capturing a reference surface of the patient at the coaching position. After the reference surface is acquired, the tracking point can be digitally placed on the reference surface and the tracking enabled. This will provide visual feedback on the SGRT system so that the clinician can determine if the breath-hold position is both reproducible and maintained for the required amount of time. The required duration of the breath-hold is dependent on the institution's protocol for breath-hold and should be determined between the physician and the physics team with an understanding of how various breath-hold durations will impact the delivery time and the imaging used for localization. The duration of the breath-hold should be on the order of 20–30 seconds if a typical chest CT is desired for SBRT planning. While this may be long enough to acquire a planning CT, this may not be long enough to make the treatment delivery as efficient as the clinical team requires. Institutional protocols that include specific selection criteria should be developed by the clinical team during breath-hold program development for SBRT so that clear and consistent criteria are used in selecting SBRT candidates for DIBH or EBH.

When a satisfactory coaching session has been completed, the patient can be scanned for treatment planning purposes in both the free-breathing and the breath-hold positions. The SGRT system can be used to ensure the patient reproduces and maintains the proper position for the breath-hold CT scan. To accomplish this, the operator would reacquire a reference surface at the starting position for the breath-hold scan. After acquiring the new SGRT reference surface the operator can once again coach the patient into the breath-hold position while actively monitoring the patient with the SGRT system. When the patient reaches the desired breath-hold position, the CT scan is initiated and completed. The operator can confirm the proper breath-hold level was maintained throughout the CT scanning process. If the two CT scans are completed in the same session, the datasets will share a common DICOM coordinate system. This is good practice when completing the breath-hold simulation process for SBRT treatments because it eliminates the potential for small misalignments between the two scans due to improper secondary registration within the treatment planning system during the planning phase. After the two CT scans are completed, the remainder of the institution's CT simulation procedures should be followed in order to complete the simulation process.

Both the free-breathing and breath-hold CT scans should be sent to the treatment planning system for SBRT. The free-breathing CT scan should only be used to generate a corresponding free-breathing body structure from within the treatment planning system while the breath-hold scan will be used for the SBRT treatment planning. The free-breathing body structure can be easily transferred to the breath-hold planning scan as a secondary structure as a result of the DICOM coordinate registration resulting from the simulation procedure outlined above. This allows the clinician to transfer both the free-breathing body structure used for initial setup and the breath-hold body structure used for intra-fractional monitoring to the SGRT system in a single plan. However, this process may vary based on the treatment planning system vendor used within the reader's clinic.

15.6 TREATMENT PLANNING CONSIDERATIONS

While there are many uses for SGRT technologies in both the CT simulation and treatment delivery phases of SBRT, there are very few SGRT-specific concerns when it comes to the treatment planning phase. The following brief discussion will highlight a few areas where the implementation of an SGRT program may impact the treatment planning portion of the reader's SBRT process. It is often dependent on an institution's process as to what impact the implementation of SGRT technology will have on the dosimetrists and the treatment planning portion of an existing SBRT program.

One of the goals of using the SGRT motion management techniques during CT simulation is to reduce motion artifacts within the planning images. This allows for accurate delineation of the targets, organs, and other normal tissues during treatment planning. The 4D-CT data facilitated by the SGRT system allows a clinician to contour and review their structures in a 4D manner by either reviewing a series of 3D datasets representing various phases of the respiratory cycle or by using those datasets to then generate synthetic datasets like the maximum intensity projection (MIP), minimum intensity projection (MinIP), and the average intensity projection (AIP). The characteristic of these datasets, when it is appropriate to use them, and for what tumor sites is beyond the scope of this chapter; however, they are mentioned here because they are made possible as a result of the data provided by the SGRT system during CT simulation. These datasets can be used to construct an internal target volume (ITV), which incorporates the 3D position of the target across all phases of respiration. The ITV can then be projected onto all phases of the 4D-CT dataset to make sure it encompasses all tumor motion prior to adding the PTV expansion. This patient-specific margin expansion is a departure from earlier SBRT planning techniques which included arbitrarily large margins (e.g., 1–2 cm margins in all directions for all cases) or margins based on population averages. In most cases, these methods produce margins that are larger than clinically required and result in the treatment of a larger volume of normal tissue.

SGRT systems can also be instrumental during SBRT delivery. SGRT systems are used as both an initial setup tool and a real-time motion management device during the delivery of the treatment. The quality of the initial setup when using SGRT technologies can be impacted by steps taken during the treatment planning phase if the external body contour is not consistently defined. Some treatment planning system vendors do not require that an external body structure be defined for dose calculation and as a result some institutions may not be accustomed to adding this structure to their plans. However, the DICOM body structure is routinely used by the SGRT system as the reference structure when setting up the patient for treatment. It is for this reason that an accurate external body contour must be generated during planning. It is also recommended to avoid manual contouring of the external body structure and instead use a consistent segmentation process such as threshold detection or some other automated method for delineating this structure.

Another area where this becomes important is when breath-hold is being used as a method of motion management for SBRT. For SBRT using a breath-hold, two scans are acquired when SGRT is to be used for setup and delivery. After a breath-hold CT simulation, both the free-breathing and breath-hold CT scans are sent to the treatment planning system. These datasets should share DICOM coordinates and be automatically registered. This allows the treatment planner to generate both a free-breathing body contour and a breath-hold body contour. The free-breathing body contour is generated from the free-breathing CT scan and can then be transferred to the breath-hold scan on which the actual SBRT treatment plan is developed. The optional process of transferring the free-breathing body structure to the breath-hold scan is made possible through the registration of the two CT scans via their shared DICOM coordinates. The specifics of the actual transfer procedure will vary based on treatment planning system vendor. This allows the treatment planner to export both body structures in the same treatment plan avoiding the potential for errors associated with having multiple plans and structure sets for the same patient in the SGRT software. To avoid the potential for initial setup errors while using breath-hold and SGRT, it is recommended that the institution adopt naming conventions for breath-hold body structures during planning. The naming convention used to distinguish the free-breathing body structure from the breath-hold body structure is left up to the reader's institutional standards.

15.7 SBRT TREATMENT DELIVERY

The use of SBRT techniques to treat targets within the lung, liver, pancreas, bone, head and neck, and others is growing in popularity with new studies suggesting an expanding role for these treatments in the future. These techniques require an accurate delivery of radiation to the target volumes to limit normal tissue toxicity. Intracranial targets and targets within the spinal column are particularly well-suited to SBRT approaches due to boney anatomy either being a reliable surrogate for the target position or, as in the latter case, the target location being within the boney tissues themselves. Reliable immobilization can help eliminate intra-fraction target motion in this group of targets; however, treatments are often complicated by sensitive OARs like the spinal cord, brain stem optic chiasm, and others being in very close proximity to the target. In contrast, there exists a second group of targets throughout the thorax and abdomen that are neither well-localized by bony anatomy nor free from the confounding effects of internal motion produced by various physiological processes like respiration. Localization and management of intra-fraction motion presents a unique challenge for this second group of potential SBRT targets. Technical advancements that help overcome this obstacle can be divided into two groups: (1) target localization and (2) tracking and motion management

In the last two decades, there has been the emergence of IGRT, which has demonstrated a high degree of accuracy in localizing targets relative to the isocenter of the treatment machines using a variety of imaging techniques. While IGRT has been shown to be successful, it does have some limitations when it comes to SBRT delivery such as:

1. the static nature of IGRT techniques,

2. the inability for IGRT to account for postural changes of the patient with its rigid linear corrections (4D or 6DOF corrections on modern linacs), and

3. IGRT's reliance on ionizing radiation makes real-time tracking and motion management dosimetrically concerning.

The dynamic real-time nature of SGRT allows for several enhancements to the standard IGRT process for SBRT. SGRT systems can complement and/ or augment an existing IGRT workflow to address the first two limitations discussed above. The real-time monitoring and 6DOF positional information provided by the SGRT system can aid in both initial setup and motion management of SBRT patients in a way that compliments an existing IGRT workflow. In a standard IGRT workflow for SBRT, images are acquired after initial positioning of the patient (either through use of SGRT-assisted setup or by utilizing surface marks on the patient). The patient is then assumed to be static until the image registration is complete and subsequent shifts are sent to the treatment machine and applied. Monitoring of SBRT patients using SGRT during the IGRT localization has shown that patients can move during the time it takes a clinician to register and apply IGRT shifts. Studies have also shown that the residual error in a patient's setup tends to increase as the treatment time increases and depends, at least in part, on the patient's performance status.[15] IGRT's dependence on ionizing radiation makes it not ideally suited to address these issues. Current rigid registration techniques used by modern IGRT systems also prevent the IGRT systems from correcting inter-fractional postural differences in the patient. When a patient is rigidly registered to the planning CT and postural changes are detected, the registration is limited to the region surrounding the target using a volume of interest. If the postural changes are significant in and around the target region, it is often required to take the patient down and set them back up again to avoid small localization errors associated with "splitting the difference" in the IGRT shifts. SGRT systems can be used to detect and correct postural changes while therapy staff are in the treatment room, prior to IGRT localization. This limits the amount of time required to register the images and potentially reduces the overall time the patient must be on the treatment machine.

With the success of IGRT for localization of SBRT targets, attention has been turned to strategies aimed at maintaining the target position throughout the remainder of the treatment. Within this group of strategies are techniques aimed at reducing target motion. Some of these include abdominal compression and breath-holds. Also, within this group are technologies that allow for well-defined target motion during delivery, having previously accounted for the motion via another technique like 4D-CT for planning. These technologies can often intervene in the delivery through vendor-provided interfaces when motion deviates from a predefined set of parameters. Technologies that provide functionality such as amplitude- and phase-based gating fit into this group. The discussion below will focus on breath-hold and respiratory gating as the two primary SGRT techniques for motion management during SBRT delivery.

From its beginnings in traditional fractionated treatments, the breath-hold method has also shown great potential for use in SBRT of the lung due to its ability to increase total lung volume, decrease the amount of lung tissue within the ITV/PTV expansions, and provide relatively reproducible target displacements. Traditionally the breath-hold position has been optically monitored using a surrogate device place on the patient. The surrogate device often must be placed in a location that exhibits a detectable level of change in the vertical position during breathing so that the tracking system is able to determine when the patient has reached maximum inspiration. Unfortunately, the location most suitable for the surrogate device is often far from the treatment area and prone to influence from other body mechanics such as arching of the back or artificially extending the abdominal wall. This can lead to poor correlation between the surrogate and the actual level of inspiration of the patient. As in all other cases, SGRT uses the reconstructed surface of the patient as the surrogate for breath-hold tracking. The surface used during treatment is pulled directly from the DICOM dataset obtained during the breath-hold CT simulation and is evaluated in all 6DOF during delivery. The first advantage of this approach is the ability to digitally place the surrogate ROI directly over the treatment site. This allows for a better correlation between the target location and the surface being monitored (in most but not all cases). The second advantage of using SGRT for breath-hold is that the selected ROI is tracked in all 6DOF during delivery. This allows the operator to detect what are referred to as "false breaths" or artificial surface excursions that are not a result of a true full inspiration breath-hold but some other combination of mechanical motions like arching of the back or shrugging of the shoulders during the breath resulting in a false vertical offset. Because all 6DOF are being monitored concurrently, it is very difficult for the patient to match the reference surface without properly breathing when proper tolerances are used.

While the general breath-hold process for SBRT is the same as in the case of a traditional breast DIBH treatment, the use of breath-hold during SBRT presents some additional challenges. The reason for this is due to the nature of SBRT targets, their varied locations throughout the body, and the imaging modalities used to localize them. As previously discussed, CBCT imaging is the most common form of imaging for SBRT localization. This differs from the planar kV or MV imaging typically used for breast setup verification. On most linacs with on-board imaging capability, a typical CBCT acquisition requires approximately 1 minute to acquire the image projections for reconstruction of the 3D dataset. Very few SBRT patients will be able to maintain a breath-hold for this amount of time. If the linac vendor allows the operator to pause image acquisition, the operator can acquire CBCT projections in discrete sectors allowing the patient to breath when necessary, while the imaging beam is held, and resuming acquisition after the patient again reaches the breath-hold position as indicated by the SGRT system. For example, a single CBCT could be acquired in three 20-second breath-holds, depending on the patient's ability. It is important to ensure that the patient returns to a consistent breath-hold level as determined by the SGRT system, otherwise artifacts may be introduced into the CBCT images.

Another challenge related to SBRT target localization using breath-hold and CBCT is the fact that many SBRT targets are found off axis within the body. Some linac vendors require a lateral couch centering to avoid collision with the X-ray source and the detector panel during CBCT. The couch centering introduces an unknown daily offset from isocenter prior to imaging that cannot be accounted for prior to the daily setup of the patient. How the institution chooses to handle this offset is important as the DICOM surface used by the SGRT system is referenced to the isocenter coordinate for treatment and is not available at the couch center position. The functional characteristics and supported workflows within the SGRT system are also critical to solving this IGRT-related problem. A potential solution to this issue is to set up the patient using the free-breathing DICOM surface, followed by verification against the DICOM surface for breath-hold at isocenter position in the same way that one would perform setup for a traditional breast patient. When the operator is ready to acquire the CBCT for localization, the patient is instructed to inhale or exhale into the desired breath-hold position and verify the match against the breath-hold DICOM reference surface displayed on the SGRT system at the console. While the patient is still holding their breath, the operator would then center the couch and capture a new reference surface at the couch-centered position and allow the patient to breath normally. This new reference surface can now be used during the CBCT acquisition following the institution's breath-hold CBCT protocol. When applying the shifts from the CBCT localization, the operator would first need to coach the patient into the breath-hold position using the new breath-hold reference image. When the patient reaches the breath-hold position, the operator would restore the couch and apply the CBCT shifts. Before allowing the patient to breathe, the operator would capture a final reference surface, which would now incorporate both the breath-hold position along with the CBCT shift. The operator can now coach the patient to the breath-hold position for treatment using the final reference surface and use kV, MV, or fluoroscopic imaging to verify final internal location prior to beam delivery. During commissioning and clinical implementation, the user should consult their vendor-specific manuals to ensure that the SGRT system supports the intended SBRT imaging protocols. Each institution should run through an end-to-end SBRT process using the SGRT system. The end-to-end testing should utilize an anthropomorphic phantom or standard phantom with a variable off-axis isocenter to simulate commonly occurring tumor sites and geometries. This process should be designed to test the workflows and identify points in the process where the various participating systems may experience difficulties in provide the required information to treat the patient. Simulating the breath-hold technique step by step will allow the physics team to map the process and identify challenges associated with the various steps in the SBRT imaging and localization workflow prior to use on a patient.

If the SGRT system is connected to the treatment machine and selected as an active device for respiratory gating, the treatment beam may be selectively turned on and off during delivery based on the output of the SGRT system. In an active gating setup, the latency of the SGRT gating signal should be evaluated during commissioning. If the SGRT system uses variable surface resolution settings (i.e., a variable number of vertices or points used to represent the surface within the software) and/or an ROI-based approach to tracking, the impact of the surface resolution and the ROI size can significantly increase the processing time and by extension the system latency. Likewise, these affects, and the combinations thereof, should be characterized under clinically relevant conditions during the commissioning of the system.

It should also be noted that the overall treatment time depends on the gating duty cycle. The wider the delivery window gets, the higher the duty cycle value will become. This results in shorter delivery times but may come at the expense of allowing for more residual motion. If the reader's SGRT system uses all 6DOF, rather than just the single vertical dimension, for the gating values the tolerances on the other dimensions will have the same impact. A compromise between treatment time and residual motion range needs to be made when implementing active gating for SBRT.

15.8 SUMMARY

The extreme demands imposed by the ablative nature of SBRT treatments amplify toxicity concerns as they relate to the volume of tissue exposed to high doses during treatment delivery. The high degree of accuracy required by these conformal and hypofractionated treatments requires the use of sophisticated motion management tools throughout the treatment development process at a level above what is routinely considered necessary for standard fractionation radiation therapy. A comprehensive approach to localization and motion management must be in place to safely and effectively treat patients using this mode of radiation therapy. SGRT systems use a 3D-reconstructed surface of the patient to facilitate 4D-CT data acquisition, 6DOF treatment positioning, respiratory phase tracking, and real-time motion management for patients undergoing radiation treatment.

KEY POINTS

- SGRT facilitates the 4D-CT data acquisition and the subsequent delineation of time-dependent target structures during treatment planning by providing high-quality respiratory trace information during the CT simulation process.

- SGRT provides a coaching and verification tool for use during CT simulation in support of breath-hold techniques such as DIBH and EBH.

- SGRT provides real-time 6DOF positional information for patient positioning prior to imaging.

- SGRT augments the IGRT localization process by providing postural information and real-time monitoring of the patient during image acquisition and registration.

- SGRT provides real-time monitoring of the patient at the approved treatment position after IGRT localization.

- SGRT provides respiratory gating functionality to the delivery platform as an active mode of respiratory motion management.

REFERENCES

1. McGarry RC, Papiez L, Williams M, Whitford T, Timmerman RD. Stereotactic body radiation therapy of early-stage non-small-cell lung carcinoma: Phase I study. *Int J Radiat Oncol Biol Phys.* 2005;63(4):1010–1015.
2. Videtic GM, Donington J, Giuliani M, et al., Stereotactic body radiation therapy for early-stage non-small cell lung cancer: Executive summary of an ASTRO evidence-based guideline. *Pract Radiat Oncol.* 2017;7:295–301.
3. Schefter TE, Kavanagh BD, Timmerman RD, et al. A phase I trial of stereotactic body radiation therapy (SBRT) for liver metastases. *Int J Radiat Oncol Biol Phys.* 2005;62:1371–1378.
4. Rwigema JC, Parikh SD, Heron DE, et al. Stereotactic body radiotherapy in the treatment of advanced adenocarcinoma of the pancreas. *Am J Clin Oncol.* 2011;34:63–69.
5. Siddiqui F, Patel M, Khan M, et al. Stereotactic body radiation therapy for primary, recurrent, and metastatic tumors in the head-and-neck region. *Int J Radiat Oncol Biol Phys.* 2009;74:1047–1053.
6. Guckenberger M, Bachmann J, Wulf J, et al. Stereotactic body radiotherapy for local boost irradiation in unfavorable locally recurrent gynecological cancer. *Radiother Oncol.* 2010;94:53–59.
7. Eldaya RW, Lo SS, Paulino AC, et al. Diagnosis and treatment options including stereotactic body radiation therapy (SBRT) for adrenal metastases. *J Radiat Oncol.* 2012;1:43–48.
8. King CR, Brooks JD, Gill H, et al. Stereotactic body radiotherapy for localized prostate cancer: Interim results of a prospective phase II clinical trial. *Int J Radiat Oncol Biol Phys.* 2009;73:1043–1048.
9. Chang EL, Shiu AS, Mendel E, et al. Phase I/II study of stereotactic body radiotherapy for spinal metastasis and its pattern of failure. *J Neurosurg Spine.* 2007;7:151–160.
10. Wulf J, Hädinger U, Oppitz U, Olshausen B, Flentje M. Stereotactic radiotherapy of extracranial targets: CT-simulation and accuracy of treatment in the stereotactic body frame. *Radiother Oncol.* 2000;57:225–236.
11. Foster R, Meyer J, Iyengar P, Pistenmaa D, Timmerman R, Choy H, Solberg T. Localization accuracy and immobilization effectiveness of a stereotactic body frame for a variety of treatment sites. *J Radiat Oncol.* 2013;87:911–916.
12. Chang J, Yenice KM, Narayana A, Gutin PH. Accuracy and feasibility of cone-beam computed tomography for stereotactic radiosurgery setup. *Med Phys.* 2007;34:2077–2084.
13. Keall PJ, Mageras GS, Balter JM, et al. The management of respiratory motion in radiation oncology report of AAPM Task Group 76. *Med Phys.* 2006;33(10):3874–3900.

14. Mutic S, Palta JR, Butker EK, Das IJ, Huq MS, Loo LN, Salter BJ, McCollough CH, Van Dyk J. Quality assurance for computed-tomography simulators and the computed-tomography-simulation process: Report of the AAPM Radiation Therapy Committee Task Group No. 66. *Med Phys.* 2003;30(10):2762–2792.

15. Li W, Purdie TG, Taremi M, et al. Effect of immobilization and performance status on intrafraction motion for stereotactic lung radiotherapy: Analysis of 133 patients. *Int J Radiat Oncol Biol Phys.* 2011;81:1568–1575.

CHAPTER 16

The Use of SGRT for SBRT without Respiratory Gating

Ryan Foster and John Heinzerling

CONTENTS

16.1 INTRODUCTION

Stereotactic body radiation therapy (SBRT), in which ablative doses are delivered to tumors while sparing surrounding normal tissues via a steep dose gradient, requires effective immobilization, motion management, and very accurate tumor localization. SBRT has been made possible largely by technological advances in in-room image guidance.[1] The primary concern is internal and external patient motion that could cause the tumor to move outside the planning target volume (PTV) and the high dose volume. SBRT requires small margins, has steep dose gradients, and is usually delivered in five fractions or less. Thus, a full or partial geometric miss of the tumor in just one fraction can have negative consequences on tumor control because unlike a more prolonged course of radiation, the effects of these misses are not likely to be averaged out.

The importance of monitoring the patient during SBRT treatments has been recognized by professional societies. The American Association of Physicists in Medicine (AAPM) Task Group 101[2] provides recommendations for performing SBRT and states "…some kind of monitoring is desirable to track patient breathing and monitor patient positioning during treatment," and "The patient position should be monitored during the entire treatment…" Likewise, the American Society for Radiation Oncology (ASTRO) white paper on quality and safety for radiosurgery and SBRT states "…the management of intra-fraction patient body movement as well as physiological motions such as breathing must be accounted for."[3]

Some strategies that have been employed to detect and correct for patient motion include intra-fraction cone-beam CT (CBCT),[1] ceiling- or floor-mounted stereoscopic kV imaging during or between beams to directly image the tumor,[4] or through electromagnetic transponders,[5] a respiratory-correlated surrogate such as fiducials on the patient's skin,[6] and more recently, the patient's surface using optical surface guidance.[7] Surface-guided radiation therapy (SGRT) has been used to set up and monitor deep inspiration breath hold (DIBH) for left breast patients[8] and intracranial stereotactic radiosurgery (SRS) patients,[9] but to date very little data exists describing the use of SGRT for setup, respiratory gating, and position monitoring of SBRT patients receiving treatment to sites in the thorax and abdomen.

16.2 BACKGROUND

The number of technologies developed and studied for inter-fraction motion management is a clear indication of the importance of understanding and eliminating or accounting for this source of localization uncertainty. For SBRT, solutions to improve immobilization and setup reproducibility include stereotactic frames,[10–13] alpha cradles,[14] and vacuum fixation.[15,16] Positioning the patient as close to the treatment position as possible before any image guidance is performed is still considered important, despite the in-room imaging that is available now. Many institutions and clinics have policies that require a repeat imaging procedure to be performed if the initial image-based shifts are greater than 1 cm. Thus, positioning the patient initially within this tolerance is highly desirable as it can reduce the need for repeat imaging and associated imaging dose.

Likewise, for intra-fraction motion management, several strategies have been employed to minimize, detect, and account for motion because intra-fraction motion drives margin size in the image-guided radiation therapy (IGRT) era and is more difficult to account for than inter-fraction setup uncertainty. Abdominal compression is frequently used to limit the magnitude of the respiration-induced tumor motion for lung and liver SBRT patients.[17,18] Respiratory gating and breath-hold are also options to manage respiration-induced tumor motion by limiting motion when the beam is on.

There are two primary methods for identifying and accounting for intra-fraction internal patient motion—respiratory gating and tracking. When respiratory gating is used, the treatment is delivered only when the patient is within a certain part of the respiratory cycle. The disadvantages of this approach is that it decreases the treatment duty cycle and increases treatment times. However, it should be pointed out that gating is very widely used and is typically performed based off the motion of a respiration-correlated external surrogate. Typically, a model must be built that correlates the internal tumor motion with the external surrogate motion. Examples of linear accelerators that utilize this technique include the Accuray CyberKnife[19] and the BrainLab Vero.[20] The correlation requires frequent verification because patient respiration has been found to vary during the course of a treatment session.[21] Malinowski et al. found that the change in tumor position from the first 10 minutes to the third 10 minutes of treatment was greater than 5 mm in 13% of lung SBRT fractions. They also reported that margins calculated from 30 minutes of tracking data were larger than margins calculated from 10 minutes of data, indicating that for long treatments, patients and tumors move farther away from the initial IGRT-determined treatment position. Other studies have also shown that for longer treatment durations, there is more intra-fraction motion;[22–24] therefore, it is beneficial to complete the treatment in as short a time as possible. A technology that continuously monitors the patient and/or tumor position and does not require an interruption in the treatment would decrease the treatment session length and provide confidence that the patient has not moved between imaging procedures.

Imaging of implanted fiducials is also commonly used for tracking intra-fraction motion during SBRT and can be used to determine the magnitude of this motion. During liver SBRT, Worm et al. observed mean 3D intra-fraction and intrafield motion of 16.6 and 11.3 mm, respectively.[25] They reported that the mean 3D intra-fractional marker displacement from the initial CBCT was 3.4 mm, with a maximum of 14.5 mm and the 3D displacements exceeded 8.8 mm 10% of the time. The clinical target volume (CTV) to PTV margins used for their patients were 5 mm in the left-right and anterior-posterior directions and 10 mm in the superior-inferior direction. They used abdominal compression on all patients but found that it did not prevent greater than 1 cm intrafield motion or baseline drift. Ge et al. imaged implanted fiducials to measure the intra-fraction motion of abdominal tumors during conventionally fractionated and SBRT patient treatments.[26] They found significant variations in patients' breathing patterns and respiration-induced tumor motion between fractions and during treatment. They observed that liver SBRT patients had statistically significant larger intra-fractional breathing amplitude changes than the conventionally fractionated patients. These changes can only be detected by a technology that continuously monitors the patient. Visual observation by the treating radiation therapists is not sensitive enough to detect small-scale (i.e., mm) but clinically relevant changes in the patient position.

Most commonly, intra-fraction position monitoring can take the form of a CBCT acquired at various time points in the treatment, usually the midpoint and/or the end of the fraction.[13,27–29] The downside of this approach is that it interrupts the treatment and lengthens the treatment session, increasing the time the patient must spend on the table. Another issue is that CBCT is only a snapshot of the patient position; nothing is known about the tumor or patient position before or after the imaging procedure. As such, these strategies have not been shown to be sensitive to intra-fraction motion of the target. Posttreatment imaging is not a sufficient indicator of intra-fraction motion, as shown by Noel et al.[30] in prostate radiotherapy. They showed that the sensitivity of pre- and post-treatment imaging at detecting 30 seconds of intra-fraction prostate motion greater than 3, 5, and 7 mm for all fractions was only 53%, 49%, and 39%, respectively. The sensitivity increased with more frequent imaging, and they make the conclusion that their findings support the value of continuous monitoring of the prostate.

16.3 USE OF SGRT FOR SBRT SETUP

Both the ASTRO and AAPM guidance documents refer to surface guidance as acceptable options for intra-fraction monitoring, provided they are evaluated and deemed to provide acceptable accuracy before implementation. Optical surface systems have been evaluated against radiographic imaging and found to have submillimeter accuracy.[31,32] The use of optical surface guidance for managing internal patient motion with respiratory gating was described in further detail in Chapter 15. This chapter focuses only on strategies and methodologies for SBRT patient setup and monitoring intra-fraction patient position for sites that do not require active management of the effects of patient respiration (gating or breath-hold).

The case for using SGRT for initial setup for thoracic and abdominal SBRT patients is driven by the accuracy of the system, which has been shown to be submillimeter, and the absence of ionizing radiation. For example, results have been presented at the 2016 ASTRO Annual Meeting that compare SGRT to planar orthogonal kV images for setup of thoracic and abdominal SBRT patients.[33] Patient characteristics for both the inter-fraction and intra-fraction parts of this study are shown in Table 16.1.

TABLE 16.1   Patient and Tumor Characteristics

	$n = 71$ subjects
Gender, *n* (%)	
Male	37 (52.1%)
Female	34 (47.9%)
Age	
Mean (SD)	71.3 (10.6)
Median (Range)	73.0 (47–87)
BMI	
Mean (SD)	26.8 (6.2)
Median (Range)	26.3 (13.8–40.3)

	$n = 85$ lesions
Site, *n* (%)	
Lung	63 (74.1%)
Liver	10 (11.8%)
Adrenal	3 (3.5%)
Spine	2 (2.4%)
Other	7 (8.2%)
Location	
Peripheral	57 (67.1%)
Central	14 (16.5%)
N/A	14 (16.5%)
4D ROM (cm)	
Mean (SD)	0.49 (0.30)
Median (Range)	0.44 (0.00–1.24)
PTV volume (cc)	
Mean (SD)	31.8 (36.7)
Median (Range)	20.3 (4.6–218.5)
ITV volume (cc)	
Mean (SD)	11.8 (18.6)
Median (Range)	5.7 (0.6–136.0)

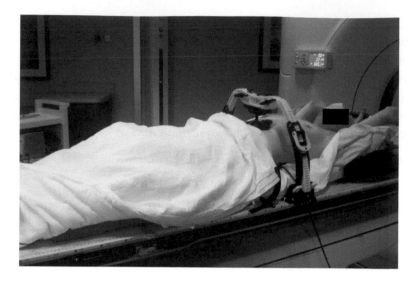

FIGURE 16.1   One type of typical immobilization equipment used for SBRT.

Lung and liver patients in this study were immobilized using a custom-molded vacuum bag and abdominal compression but were otherwise free-breathing. Patients receiving treatment to other sites were treated using the same immobilization devices but were not compressed. Typical immobilization equipment for the patients is shown in Figure 16.1.

Patients were planned with either dynamic conformal arcs or volumetric modulated arc therapy (VMAT) using either 6FFF or 10FFF photons and were treated on a 6 degrees-of-freedom couch. To account for residual respiration-induced tumor motion after compression, internal target volumes (ITV) were contoured on the maximum intensity projection (MIP) reconstructions for lung patients and on the minimum intensity projection (MinIP) reconstruction for liver patients. The ITV was uniformly expanded by 5 mm to create the planning target volumes (PTV). Dose calculation was performed on the average intensity projection (AIP) reconstruction, which was also used as the reference for image guidance.

To first quantify the accuracy of SGRT in patient positioning prior to SBRT treatments compared to orthogonal kV images, patients were initially positioned using either SGRT or planar orthogonal kV images on an alternating schedule prior to CBCT. For SGRT, patients are initially set up to the digital imaging and communications in medicine (DICOM) SGRT reference surface that was created from the body contour outlined on the AIP reconstruction CT using a density threshold of 0.6 g/cc (–350 HU).

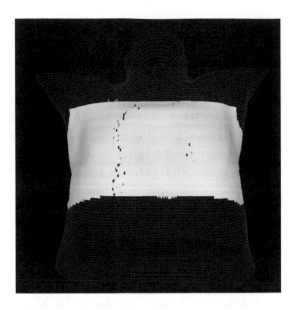

FIGURE 16.2   Typical region of interest (ROI) used for SBRT patient setup and monitoring.

The region of interest (ROI) used for tracking consisted of the patient's chest superior to the abdominal compression arch up to the clavicles and laterally down to the vacuum bag used for patient immobilization. An example ROI is shown in Figure 16.2. For this work, no issues with the compression arch interfering with the tracking were encountered.

Initial setup was performed using tolerances of 2 mm for translations and 1 degree for rotations, and adjustments were made such that offsets due to breathing motion were fluctuating about zero. On kV imaging days, patients were set up initially to skin marks, imaged with orthogonal planar kV images, and then bony anatomy in the target region was used to match the images compared to digitally reconstructed radiographs created from the treatment planning CT scan. After each setup method, CBCT was performed and additional shifts were made to match the internal target position prior to treatment. All CBCTs were evaluated by the treating physician. For lung targets, the gross tumor volume (GTV) was matched with the corresponding reference image, and the PTV was evaluated to ensure that the GTV was adequately covered. For liver targets, because the GTV is not usually visible on CBCT, the liver contour was initially matched with the reference image and then the hepatic veins and ligaments in proximity to the internal liver target were matched with the reference image to ensure accurate targeting of the area of liver containing the GTV. For adrenal targets, the adrenal gland and GTV were matched to the reference image. For spine targets, the bony anatomy was first matched to the reference image and then adjusted based on the GTV within the bone. The additional shifts from the CBCT were recorded to determine the inter-fraction setup error from the initial SGRT or kV imaging localization for comparison. The study alternated between kV imaging and surface-guided localization for initial setup of 46 SBRT patients encompassing 58 lesions and 243 fractions. The CBCT shifts provided the error in the initial localization from either the kV imaging or the surface guidance. It was found that there was no statistically significant difference between the two methods, except in the longitudinal direction, where the kV imaging produced smaller average residual CBCT shifts than SGRT, as shown in Table 16.2.

However, this difference of 1 mm is not deemed to be clinically relevant for these patients. Factors that were found to influence the difference in the shifts were body mass index (BMI) ($p = 0.001$) and increased PTV size ($p = 0.049$) as shown in Table 16.3 The localization accuracy of the SGRT system was found to be similar to that of a stereotactic body frame (SBF),[13,14,34] a device intended to position the patient close to treatment position using only external marks. Implementation of SGRT for patient setup for SBRT could eliminate the need for skin marks/tattoos or the use of a body frame.[35] Some departments have moved to using SGRT exclusively for initial setup of SBRT patients.

TABLE 16.2 Mean Quantitative Shifts and Results of Linear Mixed Models Comparisons Between Techniques for Initial Patient Setup

	Method p-Value	kV/kV Mean Shift (Standard Error)	SGRT Mean Shift (Standard Error)
Vertical shift (cm)	0.291	0.26 (0.03)	0.29 (0.03)
Longitudinal shift (cm)	0.019	0.28 (0.04)	0.41 (0.04)
Lateral shift (cm)	0.489	0.24 (0.03)	0.22 (0.02)
Translational vector	0.107	0.54 (0.05)	0.63 (0.04)
Pitch (°)	0.335	0.35 (0.09)	0.44 (0.07)
Roll (°)	0.587	0.35 (0.09)	0.40 (0.08)
Rotation (°)	0.204	0.28 (0.08)	0.39 (0.07)
Rotational vector	0.189	0.75 (0.13)	0.93 (0.12)

TABLE 16.3   Factors Influencing SGRT-kV Imaging Shift Differences

Outcome	Factor	Interaction Driver and Effect on Deltas
Vertical	BMI	Obesity (>30, $p = 0.017$); SGRT $>$ kV/kV
	AGE	Age ($p = 0.0350$); kV/kV $>$ SGRT
		Difference between methods gets smaller as age increases
Roll	Location	Non-lung targets ($p = 0.021$)
	BMI	Normal BMI (≥ 18.5 and <25, $p = 0.025$); SGRT $>$ kV/kV
Rotation	ROM	4D ROM ($p = 0.0567$); kV/kV $>$ SGRT
		Difference between methods gets smaller as ROM increases

16.4 USE OF SGRT FOR SBRT MONITORING

SGRT for intra-fraction monitoring of SBRT patients is a relatively recent development, despite the importance of assessing and accounting for intra-fraction motion during high dose per fraction treatments. To be used for monitoring in a meaningful way, the surface imaging must be reasonably accurate when compared to radiographic imaging and must represent the motion of the tumor reasonably well. Alderliesten et al. published a study where they evaluated the correlation of surface image guidance to CBCT by capturing surfaces during three separate CBCTs. The scans were performed pretreatment, after correcting for tumor misalignment, and after treatment completion.[36] They evaluated intra-fraction motion by registering the second and third CBCTs with the planning CT scan and by registering the third acquired surface with the second surface. They calculated differences between the derived intra-fraction motion determined from the CBCT and the surface imaging. They found that the differences in the 3D vector between the surface imaging and the CBCT were 0.7 and 1.3 mm for females and males, respectively, when aligning the CBCT with the tumor. Their study was performed with a single camera system, and the authors hypothesize that a three-camera system would provide better agreement between the surface imaging and the CBCT. Acknowledging this limitation, they concluded that SGRT was not suitable for monitoring of male SBRT patients, perhaps due to lack of shape on the chests of the male patients, which makes a surface registration more difficult. The authors also retrospectively calculated that tumor movement greater than 4 mm occurred in 24% of the fractions studied and the surface imaging system would have detected 20% of those incidents.

Similar studies have been performed except that SBRT patients were continuously monitored during treatment, rather than taking surface captures during intra-fraction CBCTs.[37] After the initial localization CBCT shifts were applied, a new gated reference surface is acquired by the SGRT system and used to continuously monitor the patient during treatment. The reference surface for monitoring is chosen from the gated capture to be at the 50% amplitude of the patient's respiratory cycle. For the intra-fraction motion analysis, tolerances for intra-fraction motion were defined as 2 mm translations and 1 degree rotations with a maximum of 2 seconds out of tolerance before an auto beam hold is enacted. The 2 second tolerance was used to reduce the number of beam holds that are due to the patient's normal respiratory motion. A typical monitoring example is shown in Figure 16.3.

The beam was held automatically by the SGRT system if the patient was out of tolerance for more than two seconds and the beam hold remained in effect until the patient went back in tolerance. If a patient's position as reported by the SGRT system either exceeded these thresholds three times in the fraction or went out of tolerance and stayed out of tolerance (the beam was turned off after two seconds), the SGRT-detected shifts were recorded and a repeat CBCT was performed. To determine the SGRT shifts, several (3–5) report captures were taken using the SGRT software and the average positions from these captures were used for the SGRT shifts. Any shifts that were performed after repeat CBCT with tumor volume matching were recorded as "intrafraction target motion." The shifts were then applied and a new SGRT reference image was captured. The physician

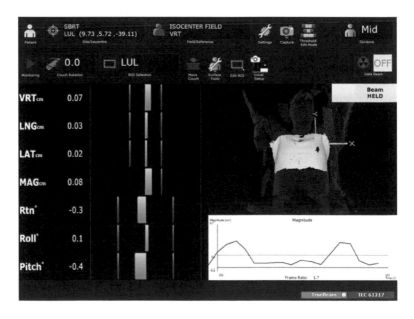

FIGURE 16.3 Example of SGRT patient monitoring during SBRT.

performed all IGRT matching. The intra-fraction shifts detected by SGRT were then compared to those determined on the subsequent CBCT. The study monitored 58 SBRT patients with abdominal or thoracic tumors during treatment delivery. The SGRT system detected motion above the threshold in 27 patients (34 individual fractions) that required an intra-fraction CBCT. Of those 34 fractions, 25 clinically meaningful shifts of at least 2 mm were performed based on the CBCT. In 10 patients, the SGRT system detected motion that resulted in shifts of at least 4 mm, and in three patients, motion was detected that resulted in shifts greater than 5 mm. No significant difference was found between the vector shifts of the SGRT system and the subsequent CBCT ($p = 0.367$). Data for the shifts is found in Table 16.4. Reasons for intra-fraction motion and repeat CBCT scans included patient coughing, patient voluntary motion, baseline drifts, and changes in respiratory amplitude. The SGRT system detected intra-fraction motion in three patients after the CBCT shifts were made, but before the first beam could be started, indicating the need for continuous monitoring of SBRT patients.

TABLE 16.4   Detected Intra-fraction Motion by SGRT and Resulting CBCT Shifts from Intra-fraction Imaging

	A	B	C	D
	$\left|Shift_{SGRT}\right|$	$\left|Shift_{CB}\right|$	Average Distance Between CBCT & SGRT Measured Shifts	Range of Distance Between CBCT & SGRT Measured Shifts
Vertical (cm)	0.135	0.144	0.165	0.003–0.610
Longitudinal (cm)	0.196	0.109	0.250	0.000–0.670
Lateral (cm)	0.100	0.158	0.169	0.000–0.440
Vector	0.332	0.314	−0.018	−0.600–0.430
Pitch (°)	0.376	0.112	0.517	0.000–2.230
Roll (°)	0.202	0.334	0.519	0.000–3.450
Rotation (°)	0.286	0.038	0.309	0.000–2.300

A. This is the average of the magnitude of the shift deltas for the SGRT method (calculated based on the **absolute value** of each shift).

B. This is the average of the magnitude of the shift deltas for the cone-beam method (calculated based on the **absolute value** of each shift).

C. This is the average of the distance between CBCT and SGRT; for the individual components, that is the average of the absolute value of the differences. For the vector, it's the average of the signed difference.

D. This is the range associated with the distances between CBCT and SGRT reported in Column C.

Mercier et al. presented results of a similar study at the 2018 European Society for Therapeutic Radiology and Oncology (ESTRO) annual meeting.[38] They monitored patients with SGRT during SBRT treatments of spine, lymph node, and non-spine bone metastases and halted treatment if intra-fraction motion was greater than 3 mm or 0.5 degrees. Their procedure was slightly different than the previous study presented; they performed three CBCT scans each fraction, similar to Alderliesten et al., and an intra-fraction CBCT was performed at the physician's discretion when the patient violated the SGRT system's tolerances. They report that out of 17 patients and 54 fractions, the treating physician deemed an intra-fraction CBCT to be necessary five times (9.2%) due to intra-fraction motion that exceeded their threshold. In those 5 instances, they found that the shifts detected by the CBCT were larger than the SGRT tolerances. For all other sessions, the shifts determined from the posttreatment CBCT were found to be within the SGRT tolerances. They conclude that with an accuracy of 99.7% surface guidance provides sufficient accuracy for monitoring intra-fraction motion in frameless SBRT of spine, nodal, and non-spine bony metastases.

16.5 CONCLUSION

SGRT has proven to be a useful technology for setup and position monitoring of patients who are receiving SBRT to the thorax and abdomen, even in the absence of respiratory gating. Using SGRT for setup can reduce the magnitude of the shifts made from the CBCT and can also monitor the patient during the imaging process to ensure that the patient is still at the same location as when the CBCT was acquired. Evaluating CBCT matches for SBRT patients can take some time due to the need for additional care and attention; however, the longer the evaluation and image match takes, the more likely the patient is no longer in the same position as when the images were acquired. Using SGRT to monitor the patient during the CBCT acquisition and evaluation provides confidence that the patient has not moved.

Studies investigating SGRT for monitoring patients during SBRT have shown that it can reliably detect intra-fraction motion, which can be clinically meaningful for treatments that involve small margins and sharp dose gradients. Advantages of SGRT over other monitoring/tracking technologies include no extra imaging dose, submillimeter accuracy, no increase in treatment times and the ability to have the system automatically hold the treatment beam if the patient's position moves out of tolerance.

KEY POINTS

- SGRT is valuable for SBRT patient setup, adds no extra imaging dose, and has accuracy comparable to stereotactic body frames for initial patient localization.

- SGRT is reliable for detecting intra-fraction external patient motion that occurs during SBRT radiographic verification imaging and treatment delivery.

- The magnitude of intra-fraction motion detected by SGRT has been shown to match well with motion determined from radiographic imaging.

- SGRT is an attractive option for intra-fraction monitoring of SBRT patients, especially for treatments that may last 30+ minutes.

REFERENCES

1. Jaffray DA, Drake DG, Moreau M, et al. A radiographic and tomographic imaging system integrated into a medical linear accelerator for localization of bone and soft-tissue targets. *Int J Radiat Oncol Biol Phys.* 1999;45:773–789.
2. Benedict SH, Yenice KM, Followill D, et al. Stereotactic body radiation therapy: The report of AAPM Task Group 101. *Med Phys.* 2010;37:4078–4101.
3. Solberg TD, Balter JM, Benedict SH, et al. Quality and safety considerations in stereotactic radiosurgery and stereotactic body radiation therapy: Executive summary. *Pract Radiat Oncol.* 2012;2:2–9.
4. Wurm RE, Gum F, Erbel S, et al. Image guided respiratory gated hypofractionated Stereotactic Body Radiation Therapy (H-SBRT) for liver and lung tumors: Initial experience. *Acta Oncol (Madr).* 2006;45:881–889.
5. Lovelock DM, Messineo AP, Cox BW, et al. Continuous monitoring and intrafraction target position correction during treatment improves target coverage for patients undergoing SBRT prostate therapy. *Int J Radiat Oncol Biol Phys.* 2015;91:588–594.
6. Herman TDLF, Vlachaki MT, Herman TS, et al. Stereotactic body radiation therapy (SBRT) and respiratory gating in lung cancer: Dosimetric and radiobiological considerations. *J Appl Clin Med Phys.* 2010;11:158–169.
7. Caillet V, Booth JT, Keall P. IGRT and motion management during lung SBRT delivery. *Phys Medica.* 2017;44:113–122.
8. Cerviño LI, Gupta S, Rose MA, et al. Using surface imaging and visual coaching to improve the reproducibility and stability of deep-inspiration breath hold for left-breast-cancer radiotherapy. *Phys Med Biol.* 2009;54:6853–6865.
9. Cerviño LI, Pawlicki T, Lawson JD, et al. Frame-less and mask-less cranial stereotactic radiosurgery: A feasibility study. *Phys Med Biol.* 2010;55:1863–1873.

10. Lax I, Blomgren H, Näslund I, et al. Stereotactic radiotherapy of malignancies in the abdomen. Methodological aspects. *Acta Oncol.* 1994;33:677–683.

11. Blomgren H, Lax I, Näslund I, et al. Stereotactic high dose fraction radiation therapy of extracranial tumors using an accelerator. Clinical experience of the first thirty-one patients. *Acta Oncol.* 1995;34:861–870.

12. Negoro Y, Nagata Y, Aoki T, et al. The effectiveness of an immobilization device in conformal radiotherapy for lung tumor: Reduction of respiratory tumor movement and evaluation of the daily setup accuracy. *Int J Radiat Oncol Biol Phys.* 2001;50:889–898.

13. Foster R, Meyer J, Iyengar P, et al. Localization accuracy and immobilization effectiveness of a stereotactic body frame for a variety of treatment sites. *Int J Radiat Oncol Biol Phys.* 2013;87:911–916.

14. Grills IS, Hugo G, Kestin LL, et al. Image-guided radiotherapy via daily online cone-beam CT substantially reduces margin requirements for stereotactic lung radiotherapy. *Int J Radiat Oncol Biol Phys.* 2008;70:1045–1056.

15. Li W, Sahgal A, Foote M, et al. Impact of immobilization on intrafraction motion for spine stereotactic body radiotherapy using cone beam computed tomography. *Int J Radiat Oncol Biol Phys.* 2012;84:520–526.

16. Hyde D, Lochray F, Korol R, et al. Spine stereotactic body radiotherapy utilizing cone-beam CT image-guidance with a robotic couch: Intrafraction motion analysis accounting for all six degrees of freedom. *Int J Radiat Oncol Biol Phys.* 2012;82:e555–e562.

17. Bouilhol G, Ayadi M, Rit S, et al. Is abdominal compression useful in lung stereotactic body radiation therapy? A 4DCT and dosimetric lobe-dependent study. *Phys Medica.* 2013;29:333–340.

18. Han K, Cheung P, Basran PS, et al. A comparison of two immobilization systems for stereotactic body radiation therapy of lung tumors. *Radiother Oncol.* 2010;95:103–108.

19. Adler Jr. JR, Chang SD, Murphy MJ, et al. The Cyberknife: A frameless robotic system for radiosurgery. *Stereotact Funct Neurosurg.* 1997;69:124–128.

20. Depuydt T, Verellen D, Haas O, et al. Geometric accuracy of a novel gimbals based radiation therapy tumor tracking system. *Radiother Oncol.* 2011;98:365–372.

21. Malinowski K, McAvoy TJ, George R, et al. Incidence of changes in respiration-induced tumor motion and its relationship with respiratory surrogates during individual treatment fractions. *Int J Radiat Oncol Biol Phys.* 2012;82:1665–1673.

22. Udrescu C, Mornex F, Tanguy R, et al. ExacTrac snap verification: A new tool for ensuring quality control for lung stereotactic body radiation therapy. *Int J Radiat Oncol.* 2013;85:e89–e94.

23. Rossi MMG, Peulen HMU, Belderbos JSA, et al. Intrafraction motion in stereotactic body radiation therapy for non-small cell lung cancer: Intensity modulated radiation therapy versus volumetric modulated arc therapy. *Int J Radiat Oncol.* 2016;95:835–843.

24. Purdie TG, Bissonnette J-P, Franks K, et al. Cone-beam computed tomography for on-line image guidance of lung stereotactic radiotherapy: Localization, verification, and intrafraction tumor position. *Int J Radiat Oncol Biol Phys.* 2007;68:243–252.

25. Worm ES, Hoyer M, Fledelius W, et al. Three-dimensional, time-resolved, intrafraction motion monitoring throughout stereotactic liver radiation therapy on a conventional linear accelerator. *Int J Radiat Oncol Biol Phys.* 2013;86:190–197.

26. Ge J, Santanam L, Noel C, et al. Planning 4-dimensional computed tomography (4DCT) cannot adequately represent daily intrafractional motion of abdominal tumors. *Int J Radiat Oncol Biol Phys.* 2013;85:999–1005.

27. Bissonnette J-P, Franks KN, Purdie TG, et al. Quantifying interfraction and intrafraction tumor motion in lung stereotactic body radiotherapy using respiration-correlated cone beam computed tomography. *Int J Radiat Oncol Biol Phys.* 2009;75:688–695.

28. Li W, Purdie TG, Taremi M, et al. Effect of immobilization and performance status on intrafraction motion for stereotactic lung radiotherapy: Analysis of 133 patients. *Int J Radiat Oncol Biol Phys.* 2011;81:1568–1575.

29. Shah C, Grills IS, Kestin LL, et al. Intrafraction variation of mean tumor position during image-guided hypofractionated stereotactic body radiotherapy for lung cancer. *Int J Radiat Oncol Biol Phys.* 2012;82:1636–1641.

30. Noel C, Parikh PJ, Roy M, et al. Prediction of intrafraction prostate motion: Accuracy of pre- and post-treatment imaging and intermittent imaging. *Int J Radiat Oncol.* 2009;73:692–698.

31. Yu AS, Fowler TL, Dubrowski P. A novel-integrated quality assurance phantom for radiographic and nonradiographic radiotherapy localization and positioning systems. *Med Phys.* 2018;45:2857–2863.

32. Oliver JA, Kelly P, Meeks SL, et al. Orthogonal image pairs coupled with OSMS for noncoplanar beam angle, intracranial, single-isocenter, SRS treatments with multiple targets on the Varian Edge radiosurgery system. *Adv Radiat Oncol.* 2017:1–9.

33. Hampton CJ, Robinson M, Foster RD, et al. Comparison of 3-dimensional surface mapping and Kv imaging technique for initial setup of stereotactic body radiation therapy treatments. *Int J Radiat Oncol Biol Phys.* 2017;96:E632.

34. Worm ES, Hansen AT, Petersen JB, et al. Inter- and intrafractional localisation errors in cone-beam CT guided stereotactic radiation therapy of tumours in the liver and lung. *Acta Oncol.* 2010;49:1177–1183.

35. Sueyoshi M, Olch AJ, Liu KX, et al. Eliminating daily shifts, tattoos, and skin marks: Streamlining isocenter localization with treatment plan embedded couch values for external beam radiation therapy. *Pract Radiat Oncol.* 2019;9:e110–e117.

36. Alderliesten T, Sonke J-J, Betgen A, et al. 3D surface imaging for monitoring intrafraction motion in frameless stereotactic body radiotherapy of lung cancer. *Radiother Oncol.* 2012;105:155–160.

37. Heinzerling JH, Hampton CJ, Robinson M, et al. Use of 3D optical surface mapping for quantification of interfraction set up error and intrafraction motion during stereotactic body radiation therapy treatments of the lung and abdomen. *Int J Radiat Oncol Biol Phys.* 2017;99:E670.

38. Mercier C, Sprangers A, Verellen D. OC-0194: Evaluation of an optical surface monitoring system for intrafractional movement during SABR. *Radiother Oncol.* 2018;127:S104–S105.

CHAPTER 17

Head and Neck

Bo Zhao

CONTENTS

17.1 INTRODUCTION

Head and neck (H&N) cancers account for 3%–5% of all cancers in the United States. According to the American Cancer Society, an estimated 55,070 people (40,220 men and 14,850 women) will develop H&N cancer.[1] Radiation treatment plays an important role in the management of H&N cancers. H&N cancers typically involve the nasopharynx, oropharynx, and/or hypopharynx. Depending on the disease involvement, it may also include neck lymph nodes. The radiation field of H&N cancers may span from the base of skull to the lower neck. Radiation treatment of H&N cancer is complex and involves many organs-at-risk (OARs). Compared with other anatomies, the H&N area is flexible and subject to deformation and postural variations. Reproducible positioning relative to simulation during H&N radiation therapy is therefore crucial to treatment quality and patient safety. Additionally, with the advent of intensity modulated radiation therapy (IMRT) techniques that employ steep dose gradients, daily setup precision and position reproducibility between fractions is particularly important in order to minimize treatment target margins. Thermoplastic masks are routinely used to immobilize the H&N during radiation treatment. The masks are shaped to the patient's external body contour prior to CT simulation. The mask is first placed in a hot water bath or oven until soft and flexible. It is then stretched over the patient and molded to fit the patient's external body contour. The mask hardens to form a reproducible shape of the patient's external body contour as it cools to room temperature. To improve setup precision of the shoulders, particularly if treating neck lymph nodes, the mask may extend from the top of the head and cover the shoulders.

The accuracy of standard mask immobilization has been studied extensively. It has been shown that standard mask immobilization provides an accuracy of 0–3.6 mm for systematic errors and 1.0–2.6 mm for random errors.[2–8] The inaccuracy is partly attributed to the highly flexible body habitus around the H&N region. Nonrigid error due to deformation accounts for up to 3–4 mm, demonstrating the contribution of the flexibility and rotations on H&N positioning uncertainty.[5,7,9] Other sources of variability include changes in primary tumor volume and deformation of normal structures resulting from tissue response as well as alteration in body habitus due to weight loss.[10,11] These factors can result in gradual development of suboptimal mask fitting throughout a course of approximately 30 fractions for conventional treatment fractionation.[12] This is manifested as changes in tumor and organ volume as well as their positional drift with subsequent fractions.[10] This can lead to changes in the dose coverage of targets as well as potential increases in dose to OARs. This highlights the importance of considering adaptive radiation therapy strategies when treating the H&N.[13–15]

The standard full-face mask forces patients to keep their eyes and mouth closed during each fraction of treatment, which can be up to 30 minutes long. Many patients find these masks constrictive, uncomfortable, and stressful. Furthermore, studies have found that these masks also possess water-equivalent density and can act like bolus.[16,17] This bolus effect increases patient skin dose and can worsen skin reactions to treatment. Studies have shown that the mask bolus effect can increase skin dose at the neck by as much as 30%.[16,18,19] Therefore, there is a clinical rationale for developing a practical alternative to the fully-enclosed masks that would improve patient comfort and/or reduce skin dose while also maintaining effectiveness at immobilizing the patient.

Previous studies have described several methods to address this clinical need by reduced mask use. Velec et al.[20] described a long mask manually cut open at the neck to reduce skin dose. Patients were initially positioned using skin marks without intra-fraction tracking. Cone-beam computed tomography (CBCT) guided setup results were then compared between 9 patients treated with modified open-neck masks and 11 patients treated with conventional long closed masks. Setup uncertainties were comparable between the two types of masks. In another study, 260 patients[21] were randomized between full-face masks and head-and-shoulder masks. The study found no statistical difference in setup reproducibility between the mask types. However, a short mask covering only the patient's head significantly reduced claustrophobic anxiety in patients. Furthermore, patients treated with the head-and-shoulder mask experienced more skin reactions.

To develop an immobilization procedure for patients who could not tolerate closed masks, Kim et al.[22] developed a mask-less technique that combined a customized head mold and a bite block plate with infrared reflectors to monitor motion. The head mold wrapped around one-third of the patient's posterior H&N. One or two straps could be used to wrap the head mold around the patient's head. They compared this setup with another bite block immobilization approach and partial mask covering the upper face. During the treatment, when the infrared marker moved beyond a threshold, the beam was interrupted. The tracking accuracy of the two methods was similar. However, the mask-less system required many more beam interruptions compared with the partial mask system. Due to oral toxicity, the biting effort on the mouthpiece, and the change in treatment positions, there were concerns as to whether this system could reduce patient comfort and tolerability.

17.1.1 Surface Guidance for H&N

Surface-guided radiation therapy (SGRT) is well established for intracranial radiation therapy and intracranial stereotactic radiosurgery (SRS).[23–30] Earlier investigations have employed a customized head cushion only to allow the entire face to be tracked.[24,26,27] Recently, there is a trend to use open-face masks that immobilize the head but allow the eyes, nose, and cheeks to be exposed.[28,29,31] Both head cushion-only and open-face mask immobilization have demonstrated an accuracy of 1–2 mm and 1°

during setup for SRS when using SGRT.[25,26] The treatment outcomes have been excellent on several diagnoses such as trigeminal neuralgia,[32] base of skull,[33] and brain metastases.[23,28,31] Outcomes for SRS and SRT using surface guidance are discussed further in Chapter 13.

With the success of intracranial SGRT, it is a logical step to apply the technique to H&N radiation therapy where a thermoplastic mask is the primary immobilization method and the exposed skin of the face is used as a surrogate for internal H&N anatomy. However, when compared with intracranial anatomy, surface-guided H&N treatments are more challenging. First, the neck is highly flexible. Even a fully enclosed mask may not provide absolute restrictive immobilization. Second, unlike intracranial lesions that move rigidly with the skull, the deformable nature of H&N anatomy may lead to poor correlation between internal lesions and the skin surface. And third, the irradiated area of the H&N typically involves a large volume and fields with long spans. Consequently, the size of the region of interest (ROI) required for surface guidance of H&N may be larger than for SRS which may increase the latency of the tracking process. These issues need to be addressed to successfully develop the application of SGRT to H&N. In short, it is essential to establish spatial and temporal correlation between the optically visible skin surface and internal anatomy.

17.1.2 Correlation between the Skin Surface and Internal Anatomy

Tumors in the H&N are usually attached to bony anatomy. Therefore, online imaging such as planar kV or CBCT is primarily used for positional verification. Several studies have examined if the skin can be reliably used as a surrogate for internal anatomy. In one retrospective study, 11 patients underwent weekly CT rescans of the head and shoulders during the course of their treatment.[34] External body contours were extracted from the CT scans, including all internal contours. The registration accuracy was categorized based on the location of two surface tracking ROIs: (1) a region for rigid setup error, defined in the CT image as all slices above the second cervical vertebra; (2) a region for nonrigid setup error, defined as the head and shoulders where the shoulders were defined on the CT image for all slices below the sixth cervical vertebra. The neck region was excluded because loose skin in this region may not be reproducible. The registration of internal bony anatomy between the weekly CT rescan and the planning CT was set as the ground truth and performed in the treatment planning system (TPS). The registration of external contours from the two CT sets was done with a proprietary algorithm embedded in a commercial SGRT system. It was found that the rigid alignment accuracy was 0.8°–2.2° in rotations and 2.4–4.5 mm in translations at a 90% confidence level. In contrast, the discrepancy for nonrigid registration was much larger at 1.9°–4.5° and 6–10 mm at a 90% confidence level. It was concluded that surface guidance is capable of detecting rigid setup errors, although there was a patient-specific time dependency on setup accuracy, likely caused by the patient's weight loss during the treatment course. Selection of surface ROIs that was less prone to weight loss improved registration accuracy. The study found that surface guidance was less capable of detecting nonrigid setup errors. Several possible reasons were: registration errors in the TPS, registration errors in the surface guidance system, and actual differences between skin and bone in the shoulder region. The correlation of internal anatomy and the patient's surface may not be consistent because the bony structures in the neck and shoulder region can move differently relative to the skin. Weight loss led to further discrepancies, which was more pronounced in the shoulder region than in the head region. The authors concluded that further measurements are needed to improve nonrigid setup accuracy.

Another similar study compared the relative accuracy of surface guidance and volumetric registration.[35] Rigid registrations between the pretreatment CBCT and planning CT based on surface and internal anatomy were retrospectively compared in 26 patients. A surface image was derived from the external body contour of the pretreatment CBCT and was registered using an ROI and third-party software to a surface image derived from the planning CT-based external body contour. The average difference in registrations was small while the standard deviations for translations were 1.4–2.2 mm and 0.5°–0.9° for rotations. The vector magnitude of the translational errors averaged 2.7 mm with a maximum of 5.2 mm. The residual error of the surface registration had an average of 0.9 mm and a maximum of 1.7 mm. Because of the large residual error, the authors concluded that surface guidance alone was not sufficient to provide accurate setup for H&N radiation therapy.

17.1.3 ROI Selection and Temporal Stability

Many studies have highlighted the importance of surface ROI selection on anatomical sites such as breast, brain, chest, and pelvis when performing surface image registration. Similar to the ROI used when matching X-ray verification films with digitally reconstructed radiographs (DRRs), the choice of ROIs on surface images can greatly influence the quality of alignment. Selection of the ROI should only include surface regions that move concurrently with the internal tumor. For example, the ROI should not contain anatomy subject to involuntary or local motion (e.g., jaw, mouth, eye, or eye brow motion) independent of net head motion. Inclusion of such regions can reduce accuracy in registration. Kang et al.[36] analyzed the surface features of the head and developed an automatic ROI selection method. Their algorithm calculated the displacements and velocities of 3D surface points between a real-time surface image acquired over a period of time and a reference surface image. The automatic ROI excluded locally varying facial motions. The tracking accuracy was better than the manually selected ROIs. Because this algorithm compared a reference image with real-time images, the automatically selected ROI could not be preset and had to be calculated online in the clinical application. This implementation is not yet commercially available.

The temporal and spatial performance are important for any guidance system. For a surface guidance system, the system response time or frame rate is dependent on the size of the ROI due to the time required to perform the calculations for automatic registration. The temporal and spatial performance of a 3D surface guidance system was measured by Wiersma et al.[37] by comparing the tracking result with a ground truth defined by an infrared marker tracker. The infrared system had a tracking resolution of less than 0.25 mm at 12 Hz. It was found that increasing the ROI size significantly decreased the frame rate. In fact, the relationship between the surface imaging system response period (1/fps) and the size of the ROI was linear. Unlike temporal performance, spatial performance, represented by the tracking accuracy, was generally independent of ROI size, provided that the chosen ROI surface was sufficiently large (greater than 1000 vertex points in this case) and contained suitable surface features for registration. For an ROI containing at least 1000 vertex points, the tracking error was constant at 0.3 mm. This provided an important guideline on the required size of the ROI and the effect of the ROI size on the surface guidance system's latency.

17.1.4 Clinical Use of SGRT in H&N

Stieler et al. evaluated several clinical cases of H&N, pelvic, and thoracic SGRT by comparing the match results based on surface and internal anatomy.[38,39] Three H&N cases were reported. The initial setup was guided by skin marks followed by a commercial surface guidance system and CBCT online imaging verification. The match between the planning CT and the CBCT was based on the treatment volume. The patient was subsequently shifted according to the CBCT match. A real-time surface image was acquired and retrospectively compared with the surface derived from the planning CT external body contour. The resulting shift vector components were reported as a way of quantifying the accuracy of SGRT. While there was no systematic error (the mean was close to zero) on translational and rotational shifts compared to CBCT, the standard deviations were approximately 3 mm and 1.1°–2.1°, respectively. The longitudinal shift was larger compared with the other two translational directions, partly due to respiration. The authors observed that match results were very different between the patient's surface and internal anatomy when patients had misaligned shoulders. This indicated the presence of body deformation.

A detailed study of surface-guided H&N radiation was reported by Li et al.[40] The authors performed an immobilization comparison between an open-face mask and a conventional closed-face mask on volunteers. The open-face mask consisted of a three locking point thermoplastic head mask with an opening for the eyes, nose, mouth, and some forehead and cheek area (Figure 17.1a). The opening also defined the surface ROI for intra-fraction motion tracking (Figure 17.1d). In contrast, the conventional three locking point full-head mask had only the nose exposed (Figure 17.1b).

The immobilization performance of the conventional mask served as a baseline for this study. The test was first performed on 10 volunteers for whom both mask types were made and tested. The head motion of the volunteers in the two masks was evaluated during mask locking and over 15 minutes lying on the treatment couch. Volunteers were asked to perform forced movements such as head rotation and chin repositioning. It was found that the forced head motion using the two masks was similar during the mask placement and continuous tracking. The ranges of forced movements in the two masks were found to be similar. Interestingly, when the forced movement was released, the head tended to fall back to its original position within 1–2 mm. The authors further investigated five patients who were simulated and treated with an open mask only, due to claustrophobia. The disease sites of these patients were close to the skin surface, including the parotids, ear, neck, nasal skin, and base of tongue. Patients

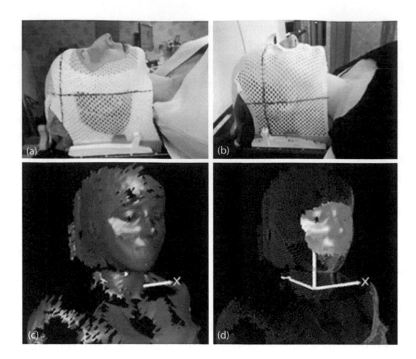

FIGURE 17.1 An open-face mask and a conventional closed-face mask molded on two volunteers. An arbitrary alignment point was marked on the masks. For the open-face mask (a), the open area was set to be the region of interest for AlignRT motion monitoring. For the conventional mask (b), the nose area was open, allowing alignment between a skin mark and the room laser in a forced motion test. A raw reference image (c), where the open area is clearly seen; the ROI (d) drawn on the reference image. (Reprinted from Li, G. et al., *J. Appl. Clin. Med. Phys.*, 14, 243–254, 2013.)

were set up initially with surface guidance. If the rotation of the head was found to be greater than 2° in any direction, the mask was repositioned to correct the head rotation. An orthogonal kV image pair was used for online image verification. A real-time surface image was acquired after setup to serve as the reference for intra-fraction motion tracking. At the end of each treatment, both kV imaging and real-time monitoring of the patient with surface imaging were repeated to evaluate patient intra-fraction motion. The difference between the verification and posttreatment kV images was obtained. This intra-fraction motion measurement by the kV imaging and surface guidance systems was found to be in a similar range of 1–2 mm, and was in agreement with the findings of the volunteer study. The authors did not report the surface guidance-based setup error compared with verification kV at the initial setup. It was concluded that the open-face mask method provided sufficient immobilization (less than 2 mm) during treatment.

17.1.5 Open-Face Long Masks

Long masks that extend from the head to the shoulders are commonly used to treat H&N cancer for sites located at the level of the neck or have involvement of the neck lymph nodes. Wiant et al.[41] conducted a prospective trial to investigate the use of open-face long masks on H&N patients. The purpose of this study was to examine if an open-face long mask would have similar levels of motion control and immobilization as closed-face masks, studied over a large group of patients throughout the course of treatment. Fifty H&N patients were included in this protocol. The patients were evenly randomized into two groups: (1) a conventional closed thermoplastic head and shoulder mask; (2) open-face thermoplastic head and shoulder mask where the face area including eyes, upper cheek, and mouth was not covered by the mask. Both groups received daily CBCT or MVCT as the image-based setup verification method. Patients with an open-face mask were initially set up and tracked during treatment with surface guidance. The setup was not conventionally reported against x-ray imaging. Instead, it was quantified by posture reproducibility on bony anatomy such as the spine and mandible. In detail, the spinal canal and mandible were both contoured on the simulation CT and verification MVCT. The contours on the simulation CT were expanded by steps of 1 mm until they covered the contours on verification CT (CBCT or MVCT). It was reported that the mean expansion thickness was approximately 2 mm for both groups. This indicates that the two patient groups had similar levels of spine flexion and mandible position change. Another evaluated metric was the similarity of the two image sets, calculated using rigid registration cost functions such as the mutual information coefficient. This metric indicates the trends of deformation over time. Both groups exhibited similar correlation with increasing fraction number, with slightly higher values on open-face mask patients. The authors noted that weight loss could be a contributor to the observed deformation. Intra-fractional motion of the open face was used as a surrogate for total body motion in this study. The motion of patients with an open-face mask was continuously monitored by surface guidance and recorded as a 6D shift based on the acquired surface compared with a reference surface. The magnitude of the 3D vector of the translational shift was reported. It was found that the average motion was 0.9 mm, with 3.7% of the fractions having movements over 2 mm. Intra-fractional motion was not correlated to patient treatment fractions or weight loss. Although it was not compared with closed-face mask patients due to a lack of tracking data, the authors compared their results with data in the literature which showed 0.5–1.8 mm of intra-fraction movement[42,43] based on a conventional mask. The authors concluded that the intra-fraction motion was similar in both groups. Both patient groups were also given a survey before simulation and then every fifth fraction during treatment. The survey rated the following items: patients' treatment-related anxiety on a scale of 0 to 10, claustrophobia on a scale of 0 to 10, and whether patients used any medication for anxiety. A score of 0 indicated no anxiety/claustrophobia, and 10 indicated extreme anxiety/claustrophobia. While two patient groups had similar survey results before simulation, the open-face mask group showed reduced mean values of anxiety, claustrophobia, and anti-anxiety drug use on the survey during treatment, although the difference compared with closed-face group was not statistically significant. The authors suggested that SGRT may be used to supplement or replace daily CT scans and that surface guidance may provide adequate real-time motion monitoring for H&N patients.

17.1.6 Minimally Open Masks

Zhao et al. performed another prospective trial using minimal mask immobilization to treat H&N cancers.[44] Twenty patients were enrolled in this protocol. The standard of care for these patients was treatment with closed-face long masks. The authors modified commercial thermoplastic masks to immobilize only the forehead and chin (Figure 17.2a and b), which has minimal open skin among all studies of this application. Shoulder movement was restricted by either a moldable cushion at the shoulders in 8 patients or shoulder retractors in 12 patients. Their standard IMRT planning employed 2–4 VMAT coplanar arcs with a matched anterior-posterior (AP) low-neck field (Figure 17.2e). Before the first day of treatment, skin surface data from the planning CT was imported into the surface guidance system. Patients were initially set up to surface markings and verified by the surface guidance system. Two ROIs for surface monitoring were then selected: (1) nose and cheeks, (2) a center neck strip

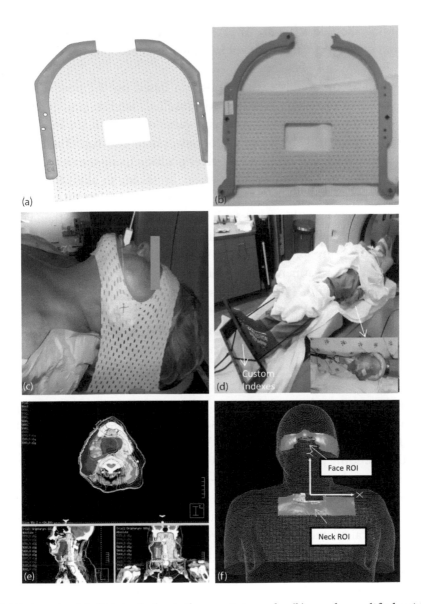

FIGURE 17.2 (a) Original short three-point mask, (b) mask modified with straight cuts at top and bottom. (c) Modified mask in place over only forehead and chin. (d) Overall patient setup. (e) Patient treatment plan with low neck coverage. (f) ROI selections on AlignRT relative to isocenter location. These two ROIs can be used to create a composite ROI for intra-fractional tracking. (Reprinted from Zhao, B. et al., *J. Appl. Clin. Med. Phys.*, 19, 17–24, 2018.)

excluding shoulders (Figure 17.2f). A mid neck ROI was not used due to motion interference from patient swallowing. A composite ROI including both ROIs was also generated.

Therapists first adjusted head position based on the nose and cheek ROI in six dimensions (vertical, longitudinal, lateral, yaw, pitch, roll). Error thresholds were set at a default value of 1.5 mm for longitudinal, lateral, and vertical shifts and 1° for yaw, pitch, and roll. The therapists then adjusted the low neck and shoulder position using the neck ROI, with 3 mm shift and 3-degree rotation error thresholds to align the AP neck field. Therapists went back to the nose and cheek ROI to ensure that the head position was within the default threshold. The therapists then verified this full setup with routine CBCT imaging. Online CBCT matching first included the bony spine and skull anatomy, followed by soft tissue matching around the treatment volume. Setup error was evaluated in 6 degrees of freedom (6DOF). The real-time surface after corrections was captured as a reference for subsequent treatments. For the version of SGRT solution being used in this study, during treatment only one ROI and one threshold could be utilized at a time. Intra-fraction motion was tracked on the composite ROIs (neck + face ROI) with default SGRT threshold settings (1.5 mm/1°). The standard clinic procedure required a beam hold for sustained movement beyond the specified error thresholds. For subsequent treatment days, therapists would use the previous day's acquired surface for initial setup. CBCT and planar kV images were acquired as prescribed by the attending physician. Beam gating was manually controlled by therapists. The authors measured surface-guided setup accuracy against CBCT. Group mean and standard deviation were calculated for all treatment fractions from all patients. Systematic setup error (Σ) and the random error (σ) were also calculated.[45] Systematic error is defined as the standard deviation of the individual patient means from his/her entire treatment fractions. Random error was defined as the root mean square of the individual patient standard deviation from treatment fractions. Similarly, these metrics were also determined from the intra-fraction motion collected by surface guidance. A total of 591 CBCTs were obtained for reference to surface guidance. The setup results based on surface guidance are tabulated in Table 17.1, including average couch shifts initiated by CBCT verification followed by surface-guided setup as well as systematic and random errors based on the two shoulder restriction methods used for this study. Overall systematic error on translational shifts was small (less than 1.4 mm) and random error varied. Vertical displacements produced the largest random error

TABLE 17.1 Summary of Setup Shifts Based on CBCT and Intra-fraction Motion

Patient Cohort		Vrt (mm) (min, max)	Lng (mm) (min, max)	Lat (mm) (min, max)	Rotation (°) (min, max)	Pitch (°) (min, max)	Roll (°) (min, max)
Setup Shift Based on CBCT (mm or °)							
Shoulder cushion	Group average	-0.73 ± 2.08	-0.21 ± 2.75	0.17 ± 2.68	-0.13 ± 0.87	-0.24 ± 0.97	0.13 ± 0.87
		(−8.0, 7.0)	(−7.0, 8.0)	(−8.0, 10.0)	(−3.6, 3.0)	(−3.0, 2.9)	(−3.0, 3.0)
	Σ	0.71	0.66	0.68	0.29	0.25	0.27
	σ	2.70	2.65	2.54	0.84	0.95	0.84
Shoulder retractors	Group average	-0.35 ± 2.60	-0.70 ± 3.65	0.28 ± 2.96	-0.16 ± 1.10	0.14 ± 1.31	0.02 ± 1.21
		(−11.1, 11.0)	(−9.0, 10.0)	(−14.0, 11.0)	(−3.0, 3.0)	(−2.8, 3.0)	(−3.0, 3.0)
	Σ	1.37	0.95	1.12	0.54	0.69	0.47
	σ	3.46	2.80	2.94	1.00	1.11	1.14
Total cohort	Group average	-0.51 ± 2.42	-0.49 ± 3.30	0.23 ± 2.58	-0.15 ± 1.01	-0.02 ± 1.19	0.06 ± 1.08
		(−11.1, 11.0)	(−9.0, 10.0)	(−14.0, 11.0)	(−3.6, 3.0)	(−3.0, 3.0)	(−3.0, 3.0)
	Σ	1.18	0.90	0.98	0.46	0.58	0.41
	σ	3.18	2.74	2.79	0.94	1.05	1.03
Intra-fraction Motion (mm or °)							
Shoulder cushion	Group average	-0.02 ± 0.74	-0.04 ± 0.87	0.04 ± 1.01	-0.05 ± 0.48	-0.03 ± 0.40	-0.05 ± 0.41
	Σ	0.15	0.15	0.13	0.08	0.09	0.11
	σ	0.69	0.81	0.82	0.44	0.35	0.40
Shoulder retractors	Group average	0.24 ± 1.00	-0.16 ± 0.78	0.10 ± 0.80	0.00 ± 0.40	-0.02 ± 0.42	-0.11 ± 0.44
	Σ	0.36	0.26	0.21	0.11	0.12	0.16
	σ	0.79	0.70	0.75	0.37	0.39	0.40
Total cohort	Group average	0.11 ± 0.89	-0.10 ± 0.83	0.07 ± 0.92	-0.02 ± 0.44	-0.02 ± 0.41	-0.08 ± 0.43
	Σ	0.30	0.24	0.18	0.10	0.11	0.14
	σ	0.75	0.75	0.78	0.40	0.38	0.40

Source: Zhao, B. et al., *J. Appl. Clin. Med. Phys.*, 19, 17–24, 2018.

Note: Σ represents systematic error, and σ represents random error.

while longitudinal errors were smallest. Approximately 5%–10% of fractions were impacted by shifts greater than 5 mm; 0%–3% fractions required greater than 7 mm shifts. Rotational errors were small (less than 1°), with few setups requiring angle correction of 2° (7%–10%) or 3° (0.3%). It was also found that translational and rotational errors settled after the first two weeks and increased to first week levels during the final week of treatment, potentially reflecting the cumulative impact of anatomical changes across the treatment course. This time-dependent setup error may be correlated with patient weight loss as nearly all patients lost weight by the end of treatment. Nonetheless, the magnitude of systematic and random errors remained small across all time points. The authors acknowledged a relatively large random error and commented that it could be due to interplay between translational shifts with angle correction when using a 6DOF couch.[2] They further analyzed the correlation between rotations and linear shifts and found that there were moderate correlations between yaw and lateral shifts and between pitch and vertical shifts. Furthermore, the location of the image registration matching box varied with location of lesion being treated, causing differences in registration that may have elevated setup error.[5,7] Another component of their study included an assessment of shoulder immobilization methods. It was found that average shifts and errors were smaller with molded shoulder cushions versus shoulder retractors (Table 17.1). This indicated that a moldable cushion provided better setup than shoulder strips. The authors speculated that with the moldable cushion, patients were able to rest their arms and elbows at anchoring points, which made shoulder and neck position more reproducible.

The authors also analyzed intra-fraction motion data including a total of 596 treatments. As shown in Table 17.1, average motion and errors were small (less than 1 mm). Differences between the two shoulder restriction methods were not significant. The magnitude of error was not time-dependent across weeks of treatment. The authors mentioned that the majority of patients completed all treatments without interruption. Nonetheless, the range of error was larger in a few cases. The authors showed examples of the intra-fraction motion pattern: (1) smooth steady sinusoidal motion where slight baseline drift can be observed; treatment was continued since motion was natural respiratory motion of the patient, and (2) acute displacement due to coughing; therapists immediately stopped treatment and waited for the patient to settle and for respiratory motion to return to baseline. The patient position was then manually checked by the therapist before treatment was restarted. It was noted that during treatment, motion tracking using a composite ROI which included the face and neck to cover the entire treatment area was applied. Contrasting reports have shown that tracking a face ROI alone detects only small positional deflections (less than 1 mm) in surface-guided SRS.[25] The authors concluded that respiratory neck motion was more likely a primary source of motion during treatment. In terms of treatment time, average total treatment time with complete surface guidance was 21.6 ± 8.4 minutes, closely matching a standard 20-minute treatment time slot for H&N IMRT. Overall treatment time with a minimal mask was comparable to standard closed mask treatment times based on their clinical practice. A patient survey was performed to determine the patient's satisfaction and comfort level with the minimal mask immobilization. Out of a maximum score of 6, with 0 being low satisfaction/comfort and 6 being highest, the average score for mask comfort was 5.11 ± 0.81, average patient perception of secure immobilization was 5.21 ± 0.71, average patient confidence of continued mask tolerability was 5.63 ± 0.60, and average overall patient satisfaction score was 5.37 ± 1.30. Although patient-reported comfort appeared to be high, the authors did not conduct a direct comparison between minimal versus standard mask comfort because the authors wished to minimize patient bias toward higher comfort scores for the minimal mask after first experiencing a standard mask fit. The authors emphasized that the primary objective of this pilot study was to confirm feasibility of proposed minimal coverage immobilization for H&N patients normally confined by large uncomfortable long masks. The authors acknowledged that surface imaging was a secondary image guidance strategy for: (1) ensuring fidelity of initial patient setup, and (2) tracking patient motion during treatment. Although the setup accuracy was comparable to that of standard masks for patients with similar treatment anatomy,[2–7,41] surface guidance alone was not sufficient to ensure reproducible daily setup given the large translational random setup errors. Therefore, radiographic imaging is recommended to confirm initial setup.

17.2 CONCLUSION

In conclusion, SGRT for H&N treatments is not yet widely used in the clinic, but the investigations to date suggest promising results with SGRT for H&N treatments and ongoing studies are being conducted. In the clinical studies performed, surface guidance provided modest help in pretreatment setup and facilitated patient motion tracking during treatment. However, reported studies have not yet proven that surface guidance techniques alone are not enough to ensure reproducible daily setup, primarily due to deformation at the neck and shoulder region. Instead, surface guidance has shown good performance for intra-fraction motion tracking. This may be beneficial for H&N radiation therapy, for example, in hypofractionated treatment schemes where intra-fractional motion control is particularly important.

KEY POINTS

- Surface-guided H&N radiation therapy can be accomplished using an open-face mask to allow tracking of the skin surface as a surrogate for internal anatomy.

- Surface guidance with an open-face mask for H&N patients provides modest help during patient setup and good intra-fractional motion tracking.

- Selection of the ROI for surface matching is important to ensure good correlation between the external skin surface and internal anatomy.

- Open-face masks have been shown to provide greater patient comfort and reduced anxiety while maintaining the same level of accuracy as closed-face masks, particularly when used in conjunction with CBCT and 2D KV.

REFERENCES

1. American Cancer Society. Cancer Facts & Figures 2014. https://www.cancer.org/research/cancer-facts-statistics/all-cancer-facts-figures/cancer-facts-figures-2014.html (accessed December 1, 2019).
2. Den RB, Doemer A, Kubicek G, et al. Daily image guidance with cone-beam computed tomography for head-and-neck cancer intensity-modulated radiotherapy: A prospective study. *Int J Radiat Oncol Biol Phys.* 2010;76(5):1353–1359.
3. Pang PPE, Hendry J, Cheah SL, et al. An assessment of the magnitude of intra-fraction movement of head-and-neck IMRT cases and its implication on the action-level of the imaging protocol. *Radiother Oncol.* 2014;112(3):437–441.
4. Vaandering A, Lee JA, Renard L, et al. Evaluation of MVCT protocols for brain and head and neck tumor patients treated with helical tomotherapy. *Radiother Oncol.* 2009;93(1):50–56.

5. van Kranen S,van Beek S, Rasch C, van Herk M, Sonke JJ. Setup uncertainties of anatomical sub-regions in head-and-neck cancer patients after offline CBCT guidance. *Int J Radiat Oncol Biol Phys.* 2009;73(5):1566–1573.

6. Velec M, Waldron JN, O'Sullivan B, et al. Cone-beam CT assessment of interfraction and intrafraction setup error of two head-and-neck cancer thermoplastic masks. *Int J Radiat Oncol Biol Phys.* 2010;76(3):949–955.

7. Zhang L, Garden AS, Lo J, et al. Multiple regions-of-interest analysis of setup uncertainties for head-and-neck cancer radiotherapy. *Int J Radiat Oncol Biol Phys.* 2006;64(5):1559–1569.

8. Zeidan OA, Langen KM, Meeks SL, et al. Evaluation of image-guidance protocols in the treatment of head and neck cancers. *Int J Radiat Oncol Biol Phys.* 2007;67(3):670–677.

9. Den RB, Doemer A, Kubicek G, et al. Daily image guidance with cone-beam computed tomography for head-and-neck cancer intensity-modulated radiotherapy: A prospective study. *Int J Radiat Oncol Biol Phys.* 2010;76(5):1353–1359.

10. Barker JL, Garden AS, Ang KK, et al. Quantification of volumetric and geometric changes occurring during fractionated radiotherapy for head-and-neck cancer using an integrated CT/linear accelerator system. *Int J Radiat Oncol Biol Phys.* 2004;59(4):960–970.

11. Lee C, Langen KM, Lu W, et al. Evaluation of geometric changes of parotid glands during head and neck cancer radiotherapy using daily MVCT and automatic deformable registration. *Radiother Oncol.* 2008;89(1):81–88.

12. Ahn PH, Chen CC, Ahn AI, et al. Adaptive planning in intensity-modulated radiation therapy for head and neck cancers: Single-institution experience and clinical implications. *Int J Radiat Oncol Biol Phys.* 2011;80(3):677–685.

13. Schwartz DL, Garden AS, Shah SJ, et al. Adaptive radiotherapy for head and neck cancer—dosimetric results from a prospective clinical trial. *Radiother Oncol.* 2013;106(1):80–84.

14. Schwartz DL, Dong L. Adaptive radiation therapy for head and neck cancer—Can an old goal evolve into a new standard? *J Oncol.* 2011;2011.

15. Schwartz DL. Current progress in adaptive radiation therapy for head and neck cancer. *Curr Oncol Rep.* 2012;14(2):139–147.

16. Lee N, Chuang C, Quivey JM, et al. Skin toxicity due to intensity-modulated radiotherapy for head-and-neck carcinoma. *Int J Radiat Oncol Biol Phys.* 2002;53(3):630–637.

17. Qi ZY, Deng XW, Huang SM, et al. *In vivo* verification of superficial dose for head and neck treatments using intensity-modulated techniques. *Med Phys.* 2009;36(1):59–70.

18. Hsu SH, Roberson PL, Chen Y, et al. Assessment of skin dose for breast chest wall radiotherapy as a function of bolus material. *Phys Med Biol.* 2008;53(10):2593.

19. Chung H, Jin H, Dempsey JF, et al. Evaluation of surface and build-up region dose for intensity-modulated radiation therapy in head and neck cancer. *Med Phys.* 2005;32(8):2682–2689.

20. Velec M, Waldron JN, O'Sullivan B, et al. Cone-beam CT assessment of interfraction and intrafraction setup error of two head-and-neck cancer thermoplastic masks. *Int J Radiat Oncol Biol Phys.* 2010;76(3):949–955.

21. Sharp L, Lewin F, Johansson H, et al. Randomized trial on two types of thermoplastic masks for patient immobilization during radiation therapy for head-and-neck cancer. *Int J Radiat Oncol Biol Phys.* 2005;61(1):250–256.

22. Kim S, Akpati HC, Li JG, et al. An immobilization system for claustrophobic patients in head-and-neck intensity-modulated radiation therapy. *Int J Radiat Oncol Biol Phys.* 2004;59(5):1531–1539.

23. Pan H, Cerviño LI, Pawlicki T, et al. Frameless, real-time, surface imaging-guided radiosurgery: Clinical outcomes for brain metastases. *Neurosurgery.* 2012;71(4):844–852.

24. Peng JL, Kahler D, Li JG, et al. Characterization of a real-time surface image-guided stereotactic positioning system. *Med Phys.* 2010;37(10):5421–5433.

25. Cervino LI, Pawlicki T, Lawson JD, Jiang SB. Frame-less and mask-less cranial stereotactic radiosurgery: A feasibility study. *Phys Med Biol.* 2010;55(7):1863.

26. Li G, Ballangrud Å, Kuo LC, et al. Motion monitoring for cranial frameless stereotactic radiosurgery using video-based three-dimensional optical surface imaging. *Med Phys.* 2011;38(7):3981–3994.

27. Cerviño LI, Detorie N, Taylor M, et al. Initial clinical experience with a frameless and maskless stereotactic radiosurgery treatment. *Pract Radiat Oncol.* 2012;2(1):54–62.

28. Lau SK, Zakeri K, Zhao X, et al. Single-isocenter frameless volumetric modulated arc radiosurgery for multiple intracranial metastases. *Neurosurgery.* 2015;77(2):233–240.

29. Lau SK, Zhao X, Carmona R, et al. Frameless single-isocenter intensity modulated stereotactic radiosurgery for simultaneous treatment of multiple intracranial metastases. *Transl Cancer Res.* 2014;3(4):383.

30. Jamshidi P, Cerviño LI, Treiber JM, et al. Frameless, real-time surface imaging-guided radiosurgery system. *J Laparoendosc Adv Surg Tech Part B Videoscop.* 2014;24(2). https://doi.org/10.1089/vor.2013.0193.

31. Pham NLL, Reddy PV, Murphy JD, et al. Frameless, real-time, surface imaging-guided radiosurgery: Update on clinical outcomes for brain metastases. *Transl Cancer Res.* 2014;3(4):351–357.

32. Paravati AJ, Manger R, Nguyen JD, et al. Initial clinical experience with surface image guided (SIG) radiosurgery for trigeminal neuralgia. *Transl Cancer Res.* 2014;3(4):333–337.

33. Lau SK, Patel K, Kim T, et al. Clinical efficacy and safety of surface imaging guided radiosurgery (SIG-RS) in the treatment of benign skull base tumors. *J Neuro Oncol.* 2017;132(2):307–312.

34. Gopan O, Wu Q. Evaluation of the accuracy of a 3D surface imaging system for patient setup in head and neck cancer radiotherapy. *Int J Radiat Oncol Biol Phys.* 2012;84(2):547–552.

35. Kim Y, Li R, Na YH, Lee R, Xing L. Accuracy of surface registration compared to conventional volumetric registration in patient positioning for head-and-neck radiotherapy: A simulation study using patient data. *Med Phys.* 2014;41(12):121701-1–121701-7.
36. Kang H, Grelewicz Z, Wiersma R. Development of an automated region of interest selection method for 3D surface monitoring of head motion. *Med Phys.* 2012;39(6Part1):3270–3282.
37. Wiersma RD, Tomarken SL, Grelewicz Z, Belcher AH, Kang H. Spatial and temporal performance of 3D optical surface imaging for real-time head position tracking. *Med Phys.* 2013;40(11):111712.
38. Stieler F, Wenz F, Shi M, Lohr F. A novel surface imaging system for patient positioning and surveillance during radiotherapy. *Strahlenther Onkol.* 2013;189(11):938–944.
39. Stieler F, Wenz F, Scherrer D, Bernhardt M, Lohr F. Clinical evaluation of a commercial surface-imaging system for patient positioning in radiotherapy. *Strahlenther Onkol.* 2012;188(12):1080–1084.
40. Li G, Lovelock DM, Mechalakos J, et al. Migration from full-head mask to "open-face" mask for immobilization of patients with head and neck cancer. *J Appl Clin Med Phys.* 2013;14(5):243–254.
41. Wiant D, Squire S, Liu H, et al. A prospective evaluation of open face masks for head and neck radiation therapy. *Pract Radiat Oncol.* 2016;6(6):e259–e267.
42. Kwa SLS, Al-Mamgani A, Osman SO, et al. Inter- and intrafraction target motion in highly focused single vocal cord irradiation of T1a larynx cancer patients. *Int J Radiat Oncol Biol Phys.* 2015;93(1):190–195.
43. Lu H, Lin H, Feng G, et al. Interfractional and intrafractional errors assessed by daily cone-beam computed tomography in nasopharyngeal carcinoma treated with intensity-modulated radiation therapy: A prospective study. *J Radiat Res.* 2012;53(6):954–960.
44. Zhao B, Maquilan G, Jiang S, Schwartz DL. Minimal mask immobilization with optical surface guidance for head and neck radiotherapy. *J Appl Clin Med Phys.* 2018;19(1):17–24.
45. Van Herk M. Errors and margins in radiotherapy. *Semin Radiat Oncol.* 2004;14(1):52–64.

CHAPTER 18

Surface Image Guidance for Treatment of Extremities

David P. Gierga

CONTENTS

18.1 INTRODUCTION

Surface image guidance has become widely utilized for a wide variety of treatment approaches and anatomical sites, including breast (e.g., accelerated partial breast irradiation and deep inspiration breath hold),[1-4] stereotactic radiosurgery (SRS),[5,6] and hypofractionated stereotactic body radiotherapy (SBRT).[7] As discussed in other chapters, surface imaging allows for three-dimensional (3D) image registration and continuous position monitoring without any additional radiation dose. This chapter will explore the utility of surface-guided radiation therapy (SGRT) for treatment of extremities.

Soft-tissue sarcomas are relatively rare, with an estimated 10,000 new cases per year in the United States.[8] Most occur in the lower extremity (45%), but about 15% are upper extremity cases. Treatment may consist of either 3D conformal radiation therapy (3D-CRT) or intensity modulated radiation therapy (IMRT), utilized in either a preoperative or postoperative setting.[9] Sarcoma extremity patients are often immobilized with custom-made vacuum bags, cushions, or thermoplastic holders. Figure 18.1 shows an example of a custom-made patient immobilization system. Efficient, accurate, and precise patient setup with these systems may be particularly challenging, given the potential for limbs to pivot about the joint, soft tissue deformation, and changes in the surface caused by tumor or normal tissue radiation response, and the lack of standardized immobilization options. Surface imaging, therefore, can be a useful imaging modality for these patients and can be used to complement standard techniques such as planar megavoltage (MV), planar kilovoltage (kV), or volumetric cone-beam CT (CBCT) imaging.

There is limited data in the literature describing the application of 3D surface imaging for extremity patient setup. However, a few published studies using both the AlignRT system (Vision RT, Inc, London, UK) and the Catalyst HD (C-RAD, Uppsala, Sweden) are available in the literature.[10-12] This chapter will provide an overview of SGRT for extremity patients in the context of assessing intra- and inter-fraction motion, calculating planning target volume (PTV) margins and aiding in patient setup. Practical considerations for clinical deployment will also be explored.

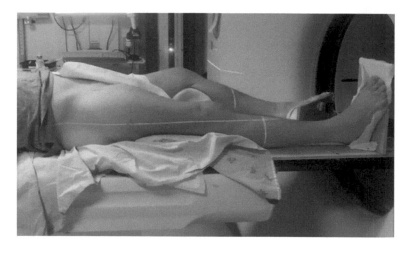

FIGURE 18.1   Example of a sarcoma extremity custom immobilization.

18.2 INTRA-FRACTION MOTION

Setup errors are often quantified in terms of random and systematic errors.[13,14] In a patient cohort, for a given translation direction, S_p is defined as the mean setup error for each patient, and the mean of all S_p values, over all patients, is the group mean, μ. The standard deviation of μ is Σ, and both μ and Σ quantify systematic errors. In addition, if σ_p is the standard deviation of the translation variations for each patient, then the average of all σ_p, or σ, quantifies the random errors.

Dickie et al.[15] examined intra-fraction errors for lower extremity soft tissue sarcomas using a combination of both pre- and postfraction CBCT scans and real-time tracking with an optical localization system (OLS). In this implementation, the OLS consisted of infrared cameras that monitored fixed reflective markers placed on the patient's surface. Using CBCT, the standard deviation of the intra-fraction systematic errors for translations was determined to be 0.2–0.6 mm, while random errors were 1.3–1.6 mm. Using the OLS, intra-fraction systematic errors for translations were determined to be 0.3–0.6 mm, while random errors were 0.5–0.6 mm. Gierga et al.[10] examined intra-fraction motion for sarcoma extremity patients using AlignRT. In this study, the group mean (μ) systematic errors ranged from −0.9 to 0.2 mm, and the standard deviation (Σ) ranged from 0.6 to 0.8 mm. Random errors (σ) were 1.1 mm in each direction.

The values from Dickie (using point-based optical surface imaging) and Gierga (using 3D surface imaging) are similar and indicate that intra-fraction errors are on the order of 1 mm or less. Both studies utilized in-house custom immobilization devices or commercially available vacuum bags customized per patient. These data indicate that the use of these techniques can limit intra-fractional motion even for sarcoma extremity patients with nonstandard or difficult setups. Since approaches to immobilization for extremity patients may vary, it is important to verify the effectiveness of a selected technique. Surface imaging is a useful tool for these studies since patients can be monitored continuously without any additional radiation dose.

18.3 INTER-FRACTION MOTION

Surface imaging has also been utilized to examine inter-fraction setup errors for extremity patients. Gierga et al.[10] quantified setup errors using surface imaging in the context of weekly image-guided radiation therapy (IGRT). The study included 236 images for 16 patients, with tumors mostly in the thigh, leg, or arms. Inter-fraction setup errors were quantified

TABLE 18.1   Inter-fraction Setup Errors for Sarcoma Extremity Patients Setup with Weekly MV Imaging

	CT-Based Reference Surface			Reference Surface at First Fraction		
	Systematic Errors (mm)		Random Errors (mm)	Systematic Errors (mm)		Random Errors (mm)
Direction	Mean (μ)	SD (Σ)	σ	Mean (μ)	SD (Σ)	σ
Ant/post	1.3	3.6	3.1	2.9	3.3	3.0
Sup/inf	−0.4	7.9	4.6	−0.6	3.7	4.3
Lateral	0.3	4.1	3.3	0.4	4.3	2.8
3D	9.5	5.1	4.1	7.6	3.9	4.2

Note: Group mean error μ, systematic error Σ, and random error σ, in mm. Data presented for both DICOM CT-based reference and reference image captured directly with surface imaging at the first fraction.

using both a digital imaging and communications in medicine (DICOM) CT-based reference image and images captured directly with surface imaging at the first fraction. The results are summarized in Table 18.1. Random and systematic errors are approximately 3–5 mm in each direction for both types of reference image. The systematic error in the longitudinal direction was 3.7 mm for the surface reference image compared to 7.9 mm for the CT DICOM reference image. The overall 3D error was 7.6 mm for the surface reference image compared to 9.5 mm for the CT DICOM reference image. Differences in setup errors depending on the type of reference surface have also been observed for other treatment sites, including brain,[5] and breast.[1] It should also be noted that these data were captured using an early version of the AlignRT software and improvements in registration relative to DICOM images may since been implemented. Systematic differences, however, could have been present at the treatment unit given the difficulty in replicating the simulation geometry for extremity setups. These may not have been detected by planar MV imaging, the "ground truth" method used in this study. Lastly, the longitudinal direction may be most susceptible to registration degeneracies, since it may be difficult to register regions of interest (ROI) for cylinder-like volumes in the leg or thigh. The design of ROIs for extremity patients will be discussed in greater detail below.

These data indicate that relatively large weekly setup variations can be present when using weekly IGRT for extremity patients, and that surface imaging can likely detect these errors.

Currently, a wider range of technologies are available for IGRT. Nabavizadeh et al.[16] reported the utilization rates of various IGRT options based on survey data, with the most common techniques including volumetric imaging (CBCT or MVCT), kV planar, or MV planar imaging. Three-dimensional surface imaging was available in 12% of the reporting treatment centers. The utilization of daily IGRT (any modality) was studied by treatment site and ranged from 18% for brain (3D or IMRT) to 96% for intact prostate IMRT. No data were reported for extremity patients, although it could be assumed that, given the complexity of extremity patient setup, the utility of daily IGRT is now relatively high.

Stanley et al.[11] compared the CBCT-reported patient setup corrections for initial positioning with either lasers and tattoos and 3D surface imaging. This study analyzed 6000 fractions for patients from four treatment sites: pelvis/lower extremities, abdomen, chest/upper extremities, and breast. Data were not reported separately for extremity patients. The distribution of initial setup corrections (3D shift vectors for pre-CBCT setup) for laser/tattoo-based setup and surface imaging-based setup for the two patient groups that included extremity patients are included in Figures 18.2 and 18.3.

Average residual CBCT corrections after setup with either method are given in Table 18.2. Statistical differences were observed for all sites. Finally, Reitz et al.[12] examined 1902 fractions for 110 patients, for a variety

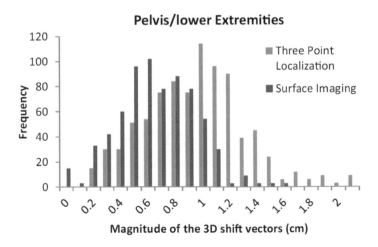

FIGURE 18.2 Cumulative histograms for pelvis and lower extremities showing 3D corrections for laser/tattoo-based setup (three-point localization) and surface imaging. (Reprinted from Stanley, D.N. et al., *J. Appl. Clin. Med. Phys.*, 18, 58–61, 2017.)

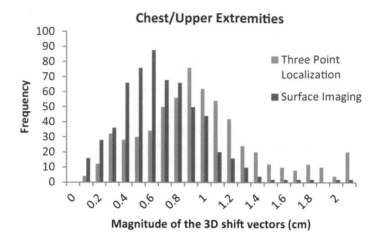

FIGURE 18.3   Cumulative histograms for chest and upper extremities showing 3D corrections for laser/tattoo-based setup (three-point localization) and surface imaging. (Reprinted from Stanley, D.N. et al., *J. Appl. Clin. Med. Phys.*, 18, 58–61, 2017.)

TABLE 18.2   Residual Setup Corrections Detected by CBCT after Either Laser/Tattoo-Based Setup (Three-Point Localization) or Surface Imaging

	Three-Point Localization		Surface Imaging	
	Average (mm)	σ (mm)	Average (mm)	σ (mm)
Pelvis/lower extremities	9	4	6	3
Chest/upper extremities	9	6	5	3

Source: Stanley, D.N. et al., *J. Appl. Clin. Med. Phys.*, 18, 58–61, 2017.

of body sites, including 123 fractions for extremity patients. In this study, patients were initially set up with skin marks and lasers, followed by surface imaging and CBCT. Setup corrections for extremity patients using laser and surface imaging generally agreed within ±2 mm.

18.4 TREATMENT PLANNING CONSIDERATIONS

Setup accuracy and precision inform the treatment planning process. The margin recipe from Van Herk et al.[17] provides a mechanism for calculating PTV margins for a given patient population. Gierga et al.[10] calculated PTV margins for sarcoma extremity patients based on inter-fraction setup errors using the reference surface captured at the first fraction. Only translational errors were considered, and rotational errors were ignored. Weekly MV images were obtained to verify patient positioning and setup corrections were not made based on surface imaging. The calculated PTV margins were 10 mm (anterior-posterior), 12 mm (superior-inferior), and 13 mm (lateral). Since clinical PTV margins likely range from 5 to 10 mm and should account for additional uncertainties in the planning or delivery process, it can be inferred that insufficient PTV margins were utilized in the context of weekly imaging and that daily image guidance is recommended. It should also be noted, however, that sarcoma extremity patients may have large GTV to CTV margins as the entire muscle compartment is included in the CTV, so an inherent safety margin may exist outside of the standard PTV expansion.

The dosimetric impact of setup errors for sarcoma extremity patients was analyzed by Arthurs et al.[18] They applied random and systematic errors to patients with lower limb sarcomas and analyzed the resulting dosimetric differences for both 3D-CRT and IMRT plans. Random errors between 1.9 and 2.9 mm, depending on direction, were applied for five fractions. Systematic errors between 1.3 and 2.0 mm were applied for three fractions, based on an offline setup protocol. Values for both random and systematic errors were based on local clinical population data. The total number of fractions per treatment course was 30–33 fractions, and CTV-PTV margin expansions of 10 mm were used, with the PTV clipped within 5 mm of the skin. Arthurs et al. concluded that systematic errors had only minimal effects on IMRT plans and no statistically significant effect on 3D-CRT plans, while random errors degraded target homogeneity for IMRT plans but had no significant effect on 3D-CRT plans. It should be noted that errors were simulated for a minimal number of fractions, and the authors recommend daily imaging to maintain target homogeneity for IMRT plans.

18.5 QUALITY ASSURANCE CONSIDERATIONS

While surface imaging has been shown to be capable of submillimeter precision in some applications,[19] it can also be a tool to detect and correct gross setup errors. Although they may be present in any patient population, these errors may be exacerbated by the unique patient geometry and immobilizations used for extremity patients. Other contributing factors could include errors in data transfer from treatment planning to delivery, unclear documentation of setup instructions, or incorrect reference images. Figure 18.4 shows histograms of the root-mean-square (RMS) errors of the ROI calculated by AlignRT of extremity sarcoma patients (data adapted from Gierga et al.[10]). The uncorrected data are for laser/tattoo setup, while the 6 degrees-of-freedom (6DOF) data show the residual RMS error after the translation and rotation corrections calculated by

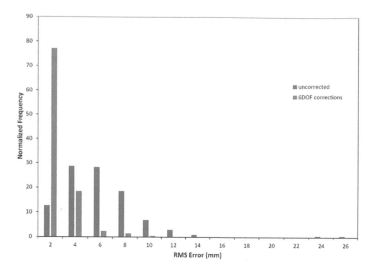

FIGURE 18.4   Root-mean-square (RMS) error histograms for uncorrected and 6 degrees-of-freedom (6DOF) corrections with surface imaging of extremity sarcoma patients.

surface imaging have been applied. Clearly, the distribution of errors is smaller for the surface imaging data. It is also quite apparent that large errors were present in the uncorrected data. Similar trends are shown in Figures 3–4 from Stanley et al.,[11] which show a significant percentage of 3D vector errors for laser-based setup approaching 2 cm.

Surface imaging may also have value in monitoring patient shape changes over the course of radiation therapy. Surface metrics (e.g., ROI RMS differences) could also be used to monitor response to therapy, detect weight loss or gain, or trigger replanning in adaptive therapy strategies. Additional studies are needed to explore these applications in greater depth.

18.6 PRACTICAL CONSIDERATIONS

It has been shown that the accuracy of surface image registration can be sensitive to the selected ROI.[5,20,21] This is also an important consideration for extremity patients, as surface fitting degeneracies may affect the calculated translations or rotations for sites that are "cylinder-like," e.g., upper or lower leg or thigh tumors, and lack distinguishing features in the long axis. Special care should be taken to ensure that the selected ROI includes enough structure or features to "anchor" the surface matching. See Figure 18.5 for example reference surface images and

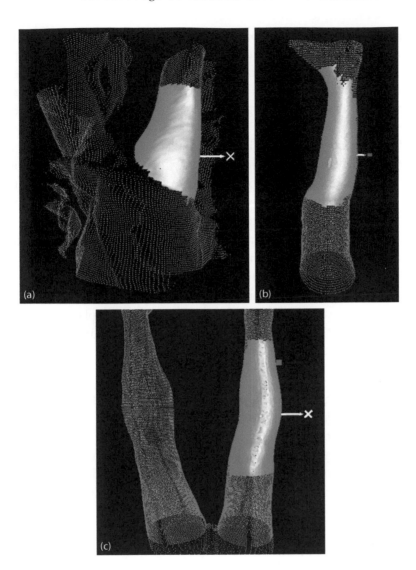

FIGURE 18.5 Examples of AlignRT reference surface images with regions of interest indicated by the solid areas. Note that (a) was captured directly at the treatment machine using the surface imaging camera system, but (b) and (c) are CT-based reference surface images.

ROI. Although not shown in the figure, multiple ROIs can also be utilized to assist with positioning limbs near pivot points, such as the upper arm and forearm around the elbow joint.

Extremity patients with superficial tumors often require bolus. Conventional SuperFlab bolus (Civco Medical Solutions, Coralville, IA) or brass mesh bolus[22] may lead to excessive reflection and degrade image quality or create image artifacts with surface imaging. Simple solutions such as spray-painting the bolus material to dull the surface can be employed. Recently, brass mesh bolus with a white paint pre-applied became commercially available (Radiation Products Design, Inc, Albertville, MN). The use of wet-towel or 3D-printed bolus materials[23] may also help minimize these effects.

The frequency of surface imaging is also an important practical consideration. For patients with weekly MV or kV IGRT, the addition of daily surface imaging is recommended to increase positioning accuracy and precision. For patients with daily CBCT, daily surface imaging can be helpful as a complement to increase the efficiency of CBCT setup (i.e., minimize CBCT shifts and reimaging) or as a safety measure (to guard against gross setup errors). In addition, a subset of extremity patients may have tumors where the soft tissue is a better surrogate than bone, which may indicate added value from surface imaging. When selecting an IGRT strategy for any clinical site, it is important to consider planning margins and techniques, clinical workload, and technology resources.

18.7 CONCLUSIONS

Surface imaging has been applied to a wide variety of anatomical sites including extremities. Surface imaging can be utilized to monitor intra- and inter-fraction setup errors and establish or confirm PTV margins. Large setup discrepancies have been observed for extremity patients set up without IGRT, and surface imaging can play a role in detecting or minimizing these errors. Daily IGRT is recommended to maximize accuracy and precision for extremity patients.

KEY POINTS

- Radiation therapy of the extremities presents a unique set of challenges to accurate and reproducible setup, including the potential for limbs to pivot about the joint, soft tissue deformation, tumor or normal tissue radiation response, and the lack of standardized immobilization options.

- Surface imaging can be utilized to monitor inter-fraction setup errors and intra-fraction motion and to establish or confirm PTV margins.

- The shape of extremities and potential absence of features can pose a challenge for surface guidance, and care must be taken when designing ROIs to account for surface fitting degeneracies.

- Surface imaging can complement standard IGRT techniques by improving initial setup and reducing the need for repeat imaging. Daily IGRT is recommended to maximize accuracy and precision for extremity patients.

ACKNOWLEDGMENTS

The author thanks Karen De Amorim Bernstein, MD; Yen-Lin E. Chen, MD; Natasha M. Colaco-Burge, RTT; Thomas F. DeLaney, MD; Drosoula Giantsoudi, PhD; Kyung-Wook Jee, PhD; Tiffany Koolakian, MBA, RTT; Long W. Tong, MS; and Julie C. Turcotte, MS, for their contributions to this work.

REFERENCES

1. Gierga DP, Riboldi M, Turcotte JC, et al. Comparison of target registration errors for multiple image-guided techniques in accelerated partial breast irradiation. *Int J Radiat Oncol Biol Phys.* 2008;70:1239–1246.
2. Chang AJ, Zhao H, Moore K, et al. Video surface image guidance for external beam partial breast irradiation. *Pract Radiat Oncol.* 2012;2:97–105.
3. Gierga DP, Turcotte JC, Sharp GC, et al. A voluntary breath hold technique for left breast with unfavorable cardiac anatomy using surface imaging. *Int J Radiat Oncol Biol Phys.* 2012;84:e663–e668.
4. Tang X, Zagar TM, Bair E, et al. Clinical experience with 3-dimensional surface matching-based deep inspiration breath hold for left-sided breast cancer radiation therapy. *Pract Radiat Oncol.* 2014;4:e151–e158.
5. Peng JL, Kahler D, Li JG, et al. Characterization of a real-time surface image-guided stereotactic positioning system. *Med Phys.* 2010;37:5421–5433.
6. Cervino LI, Pawlicki T, Lawson JD, et al. Frame-less and mask-less cranial stereotactic radiosurgery: A feasibility study. *Phys Med Biol.* 2010;55:1863–1873.
7. Alderliesten T, Sonke J, Betgen A, et al. 3D surface imaging for monitoring intrafraction motion in frameless stereotactic body radiotherapy of lung cancer. *Radiother Oncol.* 2012;105:155–160.
8. Lee B, Tsuji SY, Gottschalk AR. Soft tissue sarcoma. In: Hansen EK, Roach M, eds. *Handbook of Evidence-Based Radiation Oncology.* 2nd ed. New York: Springer; 2010: 615–628.
9. Haas RL, Delaney TF, O'Sullivan B, et al. Radiotherapy for management of extremity soft tissue sarcomas: Why, when, and where? *Int J Radiat Oncol Biol Phys.* 2012;84:572–580.
10. Gierga DP, Turcotte JC, Tong LW, Chen YL, DeLaney TF. Analysis of setup uncertainties for extremity sarcoma patients using surface imaging. *Pract Radiat Oncol.* 2014;4:261–266.
11. Stanley DN, McConnell KA, Kirby N, et al. Comparison of initial patient setup accuracy between surface imaging and three point localization: A retrospective analysis. *J Appl Clin Med Phys.* 2017;18(6):58–61.

12. Reitz D, Carl G, Schonecker S, et al. Real time optical surface IGRT: A mono-institutional prospective study of 110 patients. *Radiother Oncol.* 2018;127:S1128–S1129.

13. de Boer HC, van Sornsen de Koste JR, Creutzberg CL, et al. Electronic portal image assisted reduction of systematic set-up errors in head and neck irradiation. *Radiother Oncol.* 2001;61:299–308.

14. Bijhold J, Lebesque JV, Hart AAM, et al. Maximizing set-up accuracy using portal images as applied to a conformal boost technique for prostatic cancer. *Radiother Oncol.* 1992;24:261–271.

15. Dickie CI, Parent AL, Chung PW, et al. Measuring interfractional and intrafractional motion with cone beam computed tomography and an optical localization system for lower extremity soft tissue sarcoma patients treated with preoperative intensity-modulated radiation therapy. *Int J Radiat Oncol Biol Phys.* 2010;78:1437–1444.

16. Nabavizadeh N, Elliott DA, Chen T, et al. Image guided radiation therapy (IGRT) practice patterns and IGRT's impact on workflow and treatment planning: Results from a National Survey of American Society for Radiation Oncology Members. *Int J Radiat Oncol Biol Phys.* 2016;94:850–857.

17. van Herk M, Remeijer P, Rasch C, et al. The probability of correct target dosage: Dose-population histograms for deriving treatment margins in radiotherapy. *Int J Radiat Oncol Biol Phys.* 2000;47:1121–1135.

18. Arthurs M, Gillham C, O'Shea E, McCrickard E, Leech M. Dosimetric comparison of 3-dimensional conformal radiation therapy and intensity modulated radiation therapy and impact of setup errors in lower limb sarcoma radiation therapy. *Pract Radiat Oncol.* 2016;6:119–125.

19. Bert C, Metheany KG, Doppke K, et al. A phantom evaluation of a stereovision surface imaging system for radiotherapy patient setup. *Med Phys.* 2005;32:2753–2762.

20. Guo B, Shah CS, Magnelli A, et al. Surface guided radiation therapy (SGRT): The sensitivity of the region of interest (ROI) selection on the translational and rotational accuracy for whole breast irradiation. *Int J Radiat Oncol Biol Phys.* 2017;99(2):S1:E666–E667.

21. Zhao B, Maquilan G, Jiang SB, Schwartz DL. Minimal mask immobilization with optical surface guidance for head and neck radiotherapy. *J Appl Clin Med Phys.* 2018;19(1):17–24.

22. Healy E, Anderson S, Cui J, et al. Skin dose effects of postmastectomy chest wall radiation therapy using brass mesh as an alternative to tissue equivalent bolus. *Pract Radiat Oncol.* 2013;3:e45–53.

23. Robar JL Moran K, Allan J, et al. Intrapatient study comparing 3D printed bolus versus standard vinyl gel sheet bolus for postmastectomy chest wall radiation therapy. *Pract Radiat Oncol.* 2018;8:221–229.

CHAPTER 19

Application of SGRT in Pediatric Patients

The CHLA Experience

Arthur J. Olch, Alisha Chlebik, and Kenneth Wong

CONTENTS

19.1 INTRODUCTION

In 2011, when Children's Hospital Los Angeles (CHLA) was embarking on a linear accelerator (linac) replacement, the field of surface-guided radiation therapy (SGRT) was new and few centers had surface imaging systems. During our evaluation of the technology at that time, it became apparent that surface imaging could be particularly useful in the pediatric setting. CHLA moved forward with this technology and implemented SGRT in September 2013 with the introduction of an AlignRT (Vision RT, Ltd., London, UK) surface imaging system along with a Varian TrueBeam equipped with a PerfectPitch 6 degrees-of-freedom (6DOF) couch (Varian Medical Systems, Inc. Palo Alto, CA, USA).

The radiation therapy department of CHLA exclusively treats children, adolescents, and young adults. On average, 10 patients per day are treated with about a third requiring sedation for treatment. More than half of the patients are treated with intensity modulated radiation therapy (IMRT) or volumetric modulated arc therapy (VMAT). SGRT is currently utilized for nearly all treatment sites including brain (both stereotactic radiosurgery (SRS) and conventional fractionation), head and neck (H&N), chest (including deep inspiration breath hold), abdomen, pelvis, extremities, craniospinal irradiation (CSI), and for setup of most and monitoring of some electron treatments in cases where a region of visible skin is present around the applicator. CHLA is now a completely skin mark-less center and has been so since the fourth quarter of 2014. Although it took about a year for the SGRT process to mature, surface imaging has now come to be an essential and fully integrated part of the treatment setup, verification, and monitoring process.

19.2 BENEFITS OF SGRT FOR PEDIATRIC PATIENTS

19.2.1 Reduction of Setup Time and X-Ray Imaging

Over the past several years of clinical use, our initial vision of the potential benefits of SGRT for pediatric patients has largely been realized. With real-time information about the level of agreement between the patient's pose at treatment setup compared to simulation, many setup errors can be corrected before the first round of X-ray-based setup verification images are acquired as part of image-guided radiation therapy (IGRT). This ability has decreased the usage of ionizing radiation for patient positioning as well as the total time for setup and imaging. In particular, for treatment of long segments of the spine, the ability of SGRT to detect both translational and rotational shifts has proven to be much more useful compared with what skin marks alone provided. Quicker adjustments to the patient's position can be made due to the ability with surface imaging to adjust several directions at once, such as correcting both the lateral position, twist, and roll. These multidimensional adjustments are much more difficult when relaying on just external lasers and skin marks. Reductions in both setup time and X-ray image-related dose are also beneficial when treating children.

19.2.2 Elimination of All Skin Marks

Prior to implementing SGRT at CHLA, a three-point setup with skin marking of the isocenter to aid in patient positioning was commonly used, although CHLA practice was to never permanently tattoo patients. In addition to the three-point setup, full lines drawn on the skin were used to provide better information about rotations. CHLA setup accuracy policy requires less than 3 mm accuracy for body treatments and less than 2 mm for head and neck treatments after adjustments are made with image guidance. With the introduction of the TrueBeam linac and onboard cone-beam CT (CBCT) imaging capability, even tighter limits are achievable, especially with a 6DOF couch. With uncorrected rotations, the operator must split the difference when correcting with only couch translations, and this can make it hard to achieve the tight setup tolerances required for some treatment techniques. When SGRT was first introduced at CHLA, the use of regular skin marks was continued for a period of about 6 months while experience was gained with the surface imaging system. The initial setup was to skin marks and evaluated with SGRT but no shifts were made based on SGRT alone. Image guidance, mostly low-dose CBCT, was used to correct positional changes. After this period of comparing SGRT and X-ray-based IGRT, the transition was made to a skin mark-less setup.[1]

Prior to SGRT, keeping skin marks on children throughout the course of treatment was challenging. Permanent felt-tip markers were used along with transparent medical dressings to prolong the lifetime of marks. However, children would lose their marks due to sweating with vigorous play, bathing, and normal fading. Sometimes, children would remove marks because they did not like how visible marks on their skin indicated that they were different. Teenagers, in particular females, were the most reluctant to keep skin markings. For those patients attending school, this could be particularly stressful, and often reported being upset when they realized that the skin marks could not be removed for the duration of the treatment course. To accommodate the patients' concerns, skin marks were limited to areas that could be covered with clothing, but this was not always possible due to treatment setup requirements. For patients with central intravenous lines, weekly dressing changes could also affect nearby skin marks. In instances where the marks were lost or degraded, the patient would have to be set up with limited or no marks, requiring additional imaging and dose, as well as longer setup time. With the implementation of SGRT, the stress and difficulties caused by skin marks can be removed for both patients and staff.

19.2.3 Real-Time Patient Monitoring for Increased Safety

Real-time intra-fraction motion and position monitoring is a unique feature of SGRT systems in that it provides information about patient position that is not readily available unless using X-ray fluoroscopy. Awake children are more likely to move during treatment than adults. This makes continuous monitoring with noninvasive, nonionizing methods during treatment appealing. Without SGRT, treatment staff can only rely on closed-circuit television images where at best a human operator is capable of visually detecting patient movement during treatment on the order of 1 cm. If movement is detected, the radiation therapist must manually interrupt the beam. With a direct hardware interface between the SGRT system and the linac, treatment can be automatically paused when movement is detected that exceeds a predefined threshold. While 7 years old had been our expected sedation age prior to the introduction of SGRT, now children as young as 4 years old can be treated without anesthesia when SGRT monitoring is added to the treatment process. In many cases, children will move, pausing treatment, but will settle back down and treatment will automatically resume as soon as their position returns to within the set tolerances. Having a system to automatically detect this action gives confidence that the treatment position is correct without requiring additional imaging with X-ray-based modalities. In addition, real-time intra-fraction motion and position monitoring with SGRT permits treatment without anesthesia for patients that were otherwise deemed to be marginally compliant.

19.2.4 Safe and Efficient Treatment of Palliative Patients

Patients being treated with a palliative intent often cannot be precisely immobilized due to pain or other reasons. Furthermore, anesthesia is not always feasible with these patients. The real-time information provided by SGRT allows the patient's setup position to be more quickly approximated while therapists are in the treatment room. Additionally, if the patient moves during treatment, the beam will automatically hold until the body pose returns to the planned position within the set tolerances for the treatment technique. After the implementation of SGRT, non-immobilized palliative patients who could not be safely treated were converted into safely treatable patients, even without sedation. For example, several children under the age of 5 that were nonideal candidates for anesthesia due to large mediastinal masses were able to be treated awake with SGRT. Previously, this would not have been feasible due to the young age of the patients. With SGRT, however, careful monitoring for movement during treatment assures the treatment team that the patient remains in the correct treatment position.

19.3 EFFECTIVE APPLICATION OF SGRT BY TREATMENT SITE

Effective use of SGRT requires a sufficiently large unimpeded region of interest (ROI) with sufficient contour variation to allow unambiguous surface matching. Children are treated awake if they meet minimum age and maturity requirements. Otherwise, they are sedated. Awake children can be more resistant than adults to being uncovered for treatment, which is necessary for full skin surface viewing by the SGRT system. Sedated children often have central intravenous lines and other devices connected to their skin surface within or in close proximity to the ROI. The following sections of the chapter describe practices for applying SGRT by treatment site.

19.3.1 General Patient Setup and Monitoring

The general SGRT workflow begins with the patient being brought into the treatment room and positioned using their custom immobilization and indexing systems. The treatment couch is brought to the required position, and the SGRT reference surface obtained from the CT simulation and treatment planning system (i.e., the DICOM external body contour) is used for initial patient positioning. In some cases, an SGRT surface that was acquired during a previous treatment fraction is used for initial setup. This would occur if the patient had gained or lost weight or other features of the skin surface had changed since the CT simulation. An SGRT setup tolerance of 2 mm for vertical, longitudinal, and lateral position, 1° for rotation and roll, and 2° for pitch is typically used. The larger tolerance for pitch is because pitch is more difficult to correct in many cases. For most patients, it is attempted to get the initial setup as close as possible to having zero displacement between the patient's position as determined by SGRT and the required position. However, in most cases, some residual differences will persist but, if small enough, are acceptable. Any small residual setup differences will be corrected with additional couch shifts during the IGRT process.

After the desired setup is achieved based on SGRT, a new monitoring reference surface is captured before the therapists leave the room for verification imaging and delivery. This is because of the concern that the patient may move after staff leave the room but before commencing IGRT, leading to a posture that is uncorrectable with couch position shifts only (especially rotations) after CBCT. There is also the possibility that the patient can move between the time of the CBCT acquisition and the time when image analysis and registration is complete, and any couch shifts are executed. Acquiring a new reference image (with zero difference from the planned position) immediately before leaving the room provides the therapists an easy way to notice any such positional changes as opposed to working with various residual offsets. It should be noted that in this process, even though residual positional offsets are within tolerance when staff leave the room, patient movement during or after CBCT could move from just under tolerance on one end of the scale to just under on the opposite end of the scale, meaning a patient shift of nearly twice the tolerance could occur but go unnoticed. Before applying any IGRT-based couch shifts, the patient position is verified again with SGRT to still be within tolerance. A new monitoring reference surface is captured immediately after applying couch shifts to monitor the patient during treatment.

19.3.2 Head and Neck

For patients being treated in the H&N or brain at CHLA, a vacuum-assisted mouthpiece-based head immobilization system called HeadFIX (Elekta, Inc., Atlanta, GA, USA) is used with a custom head mold affixed to a kVue Portrait couch top (Qfix, Avondale, PA, USA), a couch extension similar to the S-frame. For sites inferior to the neck, we use a Vac-Lok bag (CIVCO Medical Solutions, Coralville, IA, USA) indexed to the kVue DoseMax (Qfix, Avondale, PA, USA) couch top. This immobilization provides daily setup reproducibility within approximately 1 mm in any direction. Because of the quality of this immobilization, SGRT is not routinely used for setup and monitoring of brain and most H&N treatments.

If the target volume comprises large regions of the neck including at the shoulder level, SGRT is used for reproducing the shoulder and neck position. If neck positioning is a concern, an ROI can also be drawn around the neck, excluding the mandible. For patients requiring accurate mandibular positioning, SGRT is a helpful tool for reproducibility. For these cases, the ROI should be drawn over the mandible. Applying these concepts should reduce the amount of imaging necessary for correcting a patient setup.

19.3.3 Brain (SRS and SRT)

For either SRS or fractionated stereotactic radiation therapy (SRT) of the brain, SGRT is used in addition to the HeadFix immobilization system due to the smaller tolerance for inter-fraction motion compared to regularly fractionated treatments. Prior to the start of treatment, a monitoring reference surface is captured, followed by CBCT acquisition and application of any IGRT-based couch shifts. A new reference surface image is captured immediately after the CBCT-based shifts are completed. The couch is rotated to each planned couch angle, and a surface is captured with the gantry at zero degrees. After all couch angles are captured, the couch is brought back to the initial CBCT position and the patient's surface is verified to still be close to zero difference from the planned setup position. During treatment, patient movement from one couch angle to the next or during any beam is assessed based on the monitoring captures taken at each couch angle. Because of the geometry of the HeadFIX head immobilization system, the mouth, chin, and the region just below the eyes is blocked from the view of the SGRT cameras. The ROI covers the forehead and temples on each side. This ROI is visible to the surface imaging cameras and is large enough to provide a reasonably stable determination of the difference between the patient's actual position and the required position, even when one camera pod is blocked by the gantry.

19.3.4 Chest

Many patients with targets in the chest have a central intravenous line or other catheters in place for various reasons including delivery of chemotherapy. This tubing and related connections can interfere with chest surface ROIs. If intravenous lines are present, the ROI is adjusted to exclude these lines (Figure 19.1). The ROI is extended laterally to at least the mid-depth of the patient, but care is taken to not include the immobilization bag. The bags only extend to mid-depth to allow more visible skin (previous to SGRT

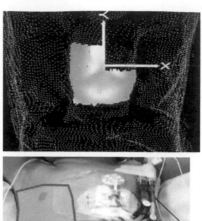
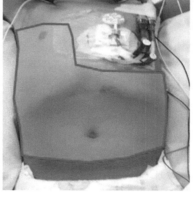

FIGURE 19.1   The surface region of interest (top) is designed to avoid tubing and central intravenous lines present on the patient's skin (bottom).

the practice was to extend them as high as possible). SGRT can be used for DIBH patients as young as 7 years old who have been carefully coached both at simulation and each treatment day. For somewhat older patients, a visual coaching device is helpful. Children who do not speak English but are old enough to understand the concept can be treated using DIBH and the coaching device. For female pre-teens or teenagers that are uncomfortable with being completely uncovered, we have used paper towels cut in small circles (about 6 cm diameter) to give more privacy. These are taped down on at least two edges to avoid migration into the ROI during setup or treatment. These must be excluded from the ROI. At CHLA, the linac room is kept warm (75°F or 24°C) to avoid needing to cover the skin surface, but this can still be a consideration for some patients.

19.3.5 Abdomen

Due to the relative lack of surface changes in the abdominal region of younger children, ROIs may have to extend outside the treatment area.

Encompassing part of the ribs increases the surface features and helps to anchor the ROI. The ROI is created to at least mid-depth to give the SGRT system as much information as possible to perform a surface match. As in adults, for very obese children, movable skin can be a problem. Other reasons for surface changes compared to CT simulation include changes in abdominal gas (bloating or constipation). Abdominal ROIs sometimes reach to the top of the pelvic area in small children and care must be taken to ensure no clothing or blankets are in the ROI.

Younger children tend to have more abdominal motion with respiration, which affects the longitudinal displacement with each breath more than the expected vertical displacement as determined by SGRT. By visually observing the respiratory pattern during simulation, it is possible to anticipate this effect. Using software features such as acquiring the reference surface image in a gated mode can mitigate the impact. If a gated capture is not supported, a 1–2 second beam hold delay may be utilized to bypass the problem. Additionally, a 1–2 mm expansion of the threshold can be used. Both options should be used with care and verified with another team member to ensure safe treatment.

19.3.6 Pelvis

Pelvis ROIs generally extend inferiorly from the umbilicus but is dependent on the child as to how inferior they extend. Diapers in very young children will limit how far inferiorly the ROI can extend; they can be unwrapped to some degree but not entirely removed. The top of the diaper can be tucked down to move as much of it out of ROI as possible. This should allow visibility down to the pelvic brim. Adolescents are more likely to resist being uncovered during treatment so efforts should be made to keep them as covered as is possible within the constraints of the setup requirements. At CHLA, the patient is undressed from the waist down and then a washcloth is used to cover them. The ROI is then able to cover more surface area including the hip bones and lateral aspects of the thighs leading to higher accuracy setup (Figure 19.2). This is especially helpful for the correction of longitudinal and pitch errors. As with abdominal setups, weight loss and constipation have been found to be factors. Typically, if consistent changes are seen over several days, new reference surface images are captured. Verification with CBCT is always performed. The fullness of the bladder or

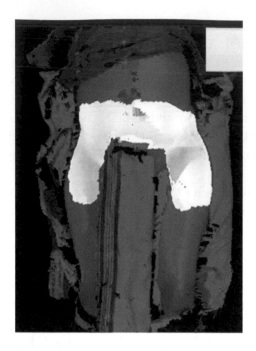

FIGURE 19.2 A pelvic surface region of interest is designed to include parts of the thighs to increase the available area for a surface match.

abdominal bloating can also affect the pitch in patients treated in the pelvis or lower abdomen. The magnitude of these changes are assessed to determine if replanning is needed.

19.3.7 Extremities

We use indexed Vac-Lok bags for extremity immobilization. For those patients with smaller extremities due to young age, we expand the ROI outside of the treatment area to capture enough surface area. The ROI must also wrap around past mid-depth to capture the contour of the extremity (Figure 19.3). The ROI should include a portion of a joint or other distinct contour to give the most accurate setup.

19.3.8 Craniospinal Irradiation

At CHLA, CSI patients are treated with VMAT.[2] They are positioned supine with an indexed Vac-Lok bag abutted to the HeadFIX system for immobilization. The transition away from skin marks took a full year as patients receiving CSI continued to be marked until the end of the

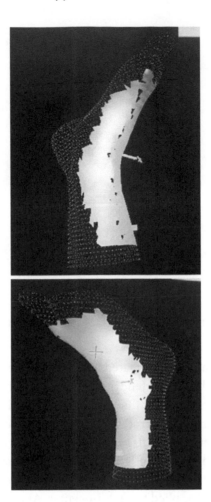

FIGURE 19.3 Two views of a surface region of interest (ROI) used for a foot treatment. The ROI includes lateral aspects of the foot to increase the surface contour variation and features available for a surface match.

transition period. For the first 3 months, full skin marks were used (horizontal lines drawn laterally and down the sagittal midline as shown in Figure 19.4) while experience was being gained with SGRT. Over the next 3 months, SGRT was used for setup with skin marks used as a backup. Intermittently during the last 6 months of the transition year, only one isocenter mark was placed on the anterior surface for each of the two spine fields but were only referred to if there was a large inconsistency between SGRT and IGRT.

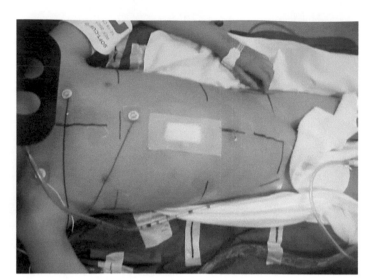

FIGURE 19.4 The system of ink pen line marking that was used for a craniospinal case prior to the introduction of surface-guided radiation therapy at CHLA.

At CHLA, CSI patients are treated with three isocenters, one positioned at mid-brain with the field extending to just above the shoulders, and the other two isocenters covering the entire thecal sac down to the S2 vertebrae. To minimize residual error, CBCT is performed at each isocenter in turn without making any shifts, all necessary shifts are evaluated simultaneously, and then one set of lateral, longitudinal, and vertical shifts is chosen and applied to all isocenters equally. Absolute couch positions are preprogrammed and applied to each isocenter. At each isocenter location, the patient is adjusted manually (e.g., not using the couch) to get agreement between SGRT and the required position. Having the 3D surface available provides the unique ability to untwist and unroll the patient along the entire spinal axis. As the patient is adjusted, vertical displacement and pitch may be marginally left out of tolerance. The ROIs used for CSI are similar to those used for chest, abdomen, or pelvis treatments. SGRT is not used for the brain isocenter due to the excellent reproducibility of the HeadFIX system. There has been a large decrease in the number of repeat CBCTs when using SGRT compared to a non-SGRT process. Visualizing and correcting the roll for the pelvis before CBCT has been the greatest help. Because VMAT is used for treatment, patient roll is important. Deviations in the setup position from the planned position of less than 3 mm for each isocenter are expected and observed. Total treatment times including setup, IGRT-based verification, and treatment have been reduced to a range of 20–40 minutes per case for VMAT treatments compared to 45–90 minutes before the use of SGRT.

19.4 CHALLENGES TO PEDIATRIC APPLICATIONS OF SGRT

There are many challenges arising from the treatment of pediatric patients using SGRT, some of which were alluded to in the previous sections. One is the limited body contour variation in children due to their small size. This gives the SGRT system less information to work with for calculating and detecting deviations in the 3D surface. The surface imaging algorithm must have enough peaks and valleys to be able to find an unambiguous 3D surface; however, younger pediatric patients tend to be more cylindrical and lack sufficient features. Adjusting the ROI so that it includes more of the patient's surface can help to mitigate the lack of contour variation. This may include extending the ROI outside of the treatment field to include more regions of variation. This issue is more pronounced in infants and toddlers.

Younger children are treated with anesthesia, which adds extra intravenous lines or tubing to the patient, interfering with the ROI and potentially leading to a less accurate 3D surface interpretation. By excluding central intravenous lines and moving other monitoring lines such as electrocardiogram (ECG) leads out of the ROI, a correct setup is possible using surface imaging. Because of this exclusion, the ROI may need to be larger than the usual ROI to provide a sufficient total area. It is recommended that the radiotherapy team work with anesthesiologists to find a way to move and/or position the lines (using paper tape) or ECG leads to a more lateral position. This can help to make more surface area available for the ROI.

If a blanket, sheet, or other covering is obscuring the ROI, it can interfere with the setup or the monitoring of the patient, giving inaccurate results. Cases under anesthesia are typically covered to keep the child's temperature above a certain threshold, so anesthesiologists may cover the child without understanding the importance of surface imaging. Parents may also cover their child to keep them warm. The child themselves may reach down and pull the blanket into the area of monitoring. It is important to always double-check and verify that a blanket or clothing item is not included within the ROI. Both staff and parents should be educated about the importance of keeping the area uncovered. Before capturing a surface or concluding that the patient is in the correct position, it should be verified that the ROI does not include any extra objects.

It should come as no surprise that children can have a hard time staying still for extended amounts of time due to several factors, including boredom, fear, pain, etc. Children in general tend to be less cooperative than older teenagers and adults. Children who have not been positioned into a comfortable position may readjust themselves. One solution is to minimize movement of the patient if they are comfortable within the SGRT setup tolerance and then use the 6DOF couch to correct the remaining differences after CBCT. Automatic gating of the beam is relied on to provide confidence that the treatment of an uncooperative or uncomfortable child can be carried out safely and accurately. Ideally, the use of anesthesia should be limited for safety reasons, but it often must be used to accurately treat the patient. This balance is more easily achieved when a clinic has SGRT available in their toolbox. It has been observed that children are more cooperative when given a fun task; they may cooperate during treatment if SGRT can be made into a game. For DIBH patients this is especially true when using a coaching device.

Explaining the treatment process at the child's level of understanding can decrease anxiety significantly. In addition, it can help to reduce the child's anxiety if a Vac-Lok is created for a stuffed animal to explain the process and show what the immobilization device or the SGRT lighting will look like during treatment (Figure 19.5). Doing this before the

FIGURE 19.5   A child's anxiety can be reduced by explaining the planning and treatment process with the help of a stuffed animal placed in an immobilization bag.

first day of treatment is helpful and typically requires less than 5 minutes. Any lighting from the SGRT camera system can sometimes cause younger children to move or became anxious. The child is instructed that they can close their eyes or watch the TV provided in the treatment room. Wash cloths can be used to cover the eyes of the patient so that they do not have to see any SGRT lighting. For those patients who fidget constantly, they are given the option as to what body parts they can move. For example, if the chest is being treated, they may be allowed to wiggle their foot. They may be given Play-Doh (Hasbro, Inc. Pawtucket, RI, USA) or a stuffed animal to hold. By decreasing the anxiety of the child, the treatment should be less stressful to all involved, including staff and family members.

KEY POINTS

- SGRT is advantageous for use with children because patient position errors can be reduced before X-ray imaging. This reduces the number of imaging studies needed for setup and associated X-ray exposure.

- Significant time savings were seen at CHLA with the addition of SGRT for patient setup due to more efficient initial patient positioning and a reduction in the number of X-ray images needed during IGRT.

- Children who for medical reasons could not be sedated can now be treated safely and confidently, even without robust immobilization. SGRT systems provide intra-fraction motion and position monitoring capability during treatment with the ability to automatically pause the treatment beam whenever the patient position is detected to exceed the set tolerances.

- Complex treatments like CSI and SRS delivered in the pediatric context can be performed more efficiently and safely with the addition of SGRT.

- Using SGRT with children presents unique challenges not seen in adults, which impact its use from a practical standpoint, including the choice of ROI.

REFERENCES

1. Sueyoshi M, Olch AJ, Liu KX, Chlebik A, Clark D, Wong KK. Streamlining isocenter localization with treatment plan embedded couch values for external beam radiation therapy. *Pract Radiat Oncol.* 2019;9(1):e110–e119. doi:10.1016/j.prro.2018.08.011.
2. Wong KK, Ragab O, Tran HN, et al. Acute toxicity of craniospinal irradiation with volumetric-modulated arc therapy in children with solid tumors. *Pediatr Blood Cancer.* 2018;65:e27050. doi:10.1002/pbc.27050.

CHAPTER 20

Skin Mark-Less Patient Setup

A Physicist's Perspective

Vanessa Panettieri, Sandra Paul,
and Catherine Russell

CONTENTS

20.1 INTRODUCTION

This chapter will describe a process for moving from radiation therapy patient setups with traditional tattoos or skin marks to a fully skin mark-less approach aided by the use of surface-guided radiation therapy (SGRT) technologies.

The chapter is organized into two main sections:

1. The first section will initially focus on the rationale of moving from a skin marks-based setup to a skin mark-less approach, providing some advice on how to implement such a change in the clinic, and including some practical examples of commissioning and end-to-end validation tests.

2. The second section will discuss the setup accuracy when moving from skin marks to skin mark-less approaches, and includes a critical examination of several studies available in the literature.

20.2 OVERVIEW OF THE TECHNICAL ASPECTS REGARDING SKIN MARK-LESS PATIENT SETUP

20.2.1 Rationale for Considering a Skin Mark-Less Setup Method

Radiation therapy has traditionally relied on skin marks, such as permanent tattoos or temporary skin marking, to reproduce the setup position and localize the isocenter placement for each fraction.[1,2]

Skin marking methods, while simple and cost effective, come with some inherent drawbacks that must be considered when selecting a patient setup technique:

- Skin marks, whether permanent tattoos or temporary skin markings, give information about limited, specific anatomical points on the patient's surface—for example, a three-point tattoo method is commonly used. By contrast, SGRT systems utilize many thousands of virtual reference points across a vastly greater surface area of interest, providing far more information with which to analyze the patient's position.

- Unlike SGRT systems, it is more difficult for skin marks to provide supplemental information about positioning discrepancies of nearby anatomy not directly in the treatment field, such as the limbs or chin. When relying on skin marks, extended reference lines must provide this information.

- As skin is mobile, skin mark points can be manipulated by the treating therapists and their anatomical relationship to each other can easily be deformed during setup, affecting rotational alignment. The skin also has the potential to be moved relative to the underlying internal anatomy, affecting the accuracy of translational alignment.

- Adverse implications of the use of radiation therapy tattoos include:

  - Risk of needle stick injury to staff;

  - A small risk of allergic reaction to the tattoo ink;[3]

  - The possibility of the ink spreading under the skin over time, resulting in a larger diameter, and less accurate marking;[2]

  - Patients' moles and skin marks have been mistaken as tattoos and used for setup;

  - Patient-related psychosocial implications of tattoos, which can be significant and which are discussed in Chapter 19 as they relate to pediatric patients, and in Chapter 21 as they relate to adult patients.[4]

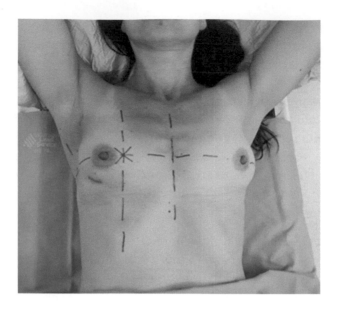

FIGURE 20.1   Example of skin marks for a patient treated with breast radiation therapy.

- Temporary skin marks, such as marker pens or adhesive marks (Figure 20.1), avoid the above limitations but introduce a different set of disadvantages:

  - They require patient education and compliance to maintain;

  - Temporary marks can fade and require reapplication, which is time consuming;

  - Each time the marks are reapplied, there is potential for inaccuracies to be introduced;

  - Adhesive tapes can irritate skin, which may already be sensitive as a result of irradiation.

At Alfred Health Radiation Oncology (AHRO) in Melbourne, Australia, the use of tattoos for breast radiation therapy patients was abandoned in 1998 in favor of temporary skin marks using marker pens, for reasons relating to the disadvantages of tattoos stated above. Particular consideration was given to the psychosocial impact of radiation therapy tattoos in the breast region specifically. An in-house analysis performed at the time determined that using temporary pen marks would provide an equivalent setup accuracy to tattoos. Daily pretreatment visual checks of the field projections on skin were performed as verification of setup in addition to routine verification imaging. These temporary skin marks required frequent reapplication by the treating therapists, and at times faded completely requiring remarking from measurements taken at simulation. To help preserve them, clear surgical tapes were placed over many of the marks, which would often become irritating to the patients' skin. However, the ability to avoid permanent tattoos was deemed to outweigh these considerations.

In 2012, after over a decade of treating breast patients without permanent tattoos, the decision was made to trial abandoning tattoos for all radiation therapy treatment sites at AHRO, instead using temporary pen marks as per the breast patients. A departmental retrospective analysis was performed, comparing the initial setup accuracy of the two skin marking methods. This study also demonstrated that there was no statistical difference in initial setup discrepancies noted on verification imaging for those patients who had tattoos compared to those who had temporary pen marks. Therefore, from 2012 onward, AHRO became the first radiation therapy department in Australia to eliminate tattoos.

Additionally, in 2009, AHRO installed Australasia's first SGRT system (AlignRT, Vision RT, London, UK). Departmental testing was performed to compare initial setup accuracy using SGRT to set up various anatomical sites, as opposed to skin marks. This analysis showed that SGRT was more accurate than skin marks for breast, supine pelvis and bellyboard sites, and comparable to skin marks for upper torso and abdominal sites. These findings, although applicable to a single institution, are supported by many other published studies that have demonstrated SGRT to be of equal or improved accuracy when positioning patients for treatment, compared to skin marks. A more detailed review of our analysis and of other published studies can be found in Sections 20.4 and 20.5.

As a result of the analysis, SGRT was initially implemented at AHRO in 2009 as a setup tool used in conjunction with skin marks for breast, supine pelvis, and bellyboard patients. Patients were grossly positioned on the treatment couch according to their skin marks, after which SGRT was utilized to fine-tune their position until the system indicated the patient was within a 1.5 mm, 1.0 degree threshold of the position from planning CT. Before verification imaging, a visual check was performed to ensure that the initial isocenter placement as indicated by SGRT was within 8 mm of the skin marks.

As the staff became more familiar with SGRT and confidence in the system increased, the workflow gradually evolved with an ever-decreasing emphasis on the skin marks for setup purposes. However, the skin marks were still maintained throughout treatment, to provide an auxiliary confirmation of initial setup position, and also as a means to set up the patient in the event that the SGRT system was unavailable or the patient needed to be transferred to a treatment machine that was not equipped with SGRT.

In 2017, the department equipped all treatment machines with SGRT technology. This eliminated transfer issues and was a key factor in being able to consider moving to a skin mark-less setup. At this time, there was also a strong push from the breast radiation oncologist to move toward eliminating all skin marks for breast patients, primarily for psychosocial reasons. In light of this, the first anatomical site to be trialed for a skin mark-less setup at AHRO was the breast. Based on our experience, the following sections provide advice on how to perform this transition.

20.3 IMPLEMENTATION

20.3.1 Physics Commissioning Tests Prior to Clinical Implementation of Skin Marks-Less Setup Technique

Before clinically implementing the use of SGRT techniques as a method for patient setup without the aid of skin marks, a series of dedicated commissioning tests should be performed to ensure that the SGRT system will provide an accurate and reproducible setup each treatment delivery. These tests should follow acceptance of the system, be performed by the radiation therapy team led by Qualified Medical Physicists and should be tailored to the characteristics of each type of SGRT system. These tests should also take into consideration the particular geometry of the treatment site for which the skin mark-less approach will be implemented.

The key component when designing these tests is to consider that, for positioning purposes, SGRT techniques will most likely complement and be coupled with radiographic (X-ray) imaging technologies. This is necessary as one of the limitations of surface guidance technology is that it can only track the surface of the patient which, depending on the treatment site and individual patient motion characteristics, might not be a good surrogate to infer the position of the internal target.[5-7]

A generic workflow for commissioning SGRT technologies for skin mark-less setup is summarized in Figure 20.2. This workflow was designed specifically for the AlignRT system but can be easily adapted for other SGRT systems.

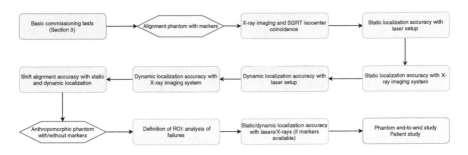

FIGURE 20.2   Basic workflow for commissioning the skin mark-less SGRT setup process.

20.3.1.1 SGRT Static Localization and X-Ray Imaging
Coincidence with an Alignment Phantom

After ensuring that the SGRT system is within recommended and established tolerances for the basic commissioning tests as explained in Chapters 6 through 8 of this book, the first step is to ensure the accuracy of the isocenter coincidence of the X-ray imaging systems used at treatment and the SGRT system. This commissioning test assumes that both the mechanical isocenter and radiographic imaging systems are maintained following a rigorous quality assurance (QA) program as described in American Association of Physicists in Medicine (AAPM) Task Group (TG) -147, TG-179, TG-142,[6,8,9] or a local QA protocol, within the tolerances used for treatment (generally 2 or 1 mm for SBRT/SRS).

To perform the X-ray imaging and coincidence test, it is advisable to use a precisely machined phantom of known size which includes an internal target of well-defined and known position (i.e., ball bearings, hollow spheres). The phantom needs to be able to be used by the X-ray imaging system as a region of interest (ROI) to calculate the position of the phantom in relation to the initial digitally reconstructed radiograph (DRR) and the machine isocenter.

Several phantoms are commercially available to be used for alignment and isocenter coincidence types of tests (some examples are shown in Figure 20.3). However, unlike X-ray systems, SGRT systems have been designed to use anatomical surface topography, texture, and color in order to be able to perform accurate tracking of the ROI defined on the patient. For this reason, it is important to select a phantom which has similar characteristics. Different groups have adapted already existing phantoms for such an application, while others have constructed custom-made

FIGURE 20.3 Example of commercial alignment phantoms used for quality assurance. From left: QUASAR™ Penta-Guide Phantom (ModusQA, London, ON, Canada), Varian MPC IsoCal Phantom, Varian OBI Phantom (Varian Medical Systems, Palo Alto, CA, USA).

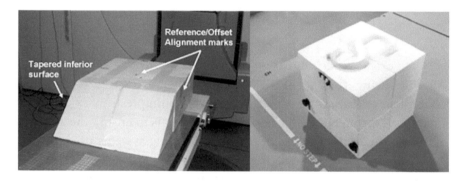

FIGURE 20.4 Custom-made phantoms to check alignment of the X-ray and SGRT imaging systems (left: styrofoam block phantom and right: Integrated S-Phantom). (From Wooten, H.O. et al., *J. Appl. Clin. Med. Phys.*, 12, 234–238, 2011 [left]; Yu, A.S. et al., *Med. Phys.*, 45, 2857–2863, 2018 [Right]. With permission.)

phantoms to be able to consistently perform coincidence of X-ray imaging systems and SGRT systems, including fully integrated solutions as shown in Figure 20.4.[10,11]

One of the key characteristics is that the phantom selected needs to have a defined feature, or suitable topography, that can be used to define the ROI by the SGRT system. An example of a well-defined ROI is shown in Figure 20.5. In this example, by using both corners of the phantom, it has been determined by our team that the SGRT system is able to track its position in all 6 degrees of freedom.

Once the phantom is selected, the following tests should be performed:

- The phantom should be CT-scanned at its isocenter and imported into the treatment planning system (TPS). An imaging field should be created and the structures sent to the SGRT system, according

FIGURE 20.5 Example of a well-defined region of interest (ROI) to be used to track the phantom's shifts in 6 degrees of freedom.

to the system's instructions. The body contour should be imported into the SGRT software to identify the ROI that should be tracked by the SGRT system on the treatment machine. (An example is shown Figure 20.5.)

- The phantom should be then positioned on the radiation therapy machine treatment couch, and, replicating a patient setup, moved to the isocenter using solely the monitoring option of the SGRT system, by observing the 6-degrees real-time displacements from the planned position (vertical, longitudinal, lateral, rotation, pitch, and roll displacements). Site-specific tolerances should be modified to be within the accepted X-ray imaging tolerances (generally 1 mm for the SRS/SBRT treatment sites). Once the tolerances have been reached, the monitoring should be stopped and a treatment capture (a surface snapshot) acquired to record the isocenter real-time position displacements.

- Once the real-time displacements are within tolerance, the phantom should be imaged with an X-ray imaging system of choice and the internal marker (e.g., the ball bearing or hollow spheres) registered to the DRR obtained from the TPS. An on-line match should be performed (manually or automatically depending on the treatment system used) and reviewed, and the couch correction should be recorded.

- The SGRT real-time displacements and the X-ray imaging couch corrections should be compared and their differences analyzed.

The differences should be within the tolerances expected for the mechanical isocenter and the X-ray imaging system, as established during the treatment machine commissioning. The phantom position should also be visually inspected and differences with the positioning lasers recorded.

• This test should be repeated several times, over several days and varying the initial position of the phantom on the treatment couch to ensure that this process provides reproducibility and repeatability of the results.

The test results should be reviewed by the radiation therapy team and the limitations of the SGRT system's static accuracy should be determined and documented.

20.3.1.2 SGRT Dynamic Localization and X-Ray Imaging Coincidence

In some cases, the SGRT technology (e.g., AlignRT) can be fully integrated with the treatment machine as a motion management device, e.g., via the Motion Management Interface (MMI) gating interface for a Varian linear accelerator (linac). This implementation may allow the use of remote couch motions that can be applied automatically to minimize the real-time displacements calculated by the SGRT software. When this option is used in the clinic, it is important to perform a series of tests similar to those designed for the static localization accuracy, described in Section 20.3.1.1. The main difference will be to apply the shifts using the couch's automatic movement function as opposed to manually moving the couch.

20.3.1.3 SGRT Static/Dynamic Localization Accuracy with an Anthropomorphic Phantom

Once the static and dynamic localization accuracy obtained by positioning a phantom of simple and known geometry have been established, it is important to test setup accuracy by using a patient-like geometry, such as the one provided by an anthropomorphic phantom. An example of phantoms used for testing the accuracy of skin mark-less setup is provided in Figure 20.6; however, its choice should be guided by the anatomical sites for which skin mark-less setup will be employed (e.g., a pelvic or thorax phantom for extracranial sites and a head phantom for intracranial sites).

FIGURE 20.6 Example of anthropomorphic phantoms used to test skin mark-less patient setup.

Several publications have also described the use of anthropomorphic phantoms for tests with SGRT systems which could be adapted for commissioning the use of skin mark-less SGRT positioning.[12-14]

An identical workflow to the one that will be used for the skin mark-less patient setup should be used on the anthropomorphic phantom. This workflow will be similar to the one developed for the simple phantom geometry; however, if the anthropomorphic phantom does not have any internal markers, it will not include a coincidence test between the SGRT and the X-ray imaging system. The workflow with the anthropomorphic phantom includes the following tests:

- The anthropomorphic phantom will be CT-scanned and the ROI prepared as previously described.

- In this test, it is important to focus on the shape of the ROI, which has been found to be one of the limiting factors in the use of SGRT systems for skin mark-less setup on patients. For example, in our experience, using a Multi-Modality Pelvic Phantom (CIRS 048A, CIRS, Norfolk, VA, USA), it has been found that a shift in the longitudinal direction of 5 cm could be underestimated by the full 5 cm if an incorrect feature-less ROI was selected, making the shift completely undetected. More details about the definition of ROI will be discussed later.

- Phantom setup accuracy should then be tested for a series of ROIs, and for different initial setup positions (different couch translations, rotations and with the gantry in different positions) to establish the limitations of the SGRT system for a wide range of initial positioning geometries, and different anatomical treatment areas for each treatment site.

- Treatment accessories, such as breast and belly boards, bolus or setup masks should also be evaluated at this stage.

Once the tests with the anthropomorphic phantoms have been completed and the limitations of the SGRT system have been evaluated, the system is now ready to be used by the radiation therapy team to begin the clinical implementation of the skin mark-less setup technique.

20.3.2 Clinical Considerations Prior to Implementation of a Skin Mark-Less Setup Technique

20.3.2.1 Adequate Training on the SGRT System

- Any radiation therapists involved in the treatment of patients using SGRT, particularly when using a skin mark-less technique, should be completely familiar and competent with the operation of the SGRT system prior to treating patients.

- The staff members responsible for delineating the surface ROI (typically either radiation therapists or physicist) must have a full understanding of the most appropriate anatomy to both include and exclude, taking care to incorporate sufficient contour changes in the ROI.

- It is advisable that any treating staff are competent, and have the user rights, to edit ROIs during the course of treatment if it is deemed that a more appropriate ROI should be utilized.

- It may be useful to provide a collection of visual examples of ideal ROIs for each anatomical site being treated with SGRT, as a guide for the treating staff.

20.3.2.2 Risk Management

The most serious risk to be considered in a skin mark-less SGRT setup is that the SGRT system will direct the user to the incorrect isocenter location, and that this will not be detected prior to treatment, resulting in a geographic miss. While this risk is very small (there have been no instances at our institution since implementing the skin mark-less technique in 2017, and to our knowledge there have been no such cases published), the consequences of such an event could have serious implications for patient outcomes; therefore, it is essential that the risk must be mitigated as much as is possible:

- It is recommended that any treatment being delivered by skin mark-less SGRT should have appropriate pretreatment verification X-ray imaging performed.

- It is recommended that in addition to pretreatment verification imaging, a robust system of cross-checking the expected versus actual couch position should be incorporated, such as requiring an override if any of the couch parameters exceed a user-defined variation tolerance.

The other significant risk to be managed is how to proceed in the event that the SGRT system fails or is unavailable for any reason:

- It is advisable that each institution should have prompt and reliable access to real-time assistance from the SGRT technical and applications support teams, as system downtime can be a significant barrier to efficient implementation of skin mark-less treatment.

- It is recommended that any patient who has a skin mark-less setup should have a full set of landmarks and reference lines recorded at simulation, and isocenter location clearly recorded in relation to these. In the event of SGRT system downtime, these landmarks and measurements should be used to temporarily mark on the approximate isocenter and assist with rotational alignment of the patient, prior to pretreatment verification imaging.

20.3.3 End-to-End Validation

It is advisable that end-to-end testing is performed by the all involved staff groups prior to implementing skin mark-less techniques.

20.4 VALIDATION OF THE USE OF SGRT TECHNIQUES FOR SKIN MARK-LESS SETUP

Once the technology has been thoroughly tested for skin mark-less setup on phantoms, it is recommended to perform an in vivo validation on a series of patients on treatment. This is a crucial step to evaluate the efficiency and the accuracy of skin mark-less setup for the treatment areas considered for this approach. In our experience, it is advisable to start with one treatment site and then extend to others once all staff have become familiar with the process.

20.4.1 Example of Validation Method: The Alfred Health Radiation Oncology Experience

As an example of the end-to-end validation, we consider the experience in our clinic. As reported in Russell et al.,[15] the first treatment technique considered for an SGRT only (skin mark-less) setup at AHRO was tangential breast radiation therapy following breast conservation, for patients not requiring nodal irradiation or deep-inspiration breath hold (DIBH). This group of patients was selected for three main reasons:

- The clinical limitations and psychological impact of having to carry and maintain temporary skin marks (as described above);

- The ability for the radiation therapists to undertake a gross visual confirmation that the treatment field coverage is correct after initial SGRT positioning and before verification imaging, via the use of the light field projection; and

- The challenges of setting up a treatment area affected by soft tissue mobility, shape changes, and respiration.

As a means of validation, we evaluated the initial setup accuracy of three different patient positioning methods that had been used at our institution. As previously described in Section 20.2.1, use of permanent tattoos for breast cancer patients had been replaced by temporary skin marks drawn by a non-permanent pen since 1998, so prior to the acquisition of the SGRT system all patients were setup by aligning the temporary skin marks to the treatment room lasers (i.e., skin marks only). After the introduction of the SGRT system in 2009, skin markings were kept as reference for an initial gross positioning followed by a "fine-tuning" positioning with SGRT (i.e., skin marks + SGRT), and in 2017 this patient group was solely setup using the SGRT system (i.e., SGRT only), with no skin markings being present.

20.4.1.1 Methodology

To validate the use of SGRT alone, a retrospective statistical analysis was performed in order to compare an initial cohort positioned with skin marks only, first with the cohort positioned with skin marks and SGRT, and then with SGRT only. For this study, 15 patients were selected for each cohort. The selection process was based on considering 15 patients, each undergoing 15–25 fractions on a linear accelerator, sequentially treated for each setup approach, in order to exclude any bias due to treatment staff, patient demographic, or clinical characteristics. For all of these patients after initial setup MV images acquired with the electronic portal imaging device (EPID) were acquired and matched to the initial planning CT DRRs. This imaging modality was selected as it was the option mandated by our clinical protocol at the time of treatment; however, it had the limitation of providing a limited view of the anatomy. So, when performing this validation, it is recommended to consider an imaging modality which provides full 3D information such as the one given by kV or MV cone-beam computed tomography (CBCT).[14,16-19] By using the acquired MV EPID images, retrospective analysis was performed, in which each image was reviewed and re-matched by two consistent independent radiation therapists. Translational displacements from the planning DRRs were calculated. These displacements were separated in each vectorial component (longitudinal, lateral, and vertical for the translations and rotation, pitch, and roll for the rotations) and the differences between the cohorts analyzed. In order to provide patients' overall displacements, translational components were also combined in an overall absolute vector (v) according to Eq. 20.1:

$$v = \sqrt{(\mathrm{Long}^2 + \mathrm{Lat}^2 + \mathrm{Vert}^2)} \qquad (20.1)$$

for which Long is the longitudinal, Lat is the lateral, and Vert is the vertical component. The median absolute overall displacements using all fractions for all patients in each cohort were compared using an unpaired Wilcoxon-signed Rank t-test ($p < 0.05$). This test was considered appropriate since in this case different patients were used in each cohort.

A summary of the workflow used in the study performed at AHRO is shown in Figure 20.7.

20.4.1.2 Findings

The results of the validation statistical analysis are shown in Table 20.1 and Figure 20.8. The estimated median absolute displacement was of the

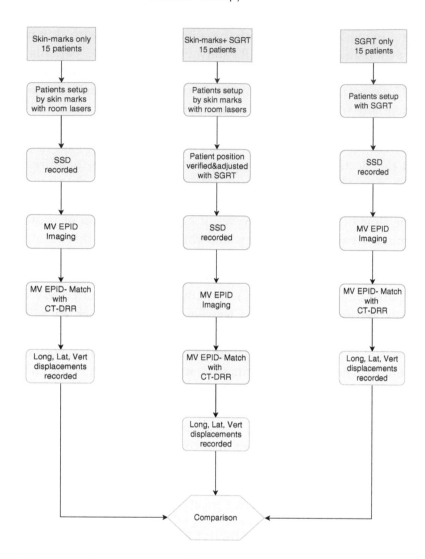

FIGURE 20.7   AHRO analysis workflow.

order of 3.8, 2.8, and 2.7 mm for skin marks only, skin marks + SGRT, and SGRT only, respectively. The *p*-values show that there was a statistically significant decrease in overall displacement when SGRT was introduced. The analysis also showed that there was a non-statistically significant difference when SGRT only was used as opposed to SGRT with skin marks ($p = 0.98$). When the displacements were separated into translational components (vertical, longitudinal, and lateral, see Figure 20.8), the main difference was found in the longitudinal component.

TABLE 20.1 Median Overall Displacements for Breast Patients with Corresponding Confidence Intervals (CI) and *p*-Values

Breast Patients	Estimated Median Overall Absolute Displacement (mm)	95% CI	*p*-Value
Skin marks only	3.8	3.3–4.4	
Skin marks + SGRT	2.8	2.2–3.4	0.029
SGRT only (skin mark-less)	2.7	2.3–3.2	0.011

Note: The comparison group is skin marks only.

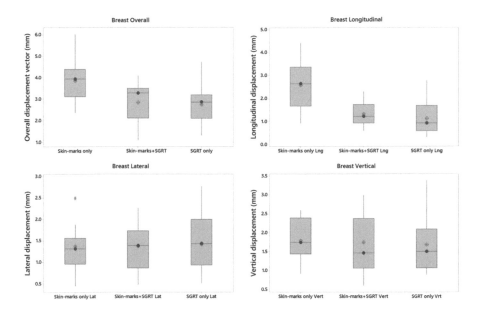

FIGURE 20.8 Box plots showing the median displacements (in mm) for skin marks only, skin marks + SGRT, and SGRT only for the overall, vertical, lateral, and longitudinal displacements.

Our experience for breast patients showed that SGRT significantly improved patient setup accuracy when compared to using skin marks alone. Standard deviations were also reduced, suggesting greater setup consistency and reliability. When SGRT was used alone (without any skin marks), there was no reduction in accuracy (Figure 20.8), indicating that a skin mark-less patient setup for breast radiation therapy was feasible and accurate.

This process of in vivo validation was an important step to give us the confidence to continue rolling out the use of SGRT-only for other treatment sites (including abdominal, pelvic, and thoracic regions). The radiation therapists had become fully familiar and expert with the skin mark-less workflow after having used it on breast patients, therefore little additional training was required. However, possible patterns of failure were still required to be identified, such as:

- Unlike tangential breast treatment, it would not generally be possible to perform a gross visual confirmation of the treatment field location accuracy before verification imaging via the light field projection, as most treatment sites were internal; and

- Greater variety of treatment sites required the skills to accurately determine a much larger range of ROIs than for breast alone.

In light of the inability to perform a gross visual confirmation of the SGRT setup prior to verification imaging without skin marks, an additional departmental requirement was that the positioning equipment must be well localized to the treatment couch (i.e., localized at two or more points), with an appropriately small table tolerance threshold. This was to safeguard against gross positioning errors. Verification imaging was required to be performed prior to every fraction and any residual displacement corrected to zero. The risk of geographic miss was deemed comparable to the one found when incorporating skin marks by ensuring these criteria were met.

With the knowledge and experience acquired for the breast site, treating abdominal was next, pelvic and thoracic patients with SGRT only setups if they fulfilled the above criteria. A review of the initial setup accuracy for each of these anatomical sites was performed by evaluating the first 15 patients for each site treated without skin marks, and comparing to a cohort of patients previously treated with skin marks and SGRT. Unlike the breast patients, a third cohort of skin mark-only patients was not included as it was the effect of removing the skin marks from an SGRT setup that was the primary interest.

The results obtained are summarized in Table 20.2 and confirmed the feasibility of skin mark-less SGRT setup for abdominal, pelvic, and thoracic sites at our institute, and support the earlier results for skin mark-less SGRT setup of breast sites.

TABLE 20.2   Median Overall Displacements for Various Treatment Sites with
Corresponding Confidence Intervals (CI) and *p*-Values

Treatment Site	Positioning Technique	Estimated Median Overall Absolute Displacement (mm)	95% CI	*p*-Value
Thoracic	Skin marks + SGRT	4.4	3.6–5.4	
	SGRT only	4.4	4.0–4.9	0.80
Pelvis	Skin marks + SGRT	4.8	4.2–5.6	
	SGRT only	4.5	3.8–5.2	0.41
Abdomen	Skin marks + SGRT	5.7	4.7–8.2	
	SGRT only	6.3	4.8–7.3	0.68

Note: The comparison group is skin marks + SGRT.

20.5 DISCUSSION ON THE ACCURACY OF SKIN MARK-LESS SETUP WITH RESPECT TO CONVENTIONAL THREE-POINT SETUPS AND IMAGING

In addition to our experience presented in the validation section, several other groups have also looked into the accuracy of positioning patients with the aid of SGRT systems, with respect to the conventional three-point setup system and imaging.

A selection of these works has been summarized in Table 20.3, which details the type of SGRT system used, the anatomical sites, the methodology, and the results of the evaluations. In contrast to the AHRO experience,[15] these works have focused on comparing patient setup using SGRT only, as opposed to SGRT with the aid of skin marks, for cohorts of patients who all had skin marks present. This chosen methodology has the advantage of allowing a comparison of the accuracy of setting up using SGRT only and SGRT with skin marks on the same patients; however, it could introduce a bias as the skin marks could potentially guide the radiation therapist setting-up toward the treatment site. Our recommendation is to start with such an approach, but then when confidence in the technique has been acquired, to move toward a fully skin mark-less SGRT-only approach and continue the evaluation for the first 10–15 patients after discontinuing the skin marks.

Similar to our results, all authors report, overall, that SGRT with skin marks can provide either superior or comparable results in comparison to skin marks alone. In one of the cases reported,[21] a decrease in accuracy was found in the longitudinal direction. In our experience, it has been found that this direction can be highly dependent on the ROI selected.

TABLE 20.3 Summary of Selected Works That Have Analyzed the Accuracy of SGRT as Opposed to Conventional Point 3D Setup Systems and Imaging

References	Treatment Site & Technique	Type of Skin Marks Used	Patients	Methodology	Findings	SGRT System
Carl et al.[16]	A number of sites: lung, prostate, head & neck, soft tissue sarcoma, intracranial	Unspecified skin marks	110	The patients were first positioned via the room lasers and skin marks. SGRT surface scan was carried out. Shift values in regard to the optimal treatment position were recorded. CBCT was acquired, in which deviations were regarded as gold standard.	1. SGRT achieved at least the same precision as positioning using spatial lasers. 2. The highest precision was achieved in scans of the head area. 3. Accuracy in the thoracic and abdominal areas decreased compared to the head area but remained comparable to laser-based positioning.	Catalyst
Cravo et al.[17]	Breast (linac)	Tattoos	20	G1: 10 patients setup with AlignRT and tattoos. G2: 10 patients setup with only tattoos. First 15 fractions analyzed by kV CBCT imaging: rotational, translational, systematic, and random errors.	1. SGRT with marks (G1) was superior to marks (G2) only in Left-Right direction. 2. Comparable in the other directions. 3. G2 higher systematic and random errors than G1. 4. SGRT should be used with another imaging method.	AlignRT

(Continued)

TABLE 20.3 (Continued) Summary of Selected Works That Have Analyzed the Accuracy of SGRT as Opposed to Conventional Point 3D Setup Systems and Imaging

References	Treatment Site & Technique	Type of Skin Marks Used	Patients	Methodology	Findings	SGRT System
Crop et al.[23]	Breast with Nodes (helical tomotherapy)	Unspecified marks	95	40 patients with Catalyst, 55 with laser-based positioning. 810 SGRT-only and 666 laser-only sessions. 31 patients had both laser-based and SGRT-based setup on different days. Analyzed by MVCT.	1. SGRT performed statistically significantly better than the laser positioning ($p<0.05$) in all directions. 2. Uncertainties were very close to MVCT imaging.	Catalyst
Herron et al.[20]	Breast (linac)	Unspecified marks	17	Half the total fractions were setup using SGRT, the others using skin marks. Couch shifts were evaluated with imaging.	SGRT setup was found to be statistically significantly more accurate and efficient than the skin mark-based technique.	AlignRT
Stanley et al.[19]	Pelvis/lower extremities/ abdomen/chest/ upper extremities/ breast (linac)	Tattoos	600– 900 fractions per site (240 patients)	All patients randomly selected to either SGRT or tattoo-based setup. A kV-CBCT was acquired immediately following to adjust for any residual corrections.	Overall 3D shift corrections for patients initially aligned with the C-RAD CatalystHD were significantly smaller than those aligned with subcutaneous tattoos.	Catalyst
Walter et al.[21]	Abdominal, thoracic, pelvic (linac)	Unspecified marks	25	Patients positioned with skin marks, then CBCT acquired. Prior to CBCT, scanned with SGRT and compared to the reference surface. Position displacements were obtained and compared to the imaging corrections. Only translations were assessed.	1. For the entire group no significant difference between SGRT and was found in any direction. 2. A significant decrease in accuracy was only found in the longitudinal direction for the pelvic region.	Catalyst

If the ROI is selected in a region without any defined feature (such as the sides of the patient for the pelvis and abdomen, particularly flat regions such as the thorax, or any structure that is cylindrical such as a limb), the SGRT system might not be able to accurately calculate the difference between the reference surface and the treatment capture surface, providing an incorrect positioning of the patient.[7,13,16,22]

Interestingly, groups that have compared positioning with tattoos to SGRT report a significant increase in accuracy with SGRT when analyzing differences with 3D imaging systems.[19,23] This behavior could be the result of the ink spreading over time, increasing the diameter of the tattoos, which is not seen for temporary pen-markings. All works use X-ray imaging as reference imaging for final patient setup before irradiation, by matching the internal anatomy. As pointed out by Cravo et al.,[17] the general advice is still to use SGRT in combination with other X-ray imaging techniques, but the introduction of SGRT has the potential to reduce repeated imaging due to the ability to perform an initial correction of the position of the patient in all 6 degrees of freedom, lowering the delivered dose and setup time. This recommendation has recently been reiterated by a study which compared DIBH breast setup retrospectively with and without X-ray imaging.[22] While for some cases SGRT has been seen to correlate with tumor motion near the surface of the patient, the use of X-ray imaging is still paramount in cases where breathing motion could affect the position of the internal organs.[16,17]

An important aspect of setup positioning is the training and experience of treatment staff as pointed out by Water et al.[21] This consideration highlights the need for the staff to be confident and well-trained on the use of SGRT before moving toward a skin mark-less patient setup.

KEY POINTS

- Traditionally, patient treatment setup has been based on the use of permanent (tattoos) or temporary (pen markings/adhesive) skin marks acquired during planning simulation. SGRT provides the ability to identify the position of the surface of the patient and provides the potential to replace skin marks and become the only initial positioning method on the treatment machine.

- In order to implement SGRT as the sole patient setup tool, a series of commissioning tests with phantoms should be initially performed in order to identify the limitations of the SGRT system, its reproducibility and accuracy. These tests should involve the physicists and radiation therapists in consultation with the radiation oncologists.

- The phantom commissioning should be followed by a validation study on patients. One treatment site should be initially selected, then extended to other sites as confidence and experience is gained.

- For all treatment sites SGRT should still be coupled to appropriate pretreatment X-ray imaging.

- Several studies published in the literature have shown that SGRT alone can provide setup accuracy comparable or superior to skin marks-only or skin marks with SGRT. These studies have also highlighted the importance of ROI definition and staff training.

REFERENCES

1. Elsner K, Francis K, Hruby G, Roderick S. Quality improvement process to assess tattoo alignment, set-up accuracy and isocentre reproducibility in pelvic radiotherapy patients. *J Med Radiat Sci.* 2014;61(4):246–252.
2. Rathod S, Munshi A, Agarwal J. Skin markings methods and guidelines: a reality in image guidance radiotherapy era. *South Asian J Cancer.* 2012;1(1):27–29.
3. Sewak S, Graham P, Nankervis J. Tattoo allergy in patients receiving adjuvant radiotherapy for breast cancer. *Australas Radiol.* 1999;43:558–561.
4. Clow B, Allen J. Psychosocial impacts of radiation tattooing for breast cancer patients: a critical review. *Can Woman Stud.* 2010;28:46–55.
5. Wikstrom K, Nilsson K, Isacsson U, Ahnesjö A. A comparison of patient position displacements from body surface laser scanning and cone beam CT bone registrations for radiotherapy of pelvic targets. *Acta Oncol.* 2014;53(2):268–277.
6. Willoughby T, Lehmann J, Bencomo JA, et al. Quality assurance for non-radiographic radiotherapy localization and positioning systems: report of Task Group 147. *Med Phys.* 2012;39(4):1728–1747.
7. Pallotta S, Vanzi E, Simontacchi G, et al. Surface imaging, portal imaging, and skin marker set-up vs. CBCT for radiotherapy of the thorax and pelvis. *Strahlenther Onkol.* 2015;191(9):726–733.
8. Klein EE, Hanley J, Bayouth J, et al. Task Group 142 report: quality assurance of medical accelerators. *Med Phys.* 2009;36(9):4197–4212.
9. Bissonnette JP, Balter PA, Dong L, et al. Quality assurance for image-guided radiation therapy utilizing CT-based technologies: a report of the AAPM TG-179. *Med Phys.* 2012;39(4):1946–1963.
10. Wooten HO, Klein EE, Gokhroo G, Santanam L. A monthly quality assurance procedure for 3D surface imaging. *J Appl Clin Med Phys.* 2011;12:234–238.

11. Yu AS, Fowler TL, Dubrowski P. A novel-integrated quality assurance phantom for radiographic and nonradiographic radiotherapy localization and positioning systems. *Med Phys.* 2018;45(7):2857–2863.

12. Bert C, Metheany KG, Doppke K, Chen GT. A phantom evaluation of a stereo-vision surface imaging system for radiotherapy patient setup. *Med Phys.* 2005;32(9):2753–2762.

13. Schoffel PJ, Harms W, Sroka-Perez G, Schlegel W, Karger CP. Accuracy of a commercial optical 3D surface imaging system for realignment of patients for radiotherapy of the thorax. *Phys Med Biol.* 2007;52(13):3949–3963.

14. Stieler F, Wenz F, Shi M, Lohr F. A novel surface imaging system for patient positioning and surveillance during radiotherapy. *Strahlenther Onkol.* 2013;189(11):938–944.

15. Russell C, Mack H, Paul S, Senthi S. OC-0190: surface guided radiation therapy for breast cancer improves accuracy without the need for skin marks. *Radiother Oncol.* 2018;127:S102.

16. Carl G, Reitz D, Schönecker S, et al. Optical surface scanning for patient positioning in radiation therapy: a prospective analysis of 1902 fractions. *Technol Cancer Res T.* 2018;17:1533033818806002.

17. Sá, CA, Fermento A, Neves D, et al. Radiotherapy setup displacements in breast cancer patients: 3D surface imaging experience. *Rep Pract Oncol Radiother.* 2018;23(1):61–67.

18. Ma Z, Zhang W, Su Y, et al. Optical surface management system for patient positioning in interfractional breast cancer radiotherapy. *Biomed Res Int.* 2018;2018:6415497.

19. Stanley DN, McConnell KA, Kirby N, Gutiérrez AN, Papanikolaou N, Rasmussen K. Comparison of initial patient setup accuracy between surface imaging and three point localization: a retrospective analysis. *J Appl Clin Med Phys.* 2017;18(6):58–61.

20. Herron E, Murray M, Hilton L, et al. Surface guided radiation therapy as a replacement for patient marks in treatment of breast cancer. *Int J Radiat Oncol Biol Phys.* 2018;102(3):e492–e493.

21. Walter F, Freislederer P, Belka C, Heinz C, Söhn M, Roeder F. Evaluation of daily patient positioning for radiotherapy with a commercial 3D surface-imaging system (Catalyst). *Radiat Oncol.* 2016;11(1):154.

22. Laaksomaa M, Sarudis S, Rossi M, et al. AlignRT® and Catalyst™ in whole-breast radiotherapy with DIBH: is IGRT still needed? *J Appl Clin Med Phys.* 2019;20(3):97–104.

23. Crop F, Pasquier D, Baczkiewic A, et al. Surface imaging, laser positioning or volumetric imaging for breast cancer with nodal involvement treated by helical TomoTherapy. *J Appl Clin Med Phys.* 2016;17(5):200–211.

CHAPTER 21

Tattoo/Skin Mark-Less Setups Using SGRT

A Radiation Therapist's Perspective

Jacqueline Dorney

CONTENTS

21.1 INTRODUCTION

During radiation therapy, patients are traditionally set up using between one to five permanent dark ink marks on the patient's skin, as show in Figure 21.1. With the introduction of surface-guided radiation therapy (SGRT) systems and techniques, there is potential for a paradigm shift to move away from these permanent marks, which can have lasting negative effects on body confidence and cosmetic outcome.[1-5]

The reproducibility of patient setup is essential to deliver highly accurate radiation therapy treatments. To consistently reproduce highly accurate set-ups, patients are traditionally marked and then set up using dark ink permanent tattoos or marks on the skin surface. Worldwide, permanent tattooing is

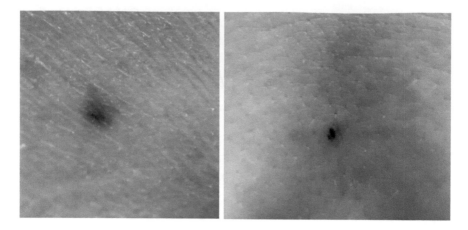

FIGURE 21.1   Examples of radiotherapy tattoos.

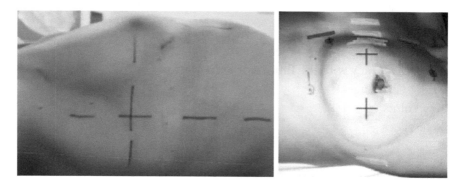

FIGURE 21.2   Examples of radiotherapy skin markings.

the most popular method for skin markings.[2] A nontoxic ink is injected into the epidermis of the skin using a disposable needle to make a 1–2 mm permanent tattoo. Practice varies across sites and even countries; however, marking of individuals either at the isocenter, field edges, or reference points of planned fields is a common practice. Both invasive (i.e. tattoos) and noninvasive (e.g., marker pen, henna, ultraviolet) skin marking techniques are available; however, all have limitations to clinical practice (Figure 12.2). Visibility or confusion of tattoo placement for radiation therapists and clinicians, social/religious beliefs, and mobility of skin are some of the main concerns relating to tattoos. Hence, additional processes are used to verify positioning prior to beam delivery. Commonly these include checking source-to-surface distances (SSDs), X-ray imaging, and, more recently, surface imaging.

For breast and chest wall patients, although not excluding other body sites, there is emerging evidence indicating that tattoos, or otherwise marking a patient's skin permanently to achieve reproducible and accurate setup for treatment, can have a significant impact on body confidence and self-esteem.[1] In one study by the National Cancer Research Institute, UK, a cohort of patients received the standard dark ink tattoo for setup while the other group received an ultraviolet "invisible" tattoo, that is, it was not visible in standard lighting. The results suggested that the permanent dark tattoos made on the skin of the women having radiotherapy reminded them of their diagnosis for years to come, reducing body confidence and self-esteem. There is a growing view that with more than half of all cancer patients surviving for 10 years and beyond,[6] it is becoming imperative to treat the whole person and reduce as much as possible the negative or psychologically damaging impacts of treatments on our patients, including cosmetic changes, where possible.

21.2 LISTENING TO OUR PATIENTS

Healthcare professionals continually strive to improve cancer services, from new innovative treatments, to improving access, to survivorship; but there is one voice that drives these changes in practice, and that is the voice of our patients. Through the emergence of patient forums and quality groups that meet and discuss services with those that have been through it firsthand, we begin to see what it is we can do as professionals to improve our care. In the discussion around tattoo or skin mark-free radiotherapy treatments, it is important to hear their voices, some of which are provided below[7-9]:

> "In the middle of everything else it is such a trivial thing, but years later they are like psychological scars on me."
>
> "I really hate tattoos and feel once my treatment is over these are going to be a constant reminder of what I have/am currently going through. I know I am lucky but I am feeling so low and dreading these tattoos. I feel I am crying more now than when I was first diagnosed. Feeling so stupid for worrying about a few dots."
>
> "I have them and didn't want a permanent reminder. When I asked to have them removed they tried to ignore my request and avoided answering, then told me it was best to leave them in case my cancer came back. You then not only have a permanent reminder, but additional concerns of my cancer returning."

> "I really do not want to have tattoo marks for my radiotherapy.
> I don't want a permanent reminder on my body... I'm so pale so
> these will really stand out."

While these skin marks may seem trivial and are essential to healthcare
professionals for highly accurate radiotherapy delivery, the duty of care
requires that patients' voices are heard and should not be ignored.

"The role of permanent radiation tattooing is neglected in the literature, in part because the marks are sometimes seen – by both patients
and practitioners – as a relatively minor issue in the context of managing a disease that is potentially life threatening."[10] Some patients have
been educated to use "reforming techniques"; that is, coaching that is
designed to train them to change their attitudes toward their bodies
and their tattoo marks, to see them as not a negative "left over" from
radiotherapy but more a reminder that they are breast cancer survivors. Sadly, for most this doesn't change the fact that they have permanent marks on their body that they did not want and do not like. When
asked, most feel that they should not have to "reframe" their attitudes
given the gravity of what they have endured during their treatment and
now their recovery.

21.3 ACCURACY WITHOUT TATTOOS OR SKIN MARKS

SGRT systems provide a safe and effective way of avoiding the use tattoos and skin marks while ensuring accurate and robust daily setup. Many
practices have adopted this technology not only for marker/tattoo-less
treatments, but for frameless SRS, and can deliver high dose radiotherapy
to extremely small targets utilizing their SGRT systems. There are many
studies that demonstrate this approach and associated accuracy.[11-20]

SGRT assesses patient positioning by comparing a real-time surface
image of the patient to a reference surface image. The reference surface is
typically imported as a DICOM data file, created from the external contour
of the patient's skin surface from their planning CT scan. Prior to each
treatment session, the patient's position is observed and compared to the
reference surface by the system's surface matching software. When patient
movement from the reference surface is detected, the software calculates
new coordinates to adjust the patient for accurate realignment back into
the required treatment position. The positional differences are displayed
as differences for three translations (vertical, longitudinal, and lateral)
and three rotations (yaw, pitch, and roll). Commercial SGRT systems have

been designed to interface directly with certain third-party couch control systems, so that table shifts can be automated.

Real-time monitoring of the patient is also possible during beam delivery, and software verification tools are provided to determine any patient movement and to hold the radiation beam should a patient move out of a desired tolerance. Commercial SGRT systems also have beam hold capabilities through dedicated hardware interfaces with the treatment delivery system.

The introduction of these systems has meant that radiation therapists who were previously solely reliant on tattoos or marks on the skin surface (and their inherent limitations) for initial patient setup can now utilize SGRT systems which gives a 3D virtual "tattoo" and allows for a much-improved setup accuracy prior to any X-ray image guidance, as well as real-time position and motion monitoring during beam delivery.

21.4 CHALLENGES

Changes to practice, emerging technologies, new hardware and software, and new techniques are all a constant in radiation oncology today, and these changes can be driven by many different forces. These include hospital management, patients, and vendors. Yet, "70 percent of all change initiatives fail"[21] and do not accomplish their intended outcomes; this may even limit the potential of an organization and its people. The consequences of not managing change effectively can be devastating and long lasting, so it is important when thinking of making a change to practice, such as the implementation of tattoo/skin mark-less treatments, that all involved understand why this change is being made and the potential issues that may be encountered. In this way, staff can equip themselves with information and techniques to support the implementation of this new approach.

In navigating an effective change event, it is imperative to have a "champion for change," a member of the radiation therapy team that will take ownership and lead the process. This clinical professional such as a medical physicist must have the appropriate skills, knowledge, and credibility within their organization to lead this change in practice. This person must lead a team that has clear roles and responsibilities to ensure that processes are followed, risk is assessed, contingency plans are made, other staff have a voice, and clear communications are appropriately and effectively addressed as part of implementation. The entirety of the multidisciplinary team including physicians, physicists, and radiation therapists must be involved from the initial planning stages to ensure the long-term sustainability as the practice changes and the techniques evolve.

KEY POINTS

- With the introduction of an SGRT system, it is entirely possible to move toward a tattoo/skin mark-less setup for all body sites.

- SGRT systems allow for improved accuracy at time of setup and real-time position and motion monitoring during beam delivery.

- Converting from a tattoo-based method for patient setups to utilizing SGRT for patient setups has the potential to relieve the negative psychological impacts that can be caused by permanent body markings.

- It is highly important to include the entirety of the multidisciplinary team in order to make safe and effective changes to clinical practice such as the introduction of SGRT for tattoo/skin mark-less patient setup.

REFERENCES

1. Landeg SJ, Kirby AM, LeeSF, et al. A randomized control trial evaluating fluorescent ink versus dark ink tattoos for breast radiotherapy. *Br J Radiol.* 2016;89:20160288.
2. Rathod S, Munshi A, Agarwal J. Skin marking methods and guidelines: a reality in image guidance radiotherapy era. *South Asian J Cancer.* 2012;1:27–29.
3. Rafi M, Tunio MA, Hashmi AH, Ahmed Z. Comparison of three methods for skin markings in conformal radiotherapy, temporary markers, and permanent Steritatt CIVCO tattooing: patients comfort and radiographers satisfaction. *SAR.* 2009;47:20–22.
4. Smolenski MC. Tattooing method for radiation therapy. 2008. Updated June 26, 2008. Available from https://www.google.com/patents/WO2008074052A1?cl=un.
5. Breshnhoi A, Haedersdal M. Q-Switched YAG laser versus punch biopsy excision for iatrogenic radiation tattoo marks—a randomized controlled trial. *J Eur Acad Dermatol Venereol.* 2010;24(10):1183–1186.
6. https://www.cancerresearchuk.org/health-professional/cancer-statistics/survival#heading-Zero. Accessed December 3, 2019.
7. https://forum.breastcancernow.org/. Accessed December 3, 2019.
8. https://community.macmillan.org.uk. Accessed December 3, 2019.

9. https://www.cancerresearchuk.org. Accessed December 3, 2019.

10. Clew B, Allen J. Psychosocial impacts of radiation tattooing for breast cancer patients. *Can Woman Stud*. 2017;28(2,3):46–52.

11. Wen N, Snyder KC, Scheib SG, et al. Technical Note: evaluation of the systematic accuracy of a frameless, multiple image modality guided, linear accelerator based stereotactic radiosurgery system. *Med Phys*. 2016;43(5):2527.

12. Smith T, Ayan A, Cochran E, Woollard J, Gupta N. Characterization of a high-definition optical patient surface tracking system across five installations. *Med Phys*. 2016;43:3648.

13. Mancosu P, Fogliata A, Stravato A, Tomatis S, Cozzi L, Scorsetti M. Accuracy evaluation of the optical surface monitoring system on edge linear accelerator in a phantom study. *Med Dosim*. 2016;41(2):173–179.

14. Wen N, Li H, Song K, et al. Characteristics of a novel treatment system for linear accelerator-based stereotactic radiosurgery. *J Appl Clin Med Phys*. 2015;16(4):5313.

15. Lau SK, Zakeri K, Zhao X, et al. Single-isocenter frameless volumetric modulated arc radiosurgery for multiple intracranial metastases. *Neurosurgery*. 2015;77(2):233–240; discussion 240.

16. Wiersma RD, Tomarken SL, Grelewicz Z, Belcher AH, Kang H. Spatial and temporal performance of 3D optical surface imaging for real-time head position tracking. *Med Phys*. 2013;40(11):111712.

17. Cervino LI, Detorie N, Taylor M, et al. Initial clinical experience with a frameless and maskless stereotactic radiosurgery treatment. *Pract Radiat Oncol*. 2012;2(1):54–62.

18. Li G, Ballangrud Å, Kuo LC, et al. Motion monitoring for cranial frameless stereotactic radiosurgery using video-based three-dimensional optical surface imaging. *Med Phys*. 2011;38(7):3981–3994.

19. Li G, Ballangrud A, Kuo L, et al. Optical surface imaging for online rotation correction and real-time motion monitoring with threshold gating for frameless cranial stereotactic radiosurgery. *Med Phys*. 2011;38:3711.

20. Cervino LI, Pawlicki T, Lawson JD, Jiang SB. Frame-less and maskless cranial stereotactic radiosurgery: a feasibility study. *Phys Med Biol*. 2010;55(7):1863–1873.

21. http://execdev.kenan-flagler.unc.edu/blog/why-change-initiatives-fail. Accessed December 3, 2019.

CHAPTER 22

Introducing SGRT into the Clinic

A Radiation Therapist's Perspective

Ellen Herron and Daniel Bailey

CONTENTS

22.1 INTRODUCTION

External beam radiation treatments require intricate collaboration between multiple disciplines, including radiation oncologists, medical dosimetrists, physicists, radiation therapists, and nurses. Each medical professional in the team provides many vital aspects of analysis, calculation, image and measurement acquisition, quality assurance (QA), or patient care to ensure safe and accurate radiotherapy treatment. The radiation therapist provides the functional link between all the preparatory work (planning, calculations, QA, etc.) and actually delivering the treatment plan to the patient, and this essential role carries very unique challenges and responsibilities.

Patient setup and immobilization are two of the most important responsibilities of the radiation therapist. Unless the patient achieves an *initial setup orientation* identical (within appropriate thresholds) to the planned treatment position, and unless the patient *maintains* this orientation throughout the course of every fraction, then all the accuracy of machine performance and fine-tuning of the treatment plan fall short of high-quality patient care.

Traditionally, every patient is given a set of semipermanent or permanent marks placed during their initial simulation (most often 3D imaging via CT and marking via laser guidance). Typically, there are three orthogonal marks: anterior (for supine setup) and bilateral locations to triangulate the isocenter within the patient. These marks become the connection between the patient's treatment planning orientation and the day-to-day clinical orientation at the treatment machine. Often, multiple sets of physical marks are required: for example, the initial marks define the triangulated position (e.g., "user origin") relating the patient's orientation to the original simulation image set, but then there is a 3D displacement between this position and the actual treatment position (e.g., the "treatment isocenter") which is also physically marked by the therapists for daily setup purposes. Orthogonal lasers, aligned to the linear accelerator (linac) isocenter, are referenced to manually set up the patient to the treatment marks, while image guidance allows the therapists and physician to fine-tune the setup position due to daily setup uncertainties (mechanical, anatomical, etc.). If the setup of the patient proves difficult for any number of reasons, more marks may be placed to assist in daily setup. Alternatively, if setup reproducibility is an ongoing concern, the patient can be resimulated and/or replanned.

Even this cursory overview of initial patient setup reveals a number of potential challenges and inefficiencies for the radiation therapist (leaving aside, for the purposes of this chapter, the mechanical and quality assurance difficulties associated with laser- and image-guidance). Marks are only as accurate as the human ability to place them, and—if not permanent—may be lost or washed away. Daily setup to physical marks is only as accurate as the human ability to visualize and manually manipulate the patient to the desired treatment position; patient compliance or problematic anatomical issues further complicate manual setup to physical marks. These marking and setup processes can be time-consuming and ultimately inexact, often requiring the therapists to split the difference between marks that do not perfectly align to the mechanical indicators. In these cases, the patient is at best set up with some inherent uncertainty, whereas daily imaging to verify and tweak patient setup requires ionizing radiation and additional dose to normal tissues not typically accounted for during treatment planning. Meanwhile, from the patient's perspective, traditional marking and setup techniques require the patient to wear semipermanent marks for at least the duration of external beam treatment, potentially creating cosmetic, emotional, and psychological concerns.[1-3] And if the marks are permanent, the patient wears these reminders of radiation treatment, and their experience with cancer, for life.

Patient immobilization and intra-fraction positional constancy are similarly challenging in the skin mark-based setup approach. After aligning the patient to marks and even fine-tuning the treatment position via image guidance, traditional treatment systems rely on closed-circuit camera monitoring—often with images distal to the patient and at nonoptimal angles—for the radiation therapist to verify that the patient is not moving during treatment beyond whatever margin is deemed allowable by the physician. While some treatment systems allow forms of motion monitoring, via surgically implanted beacons, near-continuous ionizing radiation, or an external surrogate (e.g., a trackable jig), typical treatments have no ability to gauge the magnitude or severity of patient motion while the treatment beam is on.

Surface-guided radiation therapy (SGRT) systems introduce a paradigm shift in patient setup and visualization, now enabling radiation therapists to minimize or even eliminate many of the uncertainties, inefficiencies, and other concerns associated with traditional setup marks. A typical SGRT system comprises a number of cameras that track the patient's external contour (confined to a carefully chosen region of interest (ROI), if necessary) as a 3D matrix of thousands of comparison points and refresh rates at a fraction of a second. Consequently, SGRT facilitates skin mark-less patient setup, real-time motion management options (such as deep inspiration breath hold (DIBH) and respiratory gating treatments), and near real-time intra-fraction positional monitoring of the patient's actual body contour (rather than an external surrogate) without the use of additional ionizing radiation. Depending on the anatomical target location and treatment position, SGRT may be used as a stand-alone setup and monitoring system, or in conjunction with image-guided radiation therapy (IGRT) as the gold standard of actual target location.

22.2 EDUCATION AND TRAINING

Appropriate education and training for all users before implementation of an SGRT program is vital for the safe, accurate, and efficient use of the SGRT system. For each radiation therapist, training must include at minimum:

- Functionality of all hardware, accessories, and relevant software

- Daily warm-up, quality assurance (QA), and shutdown procedures

- Rights and limitations for user profiles and groups (ultimately at the discretion of the department)

- Data import and patient management

- ROI selection and other parameters unique to each patient

- Reference and treatment image capture techniques

- Beam gating capability (if available)

- SGRT system to treatment delivery system communication and coordination (if available, e.g., shifts based on IGRT)

It is highly recommended that at least one user from each institution (and preferably from each site, if there are multiple treatment centers) receive a complete training course from the SGRT system vendor. In addition, each vendor typically includes a few days of on-site training for the treatment team with the acquisition of a new SGRT system. Functional use of the system can be readily passed on from user to user, but there is no substitute for comprehensive training of at least one active member of the treatment team.

In addition to formal training, as the SGRT user community continues to grow, the accessible online library of videos, webinars, and user forums is steadily growing, proving to be a valuable asset for initial and continued education in SGRT techniques. In our experience, it is helpful to have a follow-up round of education from the vendor—even if just an extended conference call or single day with an onsite expert—after the initial break-in period of a few weeks. As each therapist and physicist learns more about the system, questions and concerns will undoubtedly arise of a more advanced nature than were anticipated during initial training. It benefits the entire team to address these issues formally with a trainer providing the answers. In our clinic, we also recommend a weekly therapy-physics small group meeting or "huddle" once a week focused on SGRT questions, for at least one month after an SGRT system is installed. Also, it is very important to close the feedback loop: when questions arise and are answered, that vital information needs to be circulated to all the staff whose responsibilities impact the effective use of the SGRT system.

To provide a way for every team member to comprehensively learn to use a new SGRT system, while enhancing its benefits for patients, we found it advantageous to implement the system to assist in setup and for in vivo monitoring of every patient. There may be initial concerns that this additional process employed for *every patient* might require additional time and effort in the treatment workflow, and in the very beginning stage this is somewhat true. However, in our experience, the payoff of rapidly learning the SGRT system, and learning to trust the system as compared to traditional treatment aids (e.g., lasers, marks, indexing, etc.), is certainly worth the extra initial effort. Our treatment teams agree that in a relatively short uptake time (on the order of days, not weeks) the benefits of SGRT setup accuracy and motion monitoring not only ensured better treatment accuracy but made virtually every treatment faster (detailed in a subsequent section).

22.3 QUALITY ASSURANCE AND CULTURE OF SAFETY

An SGRT program requires a comprehensive QA program to ensure the performance accuracy and constancy of the monitoring system and computer control system. Such a QA program entails recommended tests conducted on regular time intervals delineated by differing levels of intensity (and is discussed elsewhere in this book).[4,5] Each radiation therapist involved in SGRT treatments must have basic knowledge of the SGRT QA program and may be required (depending on institutional policy) to participate in QA and safety checklist tasks. For example, it is often customary for therapists involved in daily warm up and QA of

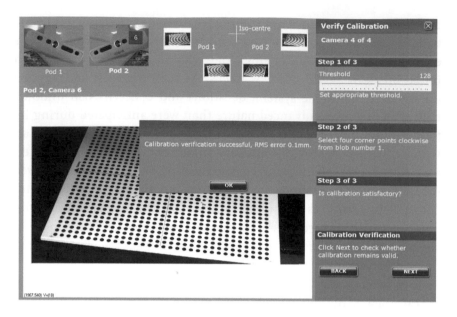

FIGURE 22.1 Example of daily SGRT QA for the AlignRT system (Vision RT Ltd., London, UK). This system's recommended daily QA involves a series of images of a calibration plate placed on the treatment couch with which the control software compares each camera to the other system cameras and to a series of baseline images of the same grid (typically captured at system commissioning and on a monthly or annual basis thereafter).

the linac to also perform daily QA of the SGRT system, typically involving a fast constancy check of the SGRT system isocenter, as shown in Figure 22.1.

The radiation therapists may also perform import tasks with patient data. For the data import process, a step-by-step procedure should be created and followed for every case, including verification of:

- Patient name and identification number

- Plan name(s), isocenter(s), and anatomical group(s)

- Patient orientation(s)

- Reference surface label(s)

- ROI

- Monitoring pass and fail thresholds in multiple dimensions

As with any complicated and sequential process, a safety checklist or questionnaire for each import session is a valuable tool to ensure that each import is performed completely and as prescribed by vendor recommendations and institutional policy.

Whereas the medical physics team usually oversees the requisite QA tests and intervals, the QA and safety program as a whole requires input from all SGRT system users. Thus, at least one representative from each discipline should be included on the QA and safety planning and review team, including a radiation therapist. User rights, training requirements, QA responsibilities, documentation, incident learning reporting, and ongoing SGRT program quality improvement are examples of matters that should be managed at the multidisciplinary team level.

22.4 CLINICAL IMPLEMENTATION

SGRT facilitates improved accuracy and efficiency during both initial patient setup and in continuous monitoring of patient positioning during treatment. While the image-guidance portion of treatments remains largely unchanged (to verify geometry of internal anatomy), the following subsections explore areas in which SGRT has proven the most beneficial in our clinical processes.

22.4.1 Initial Patient Setup

With SGRT, the initial positioning of the patient prior to treatment transitions from skin mark-based triangulation via isocentric lasers to optically tracking multiple points within an ROI on the patient's surface. The SGRT software provides shift magnitudes and directions in 6 degrees of freedom, typically including three-dimensional translation along with pitch, roll, and yaw rotations (Figure 22.2), to bring the patient into the expected position by either manual shifting or automatic couch movement (if the treatment delivery system is so equipped). SGRT is well-equipped to replace skin mark-based patient setup and complements other setup QA tools such as couch indexing, established tolerances for shifts and changes in setup parameters (e.g., couch tolerance tables), and source-to-surface distance (SSD) measurements.

The tremendous amount of data provided by the SGRT system stands in stark contrast to the limited setup information available via the traditional skin mark-based approach. Rather than using three points on the body (via triangulation) to position the patient relative to the linac isocenter, the SGRT system (calibrated to share an imaging isocenter with

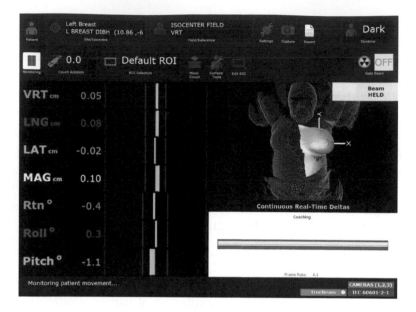

FIGURE 22.2 A screen capture of an SGRT setup for deep inspiration breath hold in which the expected patient position is compared to the current patient position and shift/rotation magnitudes are reported in real-time. In this example, the patient is coached to precisely hold her breath, enabling the therapists to compare her breath-hold thorax position to the expected external body contour from the treatment plan and evaluate achievability and uniformity.

the radiation isocenter) enables evaluation of entire portions of the body habitus and 3D alignment to the desired body position. With skin marks, the patient's body can often be manipulated such that the marks align relatively well with the wall lasers even though the desired treatment position is not achieved. By contrast, SGRT tracks many points on a patient's surface so the mathematical probability is low of having a close match to a reference surface while tracking the wrong anatomy or with the patient in the wrong setup position. As long as the patient has stable and comfortable immobilization (and relatively unchanged anatomy during the course of treatment), most body positions can be accurately and repeatedly reproduced using SGRT.

SGRT initial positioning is particularly useful in cases of difficult patient setup—for example, with rotations that are difficult to anticipate, or skin marks that are very mobile due to external patient anatomy. In the skin mark-based approach, the radiation therapists often spend a large amount of time chasing after the desired body position, sometimes requiring many radiographic images and multiple trips into the treatment room. However, even in cases that were once challenging to setup, SGRT enables the therapists to quickly achieve a treatment position that is very close to the desired position before exiting the room and proceeding with prescribed imaging procedures.

The setup process is greatly improved by SGRT for treatments involving extremities, since the mobility of the extremity makes skin mark-based setup particularly challenging. While completely encasing the extremity in an immobilization device is often required by traditional setup and monitoring methods, SGRT enables the radiation therapists to use much less immobilization material—just enough to achieve a comfortable, stable position, but not so much that the SGRT cameras are obstructed. An ROI should be carefully chosen to exclude immobilization devices and should involve a patient surface that is sufficiently nonuniform. For example, a small ROI on a uniform, flat portion of a patient's leg may not provide an accurate assessment of the position of the whole area of interest for treatment setup. Some SGRT systems provide the software capability of acquiring a static frame of the patient's external surface for real-time comparison during setup (Figure 22.3). This "treatment capture" surface is potentially very useful when positioning a body part like a breast or an extremity by enhancing the therapists' ability to visualize the desired treatment position, thereby achieving a close match before fine-tuning via the translation and rotation data provided by SGRT analysis. These techniques have proven very helpful to match arm position for breast and axilla patients, and for general positioning issues with prone breast cases.[6]

SGRT also has distinct advantages over traditional setup techniques for many electron treatments, particularly when skin marks are almost impossible to utilize or maintain. For example, open-skin wounds are very challenging for skin marks whereas SGRT matches the entire surface contour within the ROI (carefully selected to avoid areas of skin that are rapidly degrading or healing). Similarly, electron treatments of the hands and feet (such as are used to treat Ledderhose Disease or Dupuytren's Contracture) have all of the setup difficulties due to mobility (discussed above) along with impossibility of maintaining skin marks: in these cases, SGRT has proven valuable in setup efficiency and accuracy, even for treatments that are several weeks apart.

When using SGRT for electron setups, it is usually necessary to perform all required alignments and shifts before placing the electron cone in place, in order to achieve maximum visibility for the SGRT cameras. Once

FIGURE 22.3 Extremity setup assisted by "treatment capture" surface contour shown in green. In this case, the green point cloud anatomy represents the current position of the patient's arm (as seen by the visible speckle pattern tracked by the SGRT system), while the pink shows the *intended* (reference) position. With a treatment capture image acquired after setup verification on the first day of treatment (or dry run), each subsequent setup is much faster: first the treatment position is roughly achieved via comparison to this image, then fine-tuned via the SGRT shift and rotation analysis.

the patient is in position and the cone is in place, an initial photograph of the light field versus patient anatomy is acquired to compare to the treatment field each day before treatment as an additional setup tool and safety measure.

22.4.2 Intra-fraction Patient Monitoring

Traditional, nonionizing methods of patient monitoring during treatment typically consist of audio monitoring in combination with closed-circuit television (CCTV) visual monitoring. CCTV, typically with two cameras at different angles, is helpful to identify large patient movement, but is limited by the lack of quantitative shift data and limited capability to adjust viewing angles, zoom, resolution, light issues, etc. With the implementation of SGRT, CCTV monitoring is complemented with software-based matching of patient contour data that refreshes in near real-time (on the order of several times per second) and provides submillimeter and sub-degree quantitative feedback.

This new level of positional feedback has proven invaluable for accurately discerning out-of-tolerance patient movements. SGRT has demonstrated that patients move more often and with greater magnitude than observed with CCTV monitoring alone, even in situations where tighter immobilization is intended to drastically restrict patient movement, like SBRT. For example, during setup and verification imaging for an SBRT lung treatment, after successful image guidance and immediately preceding initiation of treatment, the SGRT system indicated a sizeable shift in patient position. However, the physicist and therapists present at the console were not able to discern any change via CCTV camera views. At that point, a new cone-beam CT (CBCT) was acquired, and which indicated a shift of approximately the same magnitude back to treatment position (as judged by the physician, in comparison to the planning CT). In this case, discomfort in the immobilized arm position caused the patient to adjust their shoulder and elbow. The motion was small enough that the treatment team did not see it on the visual monitor, but the SGRT system accurately indicated a shift that exceeded our clinic's SBRT tolerance levels.

SGRT is also helpful with identifying anatomy changes that could result in unacceptable delivered dosimetry. When a patient is in approximately the expected treatment position, any residual discrepancies in the SGRT match demands investigation before treatment. In one interesting case at our center, SGRT-based real-time analysis was being used to track the chest wall skin surface of a breast cancer patient. During fraction 13 in her course of radiotherapy, utilizing a DIBH technique, the SGRT system detected that the patient was unable to achieve a chest wall position that sufficiently matched the planned chest wall position. At this point, the radiation therapists acquired port films and discovered that the patient's breast-tissue expander had shifted internally: the surface image acquired at treatment indicated that the treated breast was no longer in a position similar to the original CT planning images (see Figure 22.4). Consequently, the patient was resimulated and replanned.

These are two clinical examples of the accuracy and utility of an SGRT system to reveal unacceptable changes in patient position or body habitus so that the therapists can quickly intervene. SGRT enables a level of attention and quantitative monitoring that is impossible with standard audio and visual monitoring alone. Particularly in the

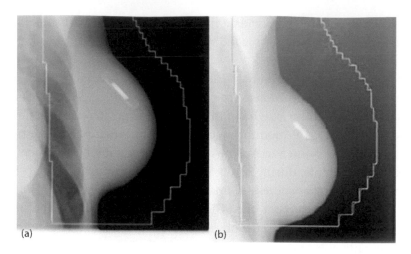

FIGURE 22.4   A breast radiotherapy case in which the patient, midway through her treatment course, was unable to achieve the required breath-hold treatment position as determined by an SGRT comparison of real-time patient position to both the most recent SGRT reference image and the original DICOM external surface contour from the treatment planning system. Figure 22.4a shows the expected portal image from the digitally reconstructed radiograph data, while Figure 22.4b shows the actual anatomy via MV portal imaging, demonstrating a breast-tissue expander that had internally shifted, altering the shape and position of the treated breast. The patient was simulated again and replanned.

realistic treatment setting in which multiple monitors require attention and sometimes interaction, and where distractions inevitably occur, the treatment team reaches the limit of human ability to constantly watch a CCTV screen and identify the subtle patient movements that could significantly and negatively impact delivered dose. These observations are particularly true in special treatment settings like the aforementioned DIBH or stereotactic procedures in which relatively small changes in position may produce dose distributions that totally miss the intended target. Since SGRT does not provide real-time positional information of internal anatomy, these systems do not substitute for IGRT. Rather, SGRT external monitoring is an excellent complement to IGRT: the desired treatment position is fine-tuned via the gold standard of IGRT, a new external contour reference image is captured, and subsequently the SGRT system provides treatment position stability information from beam on to beam off. While SGRT may potentially reduce the *frequency* of radiographic imaging, the case-by-case decision to image less frequently or extensively belongs to the physician in coordination with the treatment team.

The above case studies also demonstrate the vast improvement of surface tracking beyond the tracking of an external surrogate like a radiopaque box or jig placed on the patient's surface. SGRT enables the therapists (along with physicists and physicians if attending treatment) to monitor the actual anatomy external to the treatment volume itself, simultaneously removing the uncertainties associated with placement of an external object and providing submillimeter analysis of the patient's actual body position. The only situations in which SGRT might require monitoring of an area of the body distal to the actual treatment area are those in which some aspect of treatment obstruct the view of the SGRT cameras (e.g., an immobilization device, bolus, or electron cone). In these situations, an ROI drawn on an adjacent part of the body (e.g., the contralateral breast in an electron breast treatment) provides an approach to continue real-time SGRT intra-fraction monitoring.

Selection of an appropriate ROI is vital for all cases of SGRT setup and intra-fraction patient monitoring. Typically, the vendor of the SGRT system provides instruction and tips for drawing ROIs as part of standard training, including objects or areas to avoid and how to handle special cases. Briefly, the ROI must be specific enough to provide position and movement data relevant to the treatment position, while excluding mobile (e.g., extremities not rigidly connected to the treatment area) or changing (e.g., severe skin desquamation) anatomy that is not absolutely vital to the treatment plan, and also excluding objects like sheets, clothing, immobilization devices, etc. Closed-face masks are not compatible with SGRT, though masks can often be cut in such a way that the anatomy underneath is sufficiently trackable and potentially as immobilized as with frame-based systems.[7,8] The use of open-face masks for intracranial stereotactic treatments and for head and neck treatments is discussed in Chapters 12 and 17, respectively. A number of vendors now supply open-face masks and other head immobilization options that are SGRT compatible. SGRT is useful in treatments requiring bolus—particularly in allowing both bolus and non-bolus reference images, for ease of setup prior to placing the bolus, and intra-fraction monitoring (with bolus, or alternating bolus/non-bolus). Water-equivalent bolus, brass mesh, and wet-towel or gauze are all potentially trackable with SGRT, as long as the treatment team addresses reflection and stability issues. For example, our treatment teams observe that using polyethylene wrap around water-equivalent bolus reduces the surface glare enough to improve SGRT tracking. Similarly, brass mesh bolus with a layer of paint to eliminate reflection is compatible with SGRT tracking. Pre-painted brass mesh bolus is now commercially available (Radiation Products Design Inc., Albertville, MN, USA).

22.5 TECHNICAL CONSIDERATIONS

22.5.1 Virtual SSD Measurement

One of the SGRT system's capabilities found to be most helpful to the radiation therapist in our network is the ability to capture virtual SSD measurement, performed quickly and easily from outside the treatment vault. As long as the cameras have an unobstructed view to the patient's surface, the system uses visual surface and isocenter information to calculate the SSD at any user-defined virtual gantry angle. This procedure is faster and more accurate than reading the SSD with the treatment machine's optical distance indicator, which is sometimes difficult to read and thus often inconsistently read depending on the therapist.

22.5.2 Replication of Lost or Damaged Immobilization Devices

SGRT also provides the treatment team the capability of reproducing an immobilization device, such as a vacuum bag, if lost or damaged. The SGRT workstation has the original external surface contour from the treatment planning system, so with the assistance of radiographic imaging (typically a CBCT), the therapists can reproduce the intended patient position on the treatment table and then replicate the immobilization device configuration and setup in the treatment vault (see the schematic workflow in Figure 22.5). This technique has been used on numerous occasions, including head and neck (mask), pelvis, and breast anatomies, in which an immobilization device malfunctioned or was misplaced, even for patients planned on other similar treatment machines not equipped with an SGRT system.

22.5.3 Troubleshooting and Planning for SGRT Down Time

The vast majority of technical issues are typically solved by following the vendor's shutdown and start-up procedures, and we have empirically observed better system performance when we rigidly adhere to the vendor's recommendations for normal shutdown and start-up intervals.

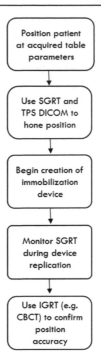

FIGURE 22.5 Workflow for the replication of a lost or damaged immobilization device in the treatment room using SGRT.

If the department employs treatment techniques and policies that *require* SGRT, a troubleshooting and down-time plan needs to be constructed in collaboration with the entire treatment team. For example, DIBH, stereotactic treatments, and skin mark-less treatments (discussed in the following section) might all require significant action if the SGRT system is rendered inoperable for an extended amount of time. In our department, this type of technological problem warrants moving patients to other treatment machines, sometimes at other treatment clinics, in order to proceed with the same treatment plan. If resources were more limited, it is possible that resimulation and replanning might be required, depending on the instructions of the physician. In any event, the occurrence of such situations can be anticipated and thus should be planned for in the early stages of commissioning an SGRT program. In addition, as new linacs are acquired, redundant SGRT capabilities is potentially an important factor to consider, similar to redundancy in treatment energy and other technological capabilities.

22.6 SGRT AND THE FEASIBILITY OF TRULY SKIN MARK-LESS RADIATION THERAPY

Skin mark-based radiotherapy may be suboptimal due to a number of setup uncertainty and logistical issues. For example, temporary marks can be washed off, lost, or cause skin reactions; permanent tattoos carry the risk of infection, allergies, and other needle-based complications; meanwhile, patients may experience social and psychological issues associated with permanent or even semipermanent skin marks for radiotherapy localization and increasingly request treatment without marks.[1-3] There are a number of potential logistical, safety, and fiscal benefits of eliminating skin marks. Transitioning to skin mark-less treatments highlighted the stress that marks, and maintaining marks, added for the patient, during an already highly stressful time of life. In the traditional approach, lost marks inherently meant additional radiographs and imaging dose and additional time on the treatment table. For all these reasons, the move to SGRT made a new way forward without treatment marks both tenable and highly desirable.

22.6.1 Step 1—Team Planning

In order for a clinic to pursue skin mark-less treatments, a number of multidisciplinary meetings were held for all team members to map out the treatment process and brainstorm potential difficulties or safety concerns. Each meeting was attended by at least a physicist, lead therapist, dosimetrist, and a clinic manager. A pilot program was crafted to collect statistical information including time spent setting up, total treatment time, number of radiographic images acquired, and shift magnitudes with skin mark versus mark-less setup, trying to minimize the number of variables (such as keeping consistency in the therapists that worked with each patient). Data was collected on all patients for three months, comparing each patient's skin mark-based setup data with respective skin mark-less setup data (adhering to the same IGRT procedures prescribed by the physician), amounting to approximately 30 patients in all.

22.6.2 Step 2—Pilot Program Results

At the conclusion of the pilot period, our data showed that skin mark-less SGRT patient setups reduced the amount of time the therapists spent in the room by about one minute per fraction, while each fractional setup time was more consistent (lower standard deviation). SSDs compared to the treatment plan remained consistent, as expected, between the two types of setups (the goal in each case being to achieve a similar accurate treatment position), but the requisite shift magnitudes (calculated via IGRT) were reduced by half on average for SGRT setups, indicating more consistently accurate initial patient position as opposed to skin mark-based triangulation. It was found that the greatest time savings were in scenarios like initial verifications, difficult/lost skin marks, and any treatment in which the angular orientation of the patient's body (e.g., roll) was difficult to achieve via skin marks. Overall, it was concluded that SGRT setup is faster, more accurate, and less variable than traditional radiotherapy marks. A sample workflow of patient setup without marks, particularly for breast treatment, is shown in Figure 22.6.

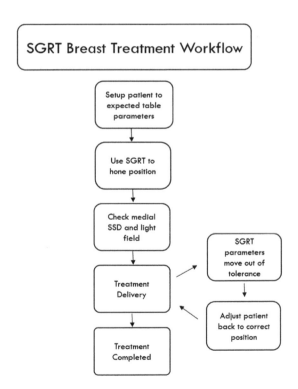

FIGURE 22.6 Workflow for the setup and treatment of a skin mark-less breast patient via SGRT and IGRT.

22.6.3 Step 3—Clinical Implementation Results

There are additional safety benefits to SGRT-based treatment setup over the skin mark-based technique. While it is possible for the wrong patient or the wrong treatment area to be aligned to the lasers via marks and subsequently treated without noticing the error, it is highly unlikely that a different patient or alternate treatment site could be aligned to the expected external contour matrix without immediately recognizing the discrepancy via the SGRT system. Meanwhile, the problem of aligning to inaccurate marks is exacerbated for patients with multiple sets of marks – a problem that becomes more acute as an increasing number of patients return for additional treatment.[9,10] SGRT eliminates these safety issues peripheral to the use of skin marks. At the same time, employing surface guidance rather than marks eliminates many potential sources of inconsistency between the setups of multiple therapists. Multiple marks, visibility of marks and precision of laser alignment, shifts from marks, and manually setting SSD are all unnecessary with SGRT-to-isocenter patient alignment, allowing for much better uniformity with all setups due to less ambiguity and uncertainty in aligning to visual indicators. Similarly, whereas skin marks allow one therapist to view only the marks visible from his or her own position with respect to the patient and the treatment machine, SGRT allows all therapists in the room to view the same monitor simultaneously with the same submillimeter quantitative indicators of patient position, thereby allowing fast, mutual verification before leaving the treatment room.

As mentioned in the previous section, skin mark-less treatment does add an additional layer of dependency on the SGRT system. In the event that an SGRT system is inoperable for an extended period of time, particularly if it is the only SGRT system available in the treatment clinic, the treatment team must have a plan for how exactly to handle treatment setups during the SGRT downtime. Resimulation and replanning is potentially a solution for some cases, while having the resource options to treat at an alternate clinic is a tremendous benefit. However, for routine treatments of patients without marks in the event of a down SGRT system, the following approach can be used:

1. Set up the patient to the most recent table indices and immobilization device settings

2. Verify SSD and light field and flash if applicable

3. Verify treatment position via image guidance in coordination with the physician

4. Place temporary skin marks for ease of subsequent treatment setups (until the SGRT system is fully functional again).

Such instances may be rare but are well within the realm of possibility and should be planned for in advance.

22.7 PATIENT SATISFACTION

When originally discussing the idea to move to skin mark-less setups, a number of treatment team members and administrators inquired as to whether or not patients would specifically choose treatment at a particular center to avoid the use of skin marks. At that initial stage, the hypothesis was that many patients would in fact make that choice. After several months of data, including patient comments and feedback via satisfaction surveys, patients do indeed value skin mark-less treatments. Many patients even choose to travel farther distances for the option of skin mark-less treatment (in addition to seeking other published benefits of treatments well-facilitated by SGRT, like DIBH and SBRT).

A number of patients have shared stories of friends, their children, or even random strangers asking them about their skin marks—and in most of these instances, the patients wished to avoid conversations and reminders like these altogether. With skin mark-less SGRT treatments, the patients gain more control of their own bodies, and can have those conversations when and if they feel ready—rather than being forced to converse about their radiotherapy skin marks due to the curiosity of other people. Allowing patients to maintain a level of normalcy as high as possible is a top priority, so that they can focus on relaxation and healing. SGRT with skin mark-less treatment setup enables them to avoid the stress of receiving and retaining skin marks.

22.8 SUMMARY

An appropriately commissioned, logically implemented SGRT program enables clinics to treat patients with a level of safety, efficiency, and accuracy previously unachievable via traditional means of radiotherapy. Ease and consistency in patient setup, and state-of-the-art intra-fraction motion monitoring are two of the most extensively improved areas of treatment quality with SGRT systems. Thorough training, including periodic refresher sessions, for all members who use the SGRT system is vital for the successful implementation of SGRT technology, as well as a detailed QA program that satisfies vendor-specific tests and current professional guidelines. During the commissioning and training period, trust in the accuracy and reliability of the SGRT system for routine treatments and special procedures like DIBH and stereotactic techniques will be developed. In addition, SGRT can enable clinics and patients to benefit from the appropriate use of a skin mark-less setup technique that improves setup accuracy and efficiency beyond traditional skin mark-based methods, simultaneously improving patient satisfaction along with their quality of care.

KEY POINTS

- With proper commissioning, quality assurance and staff training, SGRT enables wide ranging benefits to safety, efficiency, and quality in the clinic.

- In addition to stereotactic and breast applications, SGRT can be used to reduce the frequency of setup imaging, perform virtual SSD measurements, replicate lost or damaged immobilization devices, and aid in difficult electron treatment setups.

- SGRT facilitates the use of skin mark-less setups, enhancing setup accuracy and efficiency as well as patient satisfaction.

- Clinics implementing SGRT into their practice should develop contingency plans in the event the SGRT system is not available for treatment.

REFERENCES

1. Landeg S, Kirby A, Lee S, et al. A randomized control trial evaluating fluorescent ink versus dark ink tattoos for breast radiotherapy. *Br J Radiol.* 2016;89:1068.
2. Clow B, Allen J. Psychosocial impacts of radiation tattooing for breast cancer patients: a critical review. *Can Woman Stud.* 2010;28:46–52.
3. Stanley DN, Mcconnell KA, Kirby N, Gutiérrez AN, Papanikolaou N, Rasmussen K. Comparison of initial patient setup accuracy between surface imaging and three point localization: A retrospective analysis. *J Appl Clin Med Phys.* 2017;18(6):58–61. doi:10.1002/acm2.12183.
4. Paxton AB, Manger RP, Pawlicki T, Kim GY. Evaluation of a surface imaging system's isocenter calibration methods. *J Appl Clin Med Phys.* 2017;18(2):85–91. doi:10.1002/acm2.12054.

5. Willoughby T, Lehmann J, Bencomo JA, et al. Quality assurance for non-radiographic radiotherapy localization and positioning systems: Report of Task Group 147. *Med Phys.* 2012;39(4):1728–1747. doi:10.1118/1.3681967.

6. Li S, DeWeese T, Movsas B, et al. Initial validation and clinical experience with 3D optical-surface-guided whole breast irradiation of breast cancer. *Technol Cancer Res Treat.* 2012;11(1):57–68. doi:10.7785/tcrt.2012.500235.

7. Zhao B, Maquilan G, Jiang S, Schwartz DL. Minimal mask immobilization with optical surface guidance for head and neck radiotherapy. *J Appl Clin Med Phys.* 2018;19(1):17–24. doi:10.1002/acm2.12211.

8. Wiant D, Squire S, Liu H, Maurer J, Lane Hayes T, Sintay B. A prospective evaluation of open face masks for head and neck radiation therapy. *Pract Radiat Oncol.* 2016;6(6):e259–e267. doi:10.1016/j.prro.2016.02.003.

9. Nieder C, Langendijk J. *Re-Irradiation: New Frontiers.* Berlin, Germany: Springer; 2017. doi:10.1007/978-3-319-41825-4.

10. Dörr W, Stewart F. Retreatment tolerance of normal tissues. In: Joiner M, van der Kogel A, eds. *Basic Clinical Radiobiology.* 4th ed. London, UK: Hodder Arnold; 2009:259–270. doi:10.1201/9780429490606.

CHAPTER 23

SGRT for Proton Therapy

Adam B. Paxton

CONTENTS

23.1 INTRODUCTION

Proton therapy was one of the first radiation therapy modalities to utilize in-room image guidance. However, since first introducing planar kV imaging for proton therapy in 1954,[1] the continued addition of image-guided radiation therapy (IGRT) technologies has not progressed at the same pace as in linac-based radiation therapy.[2] Linac installations commonly include onboard kilovoltage (kV) imaging systems capable of 3D cone-beam computed tomography (CBCT) as well as planar kV imaging. In addition to onboard kV imaging, kV imaging in unique geometries (e.g., ExacTrac,

Brainlab, Munich, Germany), and 2D and 3D megavoltage (MV) imaging, linac-based treatments have also seen other nonionizing imaging technologies introduced and utilized for patient setup and tracking during treatment.[3] These include ultrasound, magnetic resonance imaging (MRI), and surface imaging. The use of ultrasound in proton therapy has been rare.[4] MRI guidance is an area of active interest in proton therapy due to its superior soft tissue contrast,[5] and it has been confirmed to be dosimetrically feasible to combine real-time MRI guidance with proton treatment delivery.[6] Surface imaging in proton therapy is also starting to receive more interest.[4]

As with linac-based treatments, surface-guided radiation therapy (SGRT) provides the ability to streamline the proton therapy patient setup process for a number of treatment sites, and may even afford the potential to transition to a tattoo-less setup paradigm.[7] In certain settings, SGRT also provides the potential ability to track proton therapy patients during treatment and quantify their offsets from the intended position while not delivering additional imaging dose.

However, there are fundamental differences between linac-based treatments and proton treatments. These include the dose distributions that are achievable. As protons begin traveling through tissue, they gradually lose energy and the absorbed dose is deposited nearly uniformly with depth. However, as protons near the end of their range, there is a sharp increase in energy loss. This results in a sharp increase in dose near the end of the protons' range known as the Bragg peak. After the Bragg peak, the dose drops essentially to zero. In order to cover a treatment target, the Bragg peak is "spread out" by varying the energy of the incident proton beam. Because protons deposit their dose and stop within tissue, proton treatment beams have no exit dose, whereas a large portion of MV X-ray beams from a linac pass through the patient. These differences, which are illustrated in the Figure 23.1, give proton treatments the advantage of being able to deliver the same therapeutic dose to a target with much less integral dose to the patient. However, proton treatments are more sensitive to changes in the water equivalent thickness (WET) between the incident surface and the target. A clinical scenario where a WET change is possible is weight gain in the patient. This example of patient change is also illustrated in Figure 23.1 by the curves labeled "+1 cm." These curves have an additional 1 cm of water added to the entrance of the respective beams. The X-ray curve shows a small offset in depth dose with the added water, but the proton curve is shifted by the 1 cm change in WET.

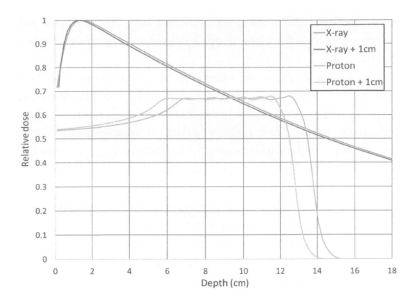

FIGURE 23.1 Plots of relative dose versus depth for proton spread out Bragg peaks and 6 MV X-ray beams in water. The curves labeled "+1 cm" show the effects of adding 1 cm of water to the beam entrance.

The previously mentioned and other fundamental differences between linac-based treatment and proton treatment lead to both unique benefits and challenges when implementing SGRT for proton therapy. This chapter discusses these benefits and challenges.

23.2 SGRT BENEFITS FOR PROTON THERAPY

23.2.1 Patient Setup

SGRT can be used for accurate patient setup and for the assessment of the posture of the patient relative to the time of CT simulation.[8-10] Chang et al.[8] compared the setup accuracy of surface imaging to laser-based alignment and kV planar imaging for partial breast patients being treated on a linac and concluded that SGRT was the most accurate patient setup method in the study. Similarly, Batin et al.[10] compared the setup accuracy of surface imaging and a traditional planar radiographic technique for postmastectomy chest wall patients treated with proton therapy. They found that surface imaging provided both more accurate and faster patient setups compared to a radiographic setup technique. This same group went on to demonstrate that surface imaging provided sufficient accuracy in post-mastectomy chest wall patient setups to allow a reduction in X-ray imaging frequency.[11]

For the previously discussed examples, the target areas were the breast or chest wall. For these treatment sites, surface imaging is able to directly visualize the treatment region of interest (ROI) and therefore provide accurate treatment setups. For target areas that are located deeper within the body, the surface of the patient may not correlate directly with the location of the target. Additional IGRT technologies are needed in these cases for an accurate patient setup. However, SGRT can still provide information on the posture of the patient prior to the application of radiographic IGRT methods, potentially allowing for patient position corrections that may not be correctable with rigid translations or rotations alone. This SGRT functionality may also reduce the need for repeated radiographic imaging, increasing setup efficiency while reducing imaging dose.

Treatment of pediatric patients is an indication for proton therapy.[12] Limiting imaging dose, particularly in this patient group where survival can extend many decades beyond treatment, is important for the potential prevention of secondary malignancies. As discussed above, SGRT may facilitate a reduction in imaging dose by providing patient position information prior to radiographic imaging. Sueyoshi et al.[7] described a workflow used at the Children's Hospital Los Angeles in which SGRT was used to make small refinement shifts of pediatric patients prior to kV image guidance. SGRT also contributed to this group's ability to transition to a workflow that eliminated the use of tattoos and skin marks. Pediatric applications of SGRT are discussed in detail in Chapter 19.

23.2.2 Patient Tracking during Treatment

In addition to patient setup, surface imaging has been used to track the position and motion of patients during treatment.[13] This monitoring can help to ensure that unintended patient motion beyond a predefined action threshold does not occur by providing a quantification of the offset of the real-time patient position relative to a reference position, typically captured immediately after IGRT-based setup. If patient motion is beyond the set threshold, the treatment can be stopped, and the patient repositioned as needed.

For intensity modulated proton therapy (IMPT), robust treatment planning methods are often used to account for setup and proton range uncertainties. In robust planning, the user enters values for setup and range uncertainty and the optimization process works to create a plan that still meets target coverage and organs-at-risk (OARs) avoidance goals with these uncertainties accounted for. If any patient motion during treatment is larger than the setup uncertainty distance entered for the robust planning optimization, the quality of the delivered treatment will decrease.

Mutter et al.[14] utilized SGRT during IMPT of postmastectomy patients with breast expanders to track the patient position. During their planning process, they used setup uncertainty values of ±5 mm. Likewise, they used a ±5 mm threshold for their SGRT tracking in any axis of movement. If a patient moved beyond this threshold, they would stop treatment and reposition the patient before continuing. SGRT tracking during treatment can provide feedback that the patient motion has not exceeded the setup uncertainty values used during robust treatment planning, helping to ensure the treatment is delivered as intended.

23.2.3 Respiratory Motion Management

Tumor motion due to respiration creates challenges in radiation therapy both in target definition and treatment delivery. Techniques such as 4D-CT are commonly used to characterize the target motion at the time of CT simulation. This helps to include the impact of motion in target definition. During treatment of moving targets on a linac, the dynamic nature of IMRT can result in interplay between the target motion and beam delivery. This can result in degraded dose distributions compared to what was intended.[15]

For proton delivery, passively scattered delivery approaches are more robust to target motion compared to a scanned beam delivery. However, scanned proton beams have the ability to modulate the dose to the target and around OARs, resulting in more conformal dose distributions. Because of these advantages, scanned delivery systems are more common in new proton facilities. As with IMRT deliveries, the dynamic delivery of scanned beams creates the potential for interplay between the beam delivery and the moving target.[16] Again, this can result in degraded dose distributions. Strategies have been implemented to limit the effects of the interplay when using a scanned beam delivery. These include rescanning of the beam and suppressing the tumor motion during beam delivery[17] (e.g., through patient compression devices, beam gating, or the use of breath-hold techniques).

Engelsman et al.[2] stated that beam tracking and minimizing target motion are the two most effective strategies for addressing interplay effects with scanned proton beams. However, dynamically tracking the proton beam with tumor motion is technically very demanding and not feasible for most proton clinics. Strategies to minimize target motion such as gating or breath-hold can be facilitated by SGRT. Freislederer et al.[18] described an analysis of gated deliveries on a linac that were facilitated by an SGRT system. As part of their analysis, they confirmed the accuracy of the machine output when the beam was gated, with the difference in dose delivery between gated and ungated treatments remaining below 1%. Additionally, they evaluated the time delay for the machine to beam back on after the SGRT system signal was sent to the machine indicating that the patient was within the intended gating window. Given that proton facilities are all fairly unique, careful analysis, such as those mentioned previously, that the fidelity of the intended treatment is maintained when delivered in a gated fashion is an important step when implementing this treatment method. SGRT systems used to facilitate gated proton delivery can also be evaluated to confirm that their system latency are within a reasonable level.

Deep inspiration breath hold (DIBH) is another form of respiratory motion management commonly used for left-sided breast cancer patients to limit heart dose. SGRT has been shown to be an effective method for confirming DIBH levels are as intended during linac-based treatments and was reviewed in Chapter 10. Proton therapy alone is another treatment technique that can spare dose to the heart because protons deposit all their dose in tissue and have no exit dose.[19] However, there can still be situations in which the position of the heart for a particular patient may be immediately adjacent to the chest wall. Due to the range uncertainties associated with proton treatment, the desired dose to the heart may not be achievable with proton therapy when treating the patient under free breathing conditions. For these situations, DIBH techniques could also be used for proton treatments. Mutter et al.[14] reported utilizing SGRT to help accommodate the DIBH delivery of a proton treatment to a chest wall patient to help displace the heart posteriorly and caudally away from the chest wall and internal mammary lymph node target.

23.2.4 Assessment of Patient Changes

As discussed in the Introduction to this chapter, proton therapy differs from linac-based treatments in that there is a strong dependence in proton therapy on the WET to the target (Figure 23.1). IGRT technologies that acquire 3D image sets provide the best option for assessment of patient changes relative to the time of the CT simulation. Reacquiring a CT of a patient undergoing proton therapy enables recalculation of the plan for the current state of the patient. This can allow clinicians to determine if the dose distribution is as intended or if patient changes (e.g., weight loss or change in the tumor size) may require replanning. However, 3D imaging modalities are not as common in proton vaults as linac vaults. Some proton treatment facilities may only have planar kV imaging for daily setup corrections. This modality gives limited information on changes in the patient that may affect the dose distribution. In these situations, facilities may adopt a workflow to acquire a weekly CT of the patient using a simulation CT (i.e., a CT not in the treatment room) to assess patient changes. Another non-imaging tool commonly used in linac-based treatments to assess patient changes is the optical distance indicator that allows therapists to measure the source to surface distance (SSD) of a particular treatment beam. An alternative method that may be used in proton treatment is to compare the measured air gap between the treatment nozzle and the patient to the planned air gap. However, both the SSD and air gap methods only provide an assessment of potential patient changes for one point on the patient and do not assess the entire area a treatment beam may cover.

SGRT can potentially provide the ability to assess patient changes in proton vaults that do not have 3D imaging capabilities. After a patient is positioned for treatment using planar kV image guidance, SGRT systems could quantify the offset of the current patient surface from the simulation CT external body reference surface. Some SGRT vendors provide heat map representations of the changes in surface anatomy from the time of simulation CT, allowing the clinical team to quickly assess surface anatomy changes. Large offsets could indicate that the patient's anatomy has changed, which may impact the intended target dose. This analysis could provide a trigger for when an additional assessment CT scan is needed. The SGRT assessment could be performed on a daily basis (i.e., more frequently than the weekly CT strategy) and would not deliver any additional imaging dose. The ROI selected in the SGRT system could encompass the entire beam entrance area. This would provide more information than an air gap measurement. However, a limitation of using SGRT for assessing patient changes is that it would only be able to directly evaluate such changes for anterior beams, where a surface would be visible to the system.

23.2.5 Evaluation of Immobilization Strategies

Similar to linac-based treatments, immobilization strategies in proton therapy aim to limit patient motion during treatment and aid in reproducing the patient's position from the time of CT simulation. However, unless repeated imaging is used during treatment, it is difficult to assess the effectiveness of a immobilization device in maintaining the patient's position. The ability of surface imaging to quantitatively track the real-time position of the patient during treatment provides the potential to evaluate immobilization effectiveness,[10] compare immobilization strategies,[20] and reduce immobilization devices.[21] Reducing immobilization may allow for more patient motion; however, it may also allow for the SGRT system to visualize an ROI (e.g., open-face masks for cranial treatments) for tracking. The potential for more motion may be mitigated by having the real-time position information of the patient. Reduced immobilization may also offer increased patient comfort, which in turn may reduce patient motion.

Batin et al.[10] discussed how using SGRT during postmastectomy chest wall treatments with protons allowed them to notice large daily variations in patient's arm and chin positions. A chinstrap, head cup, and handgrips were added to their immobilization strategy. After the change, SGRT was used daily to confirm that patient's arm and chin were in the proper location. Finally, X-ray imaging was used to confirm the patient's clavicle position, validating the effectiveness of these changes.

Proton therapy immobilization devices are often designed to limit their impact on the treatment beam. This is typically accomplished by having rounded edges (i.e., no sharp corners) that prevent drastic changes in the WET for small setup variations. As discussed in the Introduction, small changes in the WET to the target can have a large impact on the resulting dose distribution. The potential to reduce immobilization provided by SGRT could help to remove some of the uncertainties that these devices introduce.

23.2.6 Potential for Reduced Anesthesia for Pediatric Patients

Proton therapy has a strong indication for the treatment of many pediatric treatment sites.[12] Young pediatric patients (i.e., less than 7 years old) are often treated under anesthesia since otherwise they may be noncompliant with remaining still during treatment due to maturity, pain, claustrophobia, or fear of the environment.[22] However, anesthesia may not always be possible due to the current state of the patient and repeated anesthesia sessions comes with some additional risk to the patient and the need for the coordination of the radiation oncology and anesthesia teams. For some pediatric patients who may potentially be compliant or who have contraindications to anesthesia, the ability of SGRT to track a patient's position during radiographic IGRT and treatment offers a method to ensure the patient remains in the intended treatment position during treatment delivery. Rwigema et al.[23] described a case in which the patient was not a candidate for anesthesia due to superior vena cava syndrome. They discussed how SGRT was used to track a pediatric patient during treatment, which ensured the patient's position throughout the treatment process and helped them to reduce the treatment time. For this case, SGRT allowed them to circumvent the need for anesthesia. As discussed previously, SGRT may also facilitate the reduced need for some immobilization devices. Immobilization devices used for pediatric patients may create anxiety that can then require the need for anesthesia,[22] so in this way, SGRT may also provide a means to reduce the need for anesthesia for pediatric patients.

23.2.7 Robotic Couch Motion Assessment

Robotic couches in proton therapy allow for 6 degrees of freedom (6DoF) in patient setup corrections, moving the patient to imaging modalities away from treatment isocenter, and help facilitate bilateral treatment for fixed- or half-gantry installations. These systems are much more complicated than a linac treatment couch and require the coordinated movement of multiple motors to complete a desired motion. This is particularly true for isocentric rotations,[24,25] which are utilized regularly in proton therapy, especially for fixed- or half-gantry installations.

Kruse[25] reported on the experience of the Mayo Clinic in Rochester, MN, in validating isocentric couch rotations (on a half gantry) using SGRT by tracking the position of volunteers setup on the couch with various methods of immobilization. The ability of the SGRT system to detect known offsets, simulating suboptimal robotic couch performance, was also validated. It was found that the SGRT system was able to detect the known offset to within 2 mm of the actual value. This confirmation of the SGRT system to correctly detect the motion and rotation of a patient may provide the ability to replace radiographic patient position validation at every couch angle, which would reduce the imaging dose to the patient and increase the treatment efficiency.

The American Association of Physicists in Medicine Task Group 224 suggests checking the accuracy of the shifts of robotic couches used in proton therapy as part of a routine quality assurance (QA) program.[26] Previous groups have described utilizing optical tracking systems (infrared) to validate the motion of robotic couches.[24] These methods could be extended to SGRT systems. Rana et al.[27] described a comprehensive daily QA program for a modern proton therapy system. They described using a cube-shaped IGRT phantom commonly used in daily linac-based IGRT QA for the IGRT systems coupled with the proton system. Integrating surface imaging into this process would allow the SGRT system to validate the shifts indicated by the IGRT system and the ability of the couch to correctly implement the desired shifts. It would also provide a cross comparison of the SGRT system to the IGRT system. Though SGRT systems may be used to validate the shifts or rotations of patients and as a QA tool to test the performance of robotic couches, these abilities are reliant on proper isocentric calibration of the SGRT system for the results to be valid. Challenges with SGRT system isocentric calibration on a proton delivery system are discussed in a later section.

23.3 SGRT CHALLENGES FOR PROTON THERAPY

Some differences between linac-based treatments and proton treatments lead to unique challenges with the implementation of SGRT for proton therapy. This section discusses some of these challenges. If these unique challenges can be overcome, the results can sometimes lead to unique benefits.

23.3.1 Nonstandard Camera Positioning

Though there are multiple linac vault vendors available, the layout of linac vaults are relatively similar. This accommodates a standardized configuration of three-camera SGRT systems, with one camera near the foot of the couch and two cameras approximately lateral to the couch. Proton vaults that have a rotating gantry have components of the gantry that are above the ceiling and below the floor of the vault. This vault construction can limit the locations for SGRT camera placement within the treatment room. There is potential that nonstandard SGRT camera positions may lead to suboptimal operation of the system, so testing should be completed to confirm the system's operation. Mamalui-Hunter et al.[28] reported on their evaluation of an SGRT system that was installed specific to their proton treatment vault in a nonsymmetric orientation. They found the system showed satisfactory results. Batin et al.[10] also discussed the need to have the positions of their SGRT cameras dictated by their proton vault configuration. For their full gantry vault, the lateral cameras were positioned more inferiorly from isocenter compared to a typical linac vault. Batin et al. discussed that this camera configuration resulted in suboptimal images for chest wall patients. They updated their chest wall patient positioning method to help mitigate the issues due to camera position.

Another challenge in positioning SGRT cameras in proton vaults is that other imaging modalities are sometimes acquired with the patient away from the treatment isocenter. Examples of this are facilities that use CT-on-rails and some CBCT systems where the patient is moved on the treatment couch to an imaging position, the patient is imaged with the desired IGRT system, and then moved back to the treatment isocenter where the IGRT-indicated shifts can be applied. It is ideal to visualize patients with the SGRT system while they are at the treatment isocenter and also while they are being imaged with IGRT system to confirm they do not shift from their initial position during the image acquisition process (which could invalidate the IGRT session). In order to accommodate using surface imaging during other off-axis IGRT acquisitions, nonstandard SGRT camera positions may be necessary. The Maastro clinic (Maastricht, the Netherlands) is an example of a proton center utilizing nonstandard SGRT camera positions for this purpose. The Maastro clinic has a Mevion S250i Hyperscan system with the medPhoton ImagingRing CBCT system. To image with the ImagingRing system, the patient is shifted from treatment isocenter to an imaging position. The C-RAD Catalyst PT SGRT system is installed in the Maastro clinic's proton vault. To accommodate being able to image the patient with the Catalyst PT system during CBCT acquisition, a fourth Catalyst camera was added to the vault whose field of view (FOV) was more focused on the CBCT imaging position (see Figure 23.2).[29] Again, for solutions such as this that utilize a nonstandard camera arrangement, the ability of the SGRT system to correctly track the patient between the imaging and treatment locations should be validated.

23.3.2 Treatment Nozzle Positioning

Moveable treatment nozzles in proton therapy allow for a reduction in the air gap protons must traverse before reaching the patient. This reduction in the air gap leads to a sharper lateral penumbra in the dose distribution. However, the treatment nozzle location may obscure the view of SGRT cameras. Batin et al.[10] reported that when their treatment nozzle was in its extended position during the treatment of chest wall patients, it blocked their SGRT cameras view. This prevented the use of SGRT tracking during treatment.

Depending on the amount of patient surface area blocked by the nozzle, there may be an option to set the ROI to an area slightly away from the treatment area. This would allow tracking of the patient's position during treatment. In the case of a breast treatment, for example, setting the

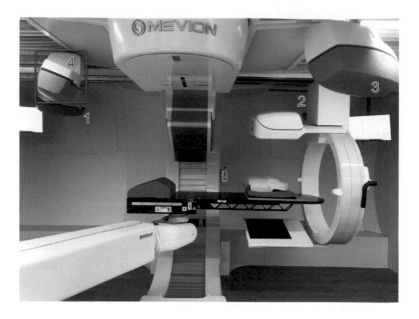

FIGURE 23.2 Photograph of the Mevion S250i Hyperscan system with the med-Photon ImagingRing CBCT system at the Maastro clinic. The C-RAD CatalystPT system is installed with a fourth camera (highlighted) to allow surface guidance during CBCT imaging away from the treatment isocenter. (Courtesy of Mattias Nilsing, C-RAD.)

ROI to the contralateral breast or the abdominal region just inferior to the treated breast may be options. If this methodology were to be employed, the validity of the surrogate ROIs would need to be confirmed.

Surface imaging may provide a means to create a model of the treatment geometry that would allow for determining a minimal air gap while preventing potential nozzle collisions with the patient or treatment accessories. Danuser et al.[30] reported on developing an in-house surface imaging system at the Paul Scherrer Institute for this purpose. They discussed how this method could save the time required to perform dry runs on the treatment machine without the patient to confirm the selected air gap would not cause a collision.

23.3.3 System Isocentric Calibration

Isocentric calibration of an SGRT system ideally places the surface imaging system's isocenter coincident with the treatment beam's isocenter. Misalignment of the two isocenters can create systematic offsets between the two systems. Additionally, couch rotations are regularly used in proton treatments, particular for fixed or half gantry installations. Even if updated reference surfaces are acquired at an initial setup couch angle, false offsets at other couch angles can be reported by the SGRT system if there is a misalignment between the surface imaging and treatment iso-centers.[31] Therefore, careful isocenter calibration is important. However, some of the surrogates commonly used for treatment isocenter in linac-based treatments, such as the light field crosshair or the front pointer, are not available in proton therapy. Care must be taken in confirming the isocenter surrogates used for SGRT calibration are an accurate representation of the treatment unit's isocenter. One SGRT vendor (Vision RT) has an isocenter calibration procedure that uses the MV images from a linac to help fine-tune the SGRT system's isocenter position. This method can resolve some of the potential false offsets seen with couch rotations.[31] Unlike linac-based treatments, it is not possible to image with the proton treatment beam, so this same procedure is not possible for a proton therapy machine. However, other radiographic imaging techniques can be used during the SGRT isocentric calibration on a proton system, thus providing a surrogate for the treatment beam isocenter.

Proton gantries are quite heavy and isocenter may drift as the gantry is rotated. Additionally, the room laser systems (i.e., a surrogate for the treatment isocenter) used in proton vaults may rotate with the gantry. Consequently, calibrating the SGRT system to be coincident with the treatment isocenter may be difficult for all gantry rotation positions. Batin et al.[10] reported having these issues. Their intended use for the SGRT system was for the treatment of chest wall patients. They decided to calibrate the SGRT system at an average gantry rotation position utilized for these patients and validated their calibration methods with a phantom study.

As discussed previously, SGRT systems may be a useful tool for confirming couch rotations or shifts have been applied correctly. Given some of the issues with the isocentric calibration of SGRT systems for proton therapy, validating the accuracy of the SGRT system's reported positions needs to be performed prior to use in proton therapy applications.

KEY POINTS

- Like linac-based treatments, proton therapy can benefit from utilizing SGRT. However, because of the fundamental differences between linac-based and proton treatment, SGRT provides both unique benefits and challenges for proton therapy.

- SGRT during proton patient setup can help to streamline the setup process and may reduce the need for radiographic imaging. Patient changes can be assessed after image-guided shifts have been performed. Additionally, robotic couch motions can be confirmed using SGRT.

- SGRT during proton treatments allows tracking of the patient's position throughout treatment delivery, can facilitate respiratory motion management, and allows assessment of immobilization effectiveness. The real-time position information provided by SGRT may also allow for the reduced need for anesthesia for pediatric patients commonly treated with proton therapy.

- Challenges for SGRT in proton vaults include the placement of the cameras, SGRT camera views being obscured by the treatment nozzle, and the isocentric calibration of the SGRT system.

REFERENCES

1. Yin F-F, Wong J, Balter J, et al. The role of in-room KV X-Ray imaging for patient setup and target localization. *AAPM Task Group Report No. 104*. 2009.
2. Engelsman M, Schwarz M, Dong L. Physics controversies in proton therapy. *Semin Radiat Oncol*. 2013;23(2):88–96. doi:10.1016/j.semradonc.2012.11.003.
3. Simpson DR, Lawson JD, Nath SK, Rose BS, Mundt AJ, Mell LK. A survey of image-guided radiation therapy use in the United States. *Cancer*. 2010;116(16):3953–3960. doi:10.1002/cncr.25129.
4. Landry G, Hua C ho. Current state and future applications of radiological image guidance for particle therapy. *Med Phys*. 2018;45(11):e1086–e1095. doi:10.1002/mp.12744.
5. Pollard JM, Wen Z, Sadagopan R, Wang J, Ibbott GS. The future of image-guided radiotherapy will be MR guided. *Br J Radiol*. 2017;90(1073):20160667. doi:10.1259/bjr.20160667.
6. Moteabbed M, Schuemann J, Paganetti H. Dosimetric feasibility of real-time MRI-guided proton therapy. *Med Phys*. 2014;41(11):1–11. doi:10.1118/1.4897570.
7. Sueyoshi M, Olch AJ, Liu KX, Chlebik A, Clark D, Wong KK. Eliminating daily shifts, tattoos, and skin marks: Streamlining isocenter localization with treatment plan embedded couch values for external beam radiation therapy. *Pract Radiat Oncol*. 2019;9(1):e110–e117. doi:10.1016/j.prro.2018.08.011.
8. Chang AJ, Zhao H, Wahab SH, et al. Video surface image guidance for external beam partial breast irradiation. *Pract Radiat Oncol*. 2012;2(2):97–105. doi:10.1016/j.prro.2011.06.013.

9. Gierga DP, Turcotte JC, Tong LW, Chen YLE, DeLaney TF. Analysis of setup uncertainties for extremity sarcoma patients using surface imaging. *Pract Radiat Oncol.* 2014;4(4):261–266. doi:10.1016/j.prro.2013.09.001.

10. Batin E, Depauw N, MacDonald S, Lu HM. Can surface imaging improve the patient setup for proton postmastectomy chest wall irradiation? *Pract Radiat Oncol.* 2016;6(6):e235–e241. doi:10.1016/j.prro.2016.02.001.

11. Batin E, Depauw N, Jimenez RB, MacDonald S, Lu HM. Reducing X-ray imaging for proton postmastectomy chest wall patients. *Pract Radiat Oncol.* 2018;8(5):e266–e274. doi:10.1016/j.prro.2018.03.002.

12. Buchsbaum JC. Pediatric proton therapy in 2015: Indications, applications and considerations. *Appl Radiat Oncol.* 2015;4:4–11.

13. Wiant DB, Wentworth S, Maurer JM, Caroline L, Terrell JA, Sintay BJ. Surface imaging-based analysis of intrafraction motion for breast radiotherapy patients. *J Appl Clin Med Phys.* 2014;15(6):12–17. doi:10.1120/jacmp.v15i6.4957.

14. Mutter RW, Remmes NB, Kahila MM, et al. Initial clinical experience of postmastectomy intensity modulated proton therapy in patients with breast expanders with metallic ports. *Pract Radiat Oncol.* 2017;7(4):e243–e252. doi:10.1016/j.prro.2016.12.002.

15. Berbeco RI, Pope CJ, Jiang SB. Measurement of the interplay effect in lung IMRT treatment using EDR2 films. *J Appl Clin Med Phys.* 2006;7(4):33–42. doi:10.1120/jacmp.v7i4.2222.

16. Dowdell S, Grassberger C, Sharp GC, Paganetti H. Interplay effects in proton scanning for lung: A 4D Monte Carlo study assessing the impact of tumor and beam delivery parameters. *Phys Med Biol.* 2013;58(12):4137–4156. doi:10.1088/0031-9155/58/12/4137.

17. Grassberger C, Dowdell S, Sharp G, Paganetti H. Motion mitigation for lung cancer patients treated with active scanning proton therapy. *Med Phys.* 2015;42(5):2462–2469. doi:10.1118/1.4916662.

18. Freislederer P, Reiner M, Hoischen W, et al. Characteristics of gated treatment using an optical surface imaging and gating system on an Elekta linac. *Radiat Oncol.* 2015;10(1):1–6. doi:10.1186/s13014-015-0376-x.

19. Shah C, Badiyan S, Berry S, et al. Cardiac dose sparing and avoidance techniques in breast cancer radiotherapy. *Radiother Oncol.* 2014;112(1):9–16. doi:10.1016/j.radonc.2014.04.009.

20. Jursinic P. Comparison of head immobilization with a metal frame and two different models of face masks. *J Cancer Cure.* 2018;1(1):36–40.

21. Cerviño LI, Detorie N, Taylor M, et al. Initial clinical experience with a frameless and maskless stereotactic radiosurgery treatment. *Pract Radiat Oncol.* 2012;2(1):54–62. doi:10.1016/j.prro.2011.04.005.

22. McMullen KP, Hanson T, Bratton J, Johnstone PAS. Parameters of anesthesia/sedation in children receiving radiotherapy. *Radiat Oncol.* 2015;10(1):10–13. doi:10.1186/s13014-015-0363-2.

23. Rwigema JCM, Lamiman K, Reznik RS, Lee NJH, Olch A, Wong KK. Palliative radiation therapy for superior vena cava syndrome in metastatic Wilms tumor using 10XFFF and 3D surface imaging to avoid anesthesia in a pediatric patient—a teaching case. *Adv Radiat Oncol.* 2017;2(1):101–104. doi:10.1016/j.adro.2016.12.007.

24. Hsi WC, Law A, Schreuder AN, Zeidan OA. Utilization of optical tracking to validate a software-driven isocentric approach to robotic couch movements for proton radiotherapy. *Med Phys.* 2014;41(8):081714. doi:10.1118/1.4890588.

25. Kruse J. Proton image guidance. In: Paganetti H, ed. *Proton Therapy Physics.* 2nd ed. Boca Raton, FL: CRC Press; 2019:615–641.

26. Arjomandy B, Taylor P, Ainsley C, et al. AAPM task group 224: Comprehensive proton therapy machine quality assurance. *Med Phys.* 2019. doi:10.1002/mp.13622.

27. Rana S, Bennouna J, Samuel EJJ, Gutierrez AN. Development and long-term stability of a comprehensive daily QA program for a modern pencil beam scanning (PBS) proton therapy delivery system. *J Appl Clin Med Phys.* 2019;20(4):29–44. doi:10.1002/acm2.12556.

28. Mamalui-Hunter M, Li Z. SU-E-T-249: Evaluation of the surface rendering patient localization system for proton therapy facility. *Med Phys.* 2011;38(6part14):3544–3544.

29. Nilsing M (C-RAD). Personal Communication. 2019.

30. Danuser S, Koschik A, Fielding A, Weber D, Lomax A, Fattori G. Surface imaging for treatment geometry virtualisation and collision detection in proton therapy. In: *Proceedings of the 58th Annual Particle Therapy Cooperative Group (PTCOG). International Journal of Particle Therapy.* Vol 6; 2019.

31. Paxton AB, Manger RP, Pawlicki T, Kim GY. Evaluation of a surface imaging system's isocenter calibration methods. *J Appl Clin Med Phys.* 2017;18(2):85–91. doi:10.1002/acm2.12054.

CHAPTER 24

Integration of Surface Imaging with Tomographic and Bore-Type Gantry Treatment Systems

Lutz Lüdemann

CONTENTS

24.1 INTRODUCTION

Optical surface imaging and registration techniques have been integrated with helical tomographic and bore-type gantry radiation delivery systems even though there exists unique challenges for the successful application of surface-guided radiation therapy (SGRT) when compared with conventional C-arm linear accelerators (linacs).[1-6] Examples of bore-type gantry systems include TomoTherapy (Accuray Inc., Sunnyvale, CA), Halcyon

(Varian Medical Systems Inc., Palo Alto, CA), and others. The aims of using surface guidance in conjunction with these systems are the same as those for C-arm linacs and include initial patient setup, intra-fraction motion monitoring, and respiratory management. Bore-type gantry systems and C-arm gantry linacs differ in terms of gantry geometry and in other important aspects of how patients are set up, imaged, and treated. For C-arm linacs, patients are generally positioned on the treatment couch while at the radiation isocenter. However for bore-type systems, access to the patient at the treatment position is limited due to the bore configuration. Initial access to the patient by the therapists and subsequent positioning takes place on the treatment couch outside of the bore (i.e., not at the radiation isocenter). A virtual isocenter that is a known distance from the radiation isocenter is used before the patient is translated longitudinally into the bore for radiographic imaging and irradiation. For example, with TomoTherapy, positioning is performed at a distance of 70 cm from the treatment isocenter, using lasers (and surface imaging if available). This presents challenges to SGRT workflows for bore-type gantries when using surface image systems and workflows that were designed for C-arm linacs. Most significant of these challenges is that the patient will be at different treatment couch positions for setup and for radiographic imaging and treatment, thus complicating the use of reference surfaces for setup and captured surface images for SGRT-based position monitoring. After setup at the virtual isocenter outside the bore, radiographic imaging is performed inside the bore using either a modified treatment beam such as megavoltage (MV) computed tomography (MVCT) as with the Accuray TomoTherapy system or a kilovoltage (kV) imaging source mounted orthogonally to the therapy beam to perform kV cone-beam computed tomography (CBCT) as in the Varian Halcyon design, for example.

In addition to the clinical challenges, having the patient at different treatment couch positions when using surface guidance for patient setup and when using it for intra-fraction motion monitoring during imaging and treatment has to be taken into account when determining optimal camera placement, performing calibration, and estimating positional accuracy. These demands require careful assessment during SGRT system commissioning and design of routine quality assurance.

24.2 CAMERA-MOUNTING POSITIONS

For optimal patient position detection in a C-arm linac, SGRT systems typically ceiling-mount three camera pods around the isocenter. When used in an enclosed bore-type radiation treatment system, the standard camera configurations used for a C-arm linac vault are suboptimal. The camera positions have to be optimized for bore-type gantry delivery systems to ensure adequate views of the patient both inside and outside the bore. Additionally, the differences in the positioning procedure between these systems and C-arm linacs have a major impact on the SGRT clinical workflow. Figure 24.1 provides an example installation of an AlignRT (Vision RT, London, UK) SGRT system in a Halcyon treatment vault and an example installation of a three-camera Catalyst (C-RAD AB, Uppsala, Sweden) SGRT system with a conventional C-arm linac.

Bore-type systems require that the patient is set up on the treatment couch while outside the bore but imaging and treatment delivery is performed with the patient inside the bore. A surface imaging camera's use for patient setup outside the bore or intra-fraction motion monitoring inside the bore will influence the optimal installation location and optical isocenter. For a two-camera installation as shown in Figure 24.1, the system is calibrated to the virtual isocenter and the system is used for patient setup only; no patient monitoring is performed inside the bore.

A three-camera system configuration was installed at Essen University Hospital, Germany, and is shown in Figure 24.2.[1] This particular configuration consists of two different subsystems—the motion detection system using only the center camera (Catalyst) and the position detection system using only the peripheral cameras (AlignRT). For such a three-camera SGRT system, the respective roles of each camera can be assigned to either patient setup outside the treatment bore or to intra-fraction patient monitoring inside the bore. Another configuration would be to move the two

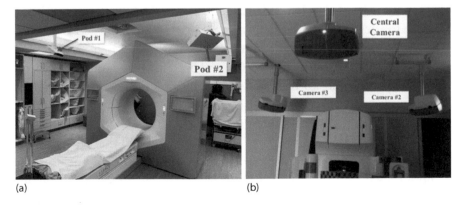

(a)   (b)

FIGURE 24.1 A two-camera pod installation of AlignRT in a Halcyon treatment vault (a) and a three-camera installation of Catalyst on a C-arm linac (b).

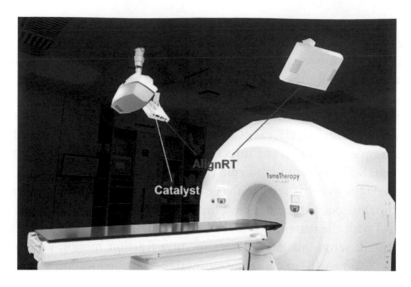

FIGURE 24.2 Illustration of the camera mounting used for evaluation of both SGRT devices–Catalyst provided by C-RAD and AlignRT from Vision RT. (From Wiencierz, M. et al., arXiv preprint arXiv:1602.03749, 2016. With permission.)

lateral cameras to minimize the impact of the occlusions by the bore when a patient is inside. The central camera can then be used for patient positioning outside the bore and additionally provides a view inside the gantry.

In the Essen installation, the peripheral cameras were directed to look upward into the bore to compensate for the fact that the center camera was not available for positioning. With this setup, after positioning, the acquisition system had to be changed manually. Figure 24.3 illustrates the difference in the detected surface using a one-camera system in central position (Figure 24.3a) versus the three-camera system (Figure 24.3b). A total of 20 surface co-registrations of the MVCT and both the one-camera and three-camera systems and an Alderson phantom on 20 different days yielded an average correction vector length of 0.91 ± 0.11 mm (mean ± standard deviation) with the three-camera system versus 1.24 ± 0.11 mm with the one-camera system. The results of this study suggested that an additional dedicated three-camera system for motion monitoring might be useful in a bore-type gantry treatment system. Alternatively, for surface detection during treatment, the central camera could be replaced by two devices – one camera with a slightly leftward view and one with a slightly rightward view into the gantry. Since the view of the patient is obscured as the

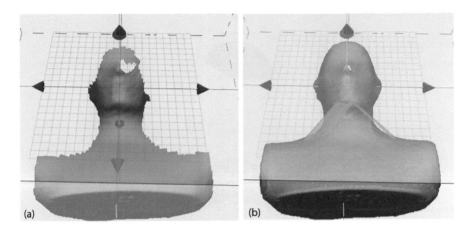

FIGURE 24.3 (a) 3D model of the Alderson phantom acquired with the one-camera system. (b) 3D-model of the phantom acquired with the three-camera system.

patient moves further into the bore, an additional camera system directed to look into the bore from the back may be useful to compensate for the loss of information. Such improvements could be added to future versions of SGRT systems intended for bore-type linacs.

The Catalyst system depicted in Figure 24.2 is a one-camera system looking into the gantry. For this Catalyst configuration, the camera was adjusted to cover the caudal direction at inclinations of 0°–120° and the cranial direction at 0°–30° (0° being orthogonal to the treatment couch). In contrast, the AlignRT camera system installation at Essen, with dedicated cameras for setup and in-bore monitoring, covers 60° in both directions. Figure 24.4 illustrates the differences in approximate coverage provided by the Catalyst and AlignRT systems for this installation. The main difference between the two systems was found in terms of the coverage provided by lateral inclination. A single camera unit ensures surface coverage of up to 60° laterally to both sides compared to a two-camera unit installation, which covers 0° to approximately 110°. Thus, the AlignRT two-camera unit solution covers nearly the complete inclination range, which would be possible to cover with a three-camera unit system.

If the SGRT system consisted of three cameras instead of a single camera, uncertainty could be reduced by approximately 36%, which was based off a set of phantom measurements comparing a three-camera unit setup with a one-camera unit setup installed on a C-arm linac. Repeated measurement at the experimental setup with an Alderson phantom demonstrated an

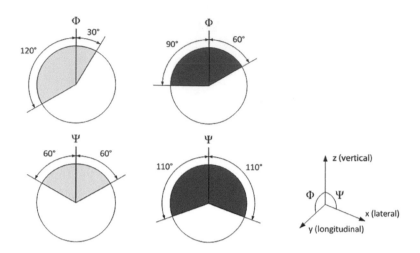

FIGURE 24.4 A comparison of the approximate coverage provided by the Catalyst (blue) and AlignRT (red) cameras installed at Essen for evaluation with a TomoTherapy unit.

overall deviation of the Catalyst system of 1.07 ± 0.26 mm (mean ± standard deviation) with a single camera. Using the known difference in positioning accuracy between a one-camera system and a three-camera system, a hypothetical three-camera Catalyst system would yield a mean deviation of approximately 0.78 ± 0.19 mm calculated by assuming a Gaussian position deviation, compared to the mean deviation of the AlignRT devices with 0.28 ± 0.22 mm. MVCT-based registration was used as the gold standard for phantom positioning, with deviations comparing the radiographic to SGRT phantom setup. The phantom surface structure data set was generated from the computed tomography (CT) scan using a CT number threshold of -350 Hounsfield Units (HU). Using an internal reference surface generated by the SGRT system yielded an average positional deviation 0.63 ± 0.42 mm with the one-camera version of Catalyst (estimated to be 0.46 ± 0.32 mm for a three-camera version) versus 0.33 ± 0.10 mm for the AlignRT system. C-RAD currently offers a three-camera system called CatalystHD but it was not available at the time this study was performed.

Optimization of surface imaging camera positioning was also studied for AlignRT and Halcyon.[3] Ceiling-mounted cameras could be used for surface guidance, but if optimized for monitoring at the radiation isocenter, surface monitoring at the virtual or pseudo-isocenter used for setup was suboptimal. Placing cameras too close to the gantry limited the visibility for head and neck, chin and upper arms of patients.

The limitations of ceiling-mounted SGRT systems have led some users of bore-type linacs to develop custom solutions for surface monitoring, for example, using a single consumer-grade camera mounted at the back end of the bore.[4] Custom surface guidance applications with consumer-grade camera are discussed in more detail in Chapter 25.

24.3 GEOMETRIC SETUP ERRORS

Since irradiation and radiographic imaging are performed inside the bore, while surface registration for initial patient setup is performed outside the bore, there is a potential for systematic setup errors. For example, the treatment couch may sag depending on the patient's weight and weight distribution. A study of surface-guided breast treatments on TomoTherapy detected a mean couch sag of 5 mm.[6] Additionally, the couch may not be at an exact right angle to the gantry rotational axis, so that the patient drifts slightly during irradiation for helical deliveries (e.g., with TomoTherapy) and continuously with the couch movement. Both effects should be corrected automatically. The first generation of SGRT systems did not allow correction for movement of the patient along with the couch. To overcome this limitation in position monitoring, couch drift for certain SGRT solutions may be eliminated by using a position-dependent calibration. Since all required parameters are available in the digital imaging and communications in medicine (DICOM) header, a weight- and patient-position-dependent calibration is feasible.

When the body surface is generated using a CT image data set, the extensions of the patient contour vary slightly with the CT number or HU threshold applied for separating patient volume from outside the body. The threshold used for detecting the patient contour is typically not adjusted to yield identical volumes enclosed by the surfaces generated by the SGRT device and the CT scan. Uncertainties in the surface image-based position may be introduced if components of the treatment delivery system occlude the camera unit from seeing the patient surface. To reduce registration uncertainties, in some instances, vendors of surface guidance systems recommend acquiring a new reference image after positioning the patient using radiographic imaging. When a bore-type linac is used, the patient has to be moved outside the gantry for acquisition of a new reference image by the surface imaging system, which would then require a correction for couch movement. In order to not disrupt the usual workflow of treatment delivery, it may be preferable to derive the new reference surface from the verification MVCT or CBCT acquired inside the bore. The HU threshold for external body contour detection will have to be optimized for these imaging modalities.

Additionally, a communication interface between the treatment console and surface registration system would be helpful. In this way, a new reference image could be acquired by using the surface cameras immediately after positioning the patient with radiographic imaging. A communication interface between the treatment console and SGRT system is available for most C-arm linacs. The addition of a communication interface between the treatment console and surface registration system can also offer additional benefits. By communicating with the treatment system, SGRT systems may be able to automatically receive information on which a patient is currently being treated and receive the actual treatment couch position at a given point in time. In the absence of automated machine communication between the SGRT system and bore-type linacs, a workaround may be improvised using a wire ruler attached to the treatment couch. This ruler allows accurate detection of the longitudinal couch position and accurate identification of the start of treatment. Additionally, a radiation detector is used to identify the start of treatment. Nevertheless, these measures account for only some of the functions provided by a communication interface between the two systems. Additional function that can be enhanced for bore-type linacs include patient management, workflow optimization, treatment couch speed, and movement. Furthermore, the DICOM-RT Plan file generated for C-arm linacs usually provides all information about the treatment parameters and is used to transfer the treatment parameters from the planning system to the treatment facility. Some vendors of bore-based solutions do not provide treatment planning information in DICOM standard format. For example, Accuray instead uses an internal standard. DICOM files are only used to provide treatment parameters to systems of other vendors, e.g., SGRT systems. The TomoTherapy unit has two laser systems, a fixed green laser to define the virtual isocenter and a movable red laser that can be programmed to identify the plan-specific patient setup marks. Unfortunately, the DICOM-RT Plan format provides no information about the position of the red and green lasers. The movable red lasers provide the actual position of a reference point in the target. By default, the green laser intersection is fixed to the virtual isocenter 70 cm outside the true isocenter, and the center of the CT image data set is adjusted by default to match the virtual isocenter.

24.4 CAMERA LIMITATIONS

In a C-arm linac, irradiation is performed in the same position as radiographic imaging. Thus, the camera calibration parameters can be adjusted based on a fixed distance. In a tomographic, or bore-type linac, positioning is performed outside the bore relative to the virtual isocenter, and for irradiation, the patient is moved from the setup location to the radiation isocenter inside the gantry. SGRT cameras are typically mounted approximately 1–2 m away from the radiation isocenter depending on the bore-type treatment unit design. Since the surface to be validated extends far beyond the true isocenter, the camera has to cover the patient's surface over a length of approximately 3 m, or additional cameras are needed to observe the patient at multiple locations. For a single-camera solution, the spatial resolution may therefore be reduced inside the bore gantry. A future improvement would be to equip the central camera with a motor-driven zoom function to achieve a spatial resolution comparable to that in the virtual isocenter. Additionally, due to self-occlusions, the view inside the bore is reduced to a small inclination. For SGRT solutions using structured light to image the surface, the intensity of this light may decrease and the light pattern may become smaller with increasing distance; therefore, the acquisition duration has to be adjusted to the actual patient position.

24.5 CLINICAL STUDIES

In clinical practice, the effective positioning accuracy yielded with SGRT devices can depend on the location of the tumor relative to the area of the body that is used for surface matching. Accuray provides a high-precision couch and an MV-imaging system using the therapy beam while Varian provides a high-precision couch and kV planar imaging as well as CBCT. Therefore, positioning accuracy is only limited by the co-registration accuracy of the MV or kV images. One study investigated the magnitude of corrections necessary for AlignRT and Catalyst (C-RAD AB, Uppsala, Sweden) integrated with a TomoTherapy unit, using MVCT as a gold standard.[1] For a one-camera Catalyst configuration, the correction vector length of the co-registrations median (M), 90% percentile (P_{90}), and 95% percentile (P_{95}) for several anatomic sites were:

- M = 2.1 mm, resp. P_{90} = 6.0 mm, P_{95} = 7.7 mm for head and neck

- M = 4.2 mm, resp. P_{90} = 10.8 mm, P_{95} = 12.4 mm for thorax

- M = 7.4 mm, resp. P_{90} = 17.2 mm, P_{95} = 18.8 mm for abdomen tumors.

The same setups were evaluated by using an AlignRT two-camera system, yielding:

- $M = 0.2$ mm, resp. $P_{90} = 0.8$mm, $P_{95} = 1.7$ mm for head and neck
- $M = 3.5$ mm, resp. $P_{90} = 10.2$ mm, $P_{95} = 12.1$ mm for thorax
- $M = 17.1$ mm, resp. $P_{90} = 15.9$ mm, $P_{95} = 18.4$ mm for abdomen tumors.

This study concluded that SGRT can improve the initial setup accuracy of thorax and abdominal patients. Improvements in setup accuracy were not significant for head and neck, as these patients were already well immobilized and could be accurately positioned with a thermoplastic mask.

Haraldsson et al. reported for a one-camera Sentinel optical surface-scanning system (C-RAD Positioning AB, Uppsala, Sweden), the correction vector length for 90% of the co-registrations was within 2.3 mm for central nervous system (CNS) tumors, 2.9 mm for head and neck, 8.7 mm for thorax, and 10.9 mm for abdomen patients.[2] Moser et al. reported similar results using a laser surface-scanning system similar to Sentinel and found that obtaining a reference surface after MVCT-based positioning on the first day of treatment was more accurate than using a planning CT-based reference surface.[5] Crop et al. investigated the setup accuracy of breast patients treated with TomoTherapy, using either lasers, Catalyst, or MVCT.[6] They found that surface guidance with Catalyst was more precise than laser-based positioning and complemented MVCT well, particularly in the correction of patient roll prior to imaging, which would otherwise require reimaging to correct. Last, Flores-Martinez et al. evaluated workflow efficiency and patient setup accuracy with SGRT for brain treatments on Halcyon.[7] Thirty-five fractions with and 35 fractions without SGRT were evaluated. AlignRT was used in a two-camera configuration (see Figure 24.1) and compared to MV-CBCT for patient setup (both being compared to the CT simulation scan). The authors found that mean treatment times were approximately 1 min shorter when SGRT was used. The mean absolute residual rotational errors were less with SGRT with head pitch having the smallest residual error of 0.5°.

24.6 CONCLUSIONS

Effective use of a surface guidance system in conjunction with a bore-type linacs relies on a communication interface between the treatment console and surface guidance system to have access to general patient information and current treatment parameters, couch position, and beam-on information. To overcome geometrical limitations, multiple cameras should be used for patient setup. Additionally, up to three cameras for intra-fraction motion monitoring may be useful. To achieve the same spatial resolution in the treatment isocenter as in the virtual setup isocenter, either an automatic zoom or multiple co-calibrated camera units would be helpful. The automatic zoom can adjust the field of view to the variable distance between camera and patient, thus compensating for the fact that patient extensions appear smaller as the distance increases, while multiple cameras can be optimally focused on different areas for setup and treatment.

KEY POINTS

- The integration of surface guidance systems with tomographic and bore-type linacs presents unique challenges including the need to set up patients outside of the bore at a virtual isocenter and moving them longitudinally to the radiation isocenter for imaging and treatment.

- Conventional surface camera configurations can be adapted to bore-type linacs by having an automatic zoom function or dedicating some cameras to the setup position and others to the treatment position; however, surface matching performance depends on the number of cameras available.

- Phantom and clinical investigations show that surface image guidance can improve the initial setup accuracy of patients that do not have additional immobilization, i.e., thermoplastic masks.

REFERENCES

1. Wiencierz M, Kruppa K, Lüdemann L. Clinical validation of two surface imaging systems for patient positioning in percutaneous radiotherapy. arXiv preprint arXiv:1602.03749. 2016.
2. Haraldsson A, Ceberg S, Crister C, Engelholm S, Bäck SÅJ, Engström PE. PO-0978 Accurate positioning with decreased treatment time using surface guided tomotherapy. *Radiother Oncol.* 2019;133:S534–S535.
3. Delombaerde L, Petillion S, Depuydt T. EP-2051: Surface scanner camera position optimization on the Varian Halcyon™ O-ring gantry linac system. *Radiother Oncol.* 2018;127:S1122–S1123.
4. Delombaerde L, Petillion S, Michiels S, Weltens C, Depuydt T. Development and accuracy evaluation of a single-camera intra-bore surface scanning system for radiotherapy in an O-ring linac. *Phys Imag Radiat Oncol.* 2019;11:21–26.

5. Moser T, Habl G, Uhl M, Schubert K, Sroka-Perez G, Debus J, Karger CP. Clinical evaluation of a laser surface scanning system in 120 patients for improving daily setup accuracy in fractionated radiation therapy. *Int J Radiat Oncol Biol Phys.* 2013;85(3):846–853.

6. Crop F, Pasquier D, Baczkiewic A, Doré J, Bequet L, Steux E, Lacour M. Surface imaging, laser positioning or volumetric imaging for breast cancer with nodal involvement treated by helical TomoTherapy. *J Appl Clin Med Phys.* 2016;17(5):200–211.

7. Flores-Martinez E, Cervino L, Pawlicki T, Kim G. Improved setup and workflow for brain treatments on the halcyon linac by using surface guided radiation therapy. *Med Phys.* 2018;45(6):E592–E592.

CHAPTER 25

Consumer-Grade Cameras and Other Approaches to Surface Imaging

Rex Cardan

CONTENTS

25.1 INTRODUCTION

Clinical surface imaging systems designed for surface-guided radiation therapy (SGRT) applications are expensive and standard configurations have a limited capacity for raw data export which can prohibit innovation, new uses, and new techniques for surface-guided applications. Fortunately, the recent emergence of low-cost consumer-grade depth cameras has enabled a new frontier of imagination and exploration of the use of three-dimensional (3D) surfaces in radiation therapy. Over the last decade, a large effort has been exerted investigating the use of both laser scanning systems and RGB-depth (RGBD) sensors, but the overwhelming majority has been a result of the introduction of Microsoft's Kinect camera (Microsoft Corporation, Redmond, WA, USA). This camera system has been used for respiratory motion management, patient identification, collision detection and avoidance, 3D printing applications, patient alignment, and image processing. This chapter reviews methods and results of consumer-grade depth camera applications for radiation therapy over recent years spanning computer vision, motion management, full room scanning, and collision avoidance applications.

25.1.1 Technology

The Kinect was introduced in 2010 for the Xbox® gaming console (Microsoft Corporation, Redmond, WA, USA), and Microsoft shortly after introduced an application programming interface (API) in 2012. Designed primarily for gaming applications, the device had embedded computer vision algorithms which parsed depth data into human skeleton models in real-time. These models provided "joint-positions," simplified bone-junction coordinates, which were made available in the API. Additionally, the API allowed access to the data streams including depth, color, and audio. In 2013, a new version of the Kinect (v2) was released with a higher resolution, different depth sensing technology, and more computer vision processing capabilities allowing the joint positions of up to six people to be tracked in real-time.

25.1.2 Hardware

The Kinect components consist of an RGB camera, infrared (IR) emitter, sensor, and a microphone array. The specifications of each Kinect device are shown in Table 25.1. The depth sensing is different between the two models. With the first Kinect, the depth was determined by means of an IR-structured light projection, which resembles a speckled dot pattern. From the distortion of the speckled pattern on the surfaces of objects, the depth could be calculated. The Kinect v2 used a time-of-flight method where the emitter varied the IR brightness with time creating an amplitude wave at the sensor. The phase shift from reflected surfaces provides the input necessary to determine depth.[1] Pöhlmann et al.[2] have provided a good overview of the technology available in both cameras.

TABLE 25.1   A Comparison of the Capabilities of the Original Microsoft Kinect, Second Generation (v2), and Azure Kinect

	Kinect	Kinect v2	Kinect Azure
Depth Sensing	Structured Light	Time-of-Flight	Time-of-Flight
IR Sensor Resolution	640 × 480	512 × 424	1024 × 1024
RGB Sensor Resolution	640 × 480	1920 × 1080	3840 × 2160 at 16:9 aspect ratio 4096 × 3072 at 4:3 aspect ratio
Frame Rate	30 fps	30 fps	30 fps
Audio Sensor	4 mic array	4 mic array	7 mic array

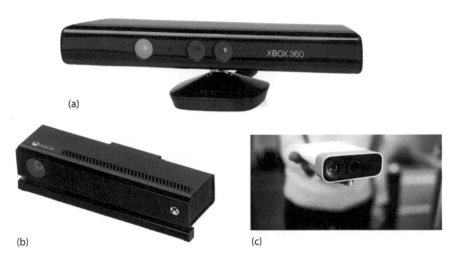

(a)

(b)   (c)

FIGURE 25.1   Examples of the now-discontinued Microsoft (a) Kinect and (b) Kinect v2 cameras. The latest version of the Azure Kinect is shown in (c).

While the gaming Kinect product has been discontinued, its capabilities now exist in both the current Microsoft HoloLens[3] and a cloud-based version of the camera called Azure Kinect released in 2019 (Figure 25.1). The Azure Kinect camera and software developer kit (SDK) consists of a 12 megapixel color imaging system, 7 audio channels, and a 1 megapixel depth camera. This is a significant improvement over the original Kinect's depth camera. The system has been designed for development into multicamera arrays or mobile applications. The Azure Kinect includes SDKs and APIs for building artificial intelligence elements such as computer vision and speech models.

25.1.3 Software

Besides the hardware components of the Kinect system, the software capabilities have been equally important in the literature. Specifically, the Kinect Fusion system outlined by Newcombe et al.[4] has allowed the creation of full three dimensional meshes using either multiple projections from a single camera or multiple cameras. Using iterative closest point (ICP), the Fusion process operates by calculating the transforms between point clouds taken at multiple projections. The transforms then allow the user to create a composite point cloud which is stored as a truncated signed distance function in a voxel grid. This surface representation can then be processed into a polygon mesh for use in collision avoidance.[5]

25.1.4 Accuracy

The accuracy of the cameras concerning radiation therapy applications has been assessed for both camera versions. The original Kinect had depth accuracies reported in the sub-2 mm range for most surfaces.[6] The structured light technique is prone to higher errors for surfaces with steep gradients (large changes in between scattered dots),[1] and errors up to 3–4 mm have been reported.[6,7] For the Kinect v2, the depth accuracy was reported to be 2.0–3.7 mm if uncorrected and 1.2–1.4 mm if calibrated prior to each scan.[8] However, the accuracy has been reported to be a function of the temperature of the sensor,[8,9] reflections of surfaces in a treatment vault,[10] but minimally affected by external electromagnetic fields created by radiation therapy devices.[8] These accuracies are for the raw depth image and not for more complex scenarios regarding depth to color registration, and Kinect Fusion. The accuracy of the complete registration chain has not been as formally assessed. It should be noted that when Kinect Fusion was used, discrepancies up to 2 cm were observed between the actual surface location and the calculated location.[5] Commercial SGRT systems have been shown to achieve accuracies on the order of 0.1 mm in comparable studies.[11]

25.2 A REVIEW OF DEVELOPMENT AND APPLICATIONS

25.2.1 Computer Vision

The Kinect API allows for more than just raw point cloud data to be acquired. Computer vision capabilities are also embedded in both the hardware and the software system which allows the parsing of estimated skeletal positions and facial features. Both of these functions have been exploited and applied to radiation therapy needs. Silverstein et al.

demonstrated a rudimentary facial recognition system for verifying patient identity prior to treatment.[12] The system relied on the Kinect API to provide 31 unique facial points from the subject's surface scan. From these points, the distance between each pair of points was calculated to create a vector array unique to each patient's face. Using just the mean and median differences between two scanned vector arrays, the discrimination of identity could be obtained (sensitivity 96.5%, specificity 96.7%). The authors did note that the results were highly dependent on lighting conditions. Another use of the Kinect was by Mullaney et al. in which they demonstrated patients could be allowed to participate in their positioning.[13] In their prototype example, the depth sensors parsed the patient joint locations and the patients watched a display showing the desired and current position of their joints. A feedback system, consisting of a simple user interface, was used to positively validate the correct position in a user interface.

25.2.2 Full Room Scanning

The exact positions of multiple equipment components in a treatment vault are extremely important to a safe and high-quality treatment. The accessories must be correct, the gantry and patient support must be at the correct angle, and the patient must be set up in the same position as in the simulation. The basic framework for verifying the position of these components by performing full room scanning using multiple Kinect cameras was originally outlined by Santhanam et al.[14] In the original proposal, it was envisioned to use the sensors to create a 3D live view of a treatment vault streamed over a network connection so experts could evaluate the setup remotely. In a later publication, the same group outlined a process for the specific calibration and registration of multiple cameras using a custom registration jig.[15]

25.2.3 Motion Management

One of the most published areas that has been explored with the Kinect is its use in respiratory motion management.[7,10,16-22] Conceptually, each method involved tracking an array of points across the surface of a phantom or patient as a surrogate for internal motion of a volume. Techniques have included tracking a flat surface placed on the thorax, a custom t-shirt with an optical pattern,[16] and direct imaging of the actual patient surface.[19-21] Analysis has included dividing the motion into abdominal and thoracic motions using principle component analysis (PCA) of the data.[21]

This method allowed for elucidation of the most prominent surface features that are correlated with internal motion. Accuracy of the relative motion compared to commercial respiratory gating systems was between 2 and 3 mm,[19] similar to the overall accuracy reported from the sensors.

Aside from respiratory monitoring, the Kinect has been used to track head motion. In one case, Noonan et al. demonstrated its potential in improving high-resolution positron emission tomography (PET) scans, proposing to use tracking during signal acquisition to reduce motion artifacts during the image processing.[9] Another novel use for the Kinect has been in validating and controlling a motion "soft robot" for positioning head and neck (H&N) patients without a thermoplastic mask. Using both the original Kinect[23] and Kinect v2,[24] Ogunmolu et al. demonstrated a unique liquid mechanical system capable of positioning a phantom head within 2.5 mm using the Kinect sensor data in a feedback loop.

25.2.4 Collision Avoidance

Collisions of in-room mechanical components and patients can be catastrophic and result in injury, loss of patient life, equipment damage, and financial costs. A few groups have attempted to predict the collision space using data from Kinect cameras.[5,25,26] Cardan et al. first demonstrated the framework using a generalized polygon interference algorithm on a phantom[25] and later refined the technique using four patients.[5] This technique iterated a gantry polygon model, couch model, and Kinect scanned patient model in many orientations to determine the entire collision free space for treatment as seen in Figure 25.2. It also included a process in which the mesh was oriented to IEC coordinates using optical markers during simulation. Padilla et al. used specialized software to scan a polygon mesh and demonstrated a technique to determine if parts of the mesh were outside of axially oriented bounding cylinders for a head first supine treatment.[26]

25.2.5 Other Methods

It should be noted that other consumer or at least nonclinical level devices beyond the Kinect have been used for similar research. Laser scanner systems and industrial scanners have been used in multiple works for both patient modelling,[27] patient setup,[28] collision avoidance,[29,30] and respiratory gating.[31] Also, 3D surfaces have been created for use in radiation therapy applications by consumer level 2D cameras using photogrammetry.[32] This technique is available with many low-cost cameras and can provide much of the same data and accuracy as 3D surface scanning systems.[33]

FIGURE 25.2 A detected collision state of the polygon meshes for a scanned patient. In the collision simulation, the scanned gantry, patient, and modeled couch are positioned in 64,800 configurations to determine the feasible space for treatment.

Nonetheless, most research and development performed on nonclinical systems has made the use of the Microsoft Kinect, likely due to its low cost, wide availability, ease of use, and availability of third-party software libraries. Several other manufacturers have released similar products at the consumer level including the Structure by Occipital,[34] Orbbec Astra,[35] the PrimeSense Carmine,[36] and more. Additionally, the ability to create 3D point clouds with 2D images through photogrammetry[37] has been stream-lined and multiple software systems are available to perform this task.

25.3 SUMMARY

In summary, there are numerous commercially available low-cost consumer-grade depth imaging devices for surface imaging applications. These devices, such as the Kinect, have produced a rich landscape of RT research and development related to image guidance, computer vision, and motion management. As discussed in this chapter, there have been many innovating uses of depth scanning technologies over the last decade. As imaging hardware continues to advance and become more widely available, the improvement of patient care from its experimental use will surely continue.

KEY POINTS

- Consumer-grade laser scanning and depth camera technologies can allow innovative surfacing imaging solutions for a variety of applications in radiation therapy.

- The majority of custom surface imaging solutions have used the Microsoft Kinect, which has evolved through multiple versions since first being introduced in 2010.

- Consumer grade camera systems are relatively low cost and highly customizable through the availability of application programming interfaces and software development kits.

- Applications of surfacing imaging applications using consumer grade camera systems include patient identification, collision detection and avoidance, respiratory motion management, 3D printing of treatment accessories, patient alignment, and image processing.

REFERENCES

1. Zhang Y, Xiong Z, Yang Z, Wu F. Real-time scalable depth sensing with hybrid structured light illumination. *IEEE Trans Image Process.* 2014;23(1):97–109.
2. Pöhlmann ST, Harkness EF, Taylor CJ, Astley SM. Evaluation of Kinect 3D sensor for healthcare imaging. *J Med Biol Eng.* 2016;36(6):857–870.
3. Cardan R, Covington E, Popple R. A holographic augmented reality guidance system for patient alignment: a feasibility study: We-ram2-gepd-t-01. *Med Phys.* 2017;44(6):3215.
4. Newcombe RA, Izadi S, Hilliges O, et al. KinectFusion: real-time dense surface mapping and tracking. in Mixed and augmented reality (ISMAR), 2011. *10th IEEE International Symposium On.* 2011. IEEE.
5. Cardan RA, Popple RA, Fiveash J. A priori patient-specific collision avoidance in radiotherapy using consumer grade depth cameras. *Med Phys.* 2017;44(7):3430–3436.
6. Shin B, Venkatramani R, Borker P, et al. Spatial accuracy of a low cost high resolution 3D surface imaging device for medical applications. *Int J Med Phys Clin Eng Radiat Oncol.* 2013;2(2):45.
7. Alnowami M et al. A quantitative assessment of using the Kinect for Xbox 360 for respiratory surface motion tracking. in *Medical Imaging 2012: Image-Guided Procedures, Robotic Interventions, and Modeling.* 2012. International

Society for Optics and Photonics. https://www.spiedigitallibrary.org/conference-proceedings-of-spie/8316/83161T/A-quantitative-assessment-of-using-the-Kinect-for-Xbox-360/10.1117/12.911463.short?SSO=1

8. Edmunds DM, Bashforth SE, Tahavori F, et al. The feasibility of using Microsoft Kinect v2 sensors during radiotherapy delivery. *J Appl Clin Medical Phys.* 2016;17(6):446–453.

9. Noonan PJ, Howard J, Hallett WA, Gunn RN. Repurposing the Microsoft Kinect for Windows v2 for external head motion tracking for brain PET. *Phys Med Biol.* 2015;60(22):8753.

10. Nazir S, Rihana S, Visvikis D, Fayad H. Kinect v2 surface filtering during gantry motion for radiotherapy applications. *Med Phys.* 2018;45(4):1400–1407.

11. Wen N, Snyder KC, Scheib SG, et al. Evaluation of the systematic accuracy of a frameless, multiple image modality guided, linear accelerator based stereotactic radiosurgery system. *Med Phys.* 2016;43(5):2527–2537.

12. Silverstein E, Snyder M. Implementation of facial recognition with Microsoft Kinect v2 sensor for patient verification. *Med Phys.* 2017;44(6):2391–2399.

13. Mullaney T, Yttergren B, Stolterman E. Positional acts: using a Kinect™ sensor to reconfigure patient roles within radiotherapy treatment. in *Proceedings of the 8th International Conference on Tangible, Embedded and Embodied Interaction.* 2014. ACM.

14. Santhanam AP, Min Y, Dou HT, Kupelian P, Low D. A client–server framework for 3D remote visualization of radiotherapy treatment space. *Front Oncol.* 2013;3:18.

15. Santhanam AP, Min Y, Kupelian P, Low D. Multi-Kinect v2 camera based monitoring system for radiotherapy patient safety. *MMVR.* 2016.

16. Ernst F, Saß P. Respiratory motion tracking using Microsoft's Kinect v2 camera. *Curr Dir Biomed Eng.* 2015;1(1):192–195.

17. Lim SH, Golkar E, Rahni AAA. Respiratory motion tracking using the Kinect camera. in *Biomedical Engineering and Sciences (IECBES), 2014 IEEE Conference on.* 2014. IEEE.

18. Samir M, Golkar E, Rahni AAA. Development of a respiratory motion tracking system using a distance camera for diagnostic imaging and external beam radiotherapy delivery. in *Biomedical Engineering and Sciences (IECBES), 2014 IEEE Conference on.* 2014. IEEE.

19. Silverstein E, Snyder M. Comparative analysis of respiratory motion tracking using Microsoft Kinect v2 sensor. *J Appl Clin Med Phys.* 2018;19(3):193–204.

20. Tahavori F, Adams E, Dabbs M, et al. Combining marker-less patient setup and respiratory motion monitoring using low cost 3D camera technology. in *Medical Imaging 2015: Image-Guided Procedures, Robotic Interventions, and Modeling.* 2015. International Society for Optics and Photonics.

21. Tahavori F, Alnowami M, Wells K. Marker-less respiratory motion modeling using the Microsoft Kinect for Windows. in *Medical Imaging 2014: Image-Guided Procedures, Robotic Interventions, and Modeling.* 2014. International Society for Optics and Photonics.

22. Xia J, Siochi RA, A real-time respiratory motion monitoring system using KINECT: proof of concept. *Med Phys.* 2012;39(5):2682–2685.

23. Ogunmolu OP, Gu X, Jiang S, Gans NR. A real-time, soft robotic patient positioning system for maskless head-and-neck cancer radiotherapy: an initial investigation. in *Automation Science and Engineering (CASE), 2015 IEEE International Conference on*. 2015. IEEE.

24. Ogunmolu OP, Gu X, Jiang S, Gans NR. Vision-based control of a soft robot for maskless head and neck cancer radiotherapy. in *Automation Science and Engineering (CASE), 2016 IEEE International Conference on*. 2016. IEEE.

25. Cardan R, Popple R, Duan J, et al. MO-D-BRB-06: junior investigator winner-fast and accurate patient specific collision detection for radiation therapy. *Med Phys*. 2012;39(6Part21):3867–3867.

26. Padilla L, Pearson EA, Pelizzari CA. Collision prediction software for radiotherapy treatments. *Med Phys*. 2015;42(11):6448–6456.

27. Singh V, Chang Y-J, Ma K, et al. Estimating a patient surface model for optimizing the medical scanning workflow. in *Med Image Comput Comput Assist Interv*. 2014;17(Pt 1):472–479.

28. McKernan B, Bydder SA, Deans T, et al. Surface laser scanning to routinely produce casts for patient immobilization during radiotherapy. *Australa Radiol*. 2007;51(2):150–153.

29. Jung H, Kum O, Han Y, et al. A virtual simulator designed for collision prevention in proton therapy. *Med Phys*. 2015;42(10):6021–6027.

30. Yu VY, Tran A, Nguyen D, et al. The development and verification of a highly accurate collision prediction model for automated noncoplanar plan delivery. *Med Phys*. 2015;42(11):6457–6467.

31. Schaller C, Penne J, Hornegger J. Time-of-flight sensor for respiratory motion gating. *Med Phys*. 2008;35(7Part1):3090–3093.

32. Popple R, Cardan R. SU-E-T-754: Three-dimensional patient modeling using photogrammetry for collision avoidance. *Med Phys*. 2015;42(6Part24):3510–3510.

33. Jiang R, Jáuregui DV, White KR. Close-range photogrammetry applications in bridge measurement: Literature review. *Measurement*. 2008;41(8):823–834.

34. Occipital I. Structure Sensor—3D scanning, augmented reality, and more. 2018 [cited 2018 12/4/2018]; Available from: https://structure.io/.

35. Giancola S, Valenti M, Sala R. Metrological qualification of the Orbbec Astra S™ structured-light camera, in *A Survey on 3D Cameras: Metrological Comparison of Time-of-Flight, Structured-Light and Active Stereoscopy Technologies*. 2018, London, Springer. pp. 61–69.

36. Bandini A, Ouni S, Orlandi S, Manfredi C. Evaluating a markerless method for studying articulatory movements: Application to a syllable repetition task. in *Proceedings of the 9th International Workshop on Models and Analysis of Vocal Emissions for Biomedical Applications MAVEBA*. 2015.

37. Mikhail EM, Bethel JS, McGlone JC. *Introduction to Modern Photogrammetry*. New York, Wiley, 2001.

CHAPTER 26

Future Directions for SGRT

David Wiant

CONTENTS

26.1 INTRODUCTION

The first radiation therapy-specific commercial surface imaging system was installed at Massachusetts General Hospital in 2006. Since then almost 1000 commercial and numerous home-grown surface imaging solutions have been put in place around the world. This proliferation of surface imaging has sparked many new ideas and insights around the delivery of what is now called surface-guided radiation therapy (SGRT).

Before SGRT, patients receiving linear accelerator (linac)-based radiotherapy were typically set up with a combination of tattoos, lasers, and radiographic images. Any information on inter- and intra-fraction motion was generated from radiographic imaging. However, the number and frequency of these images were necessarily limited by the dose delivered from the ionizing radiation. The limitations of radiographic imaging only allowed patient setups to be evaluated at single points or "snapshots" in time. These snapshots occurred once per week (e.g., breast), once per fraction (e.g., prostate or head and neck), or several times per fraction (e.g., brain and intra-fraction motion studies). However, even the most image-intensive sessions would end with 3–4 image sets per 15 minute treatment. Treatment sessions with a finite number of images are very easy to characterize and understand, as the motion can be clearly described by the displacement or difference between the images.

Surface imaging does not suffer from the same dose limitations as radiographic imaging. Since surface imaging does not use ionizing radiation, it allows for continuous imaging on a daily basis without the risks from ionizing radiation. In contrast to the 3–4 motion data points generated by radiographic imaging, surface imaging can produce thousands of data points per treatment session. In fact, surface imaging produces so much information that it has become a challenge to interpret the data in a meaningful way. The primary challenge is how to incorporate time into the motion analysis. As shown in Figure 26.1, motion between radiographic images is easily represented from a single displacement number, while the surface imaging motion data needs a time component to be fully described. The methods used to interpret surface imaging and its clinical applications are rapidly evolving and will continue to evolve for the near future.

Surface imaging has found success in many clinical applications up to this time, as described throughout this book. Prior to SGRT, it was common practice to set up breast patients with lasers and tattoos and to monitor with radiographic images once per week. The daily use of surface imaging has led to more consistent inter-fraction positioning[1-3] and more accurate intra-fraction positioning for applications such as deep inspiration breath hold.[4-6] Intracranial treatments, in particular stereotactic radiosurgery (SRS), have also benefited from SGRT. The adoption of SGRT-SRS has allowed for less restrictive immobilization (see Figure 26.2) and a reduced need for intra-fraction radiographic imaging. Finally, regular use of SGRT

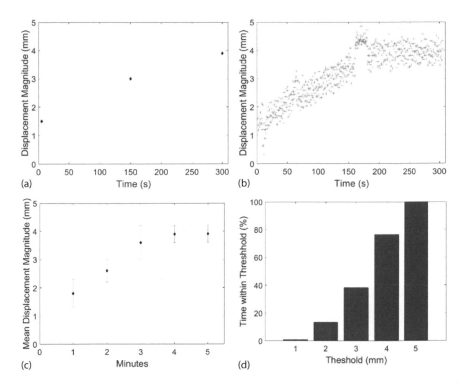

FIGURE 26.1 Typical breast patient displacement for a fraction as measured by radiographic imaging and surface imaging. (a) Radiographic images were acquired at the beginning of treatment, mid-treatment, and at the end of treatment for a total of 3 data points. This data is completely described by the differences between the points. (b) Surface imaging was continuously acquired at about 2 frames per second giving 588 data points. Clearly, this data cannot be described by the differences between the points. Different methods to present surface imaging data are (c) the mean displacement for each minute, and the (d) percent of time within a given threshold.

for other disease sites can lead to improved setup and fewer gross errors by eliminating the effects of misplaced tattoos and incorrect shifts from tattoos.

The remainder of this chapter will delve into the future of SGRT. The first part of the chapter will look into the near future at several ideas that are likely to arrive soon in the clinic. In the final part of the chapter, we will indulge in some speculative thinking and imagine how surface imaging technology and the practice of SGRT might evolve in the distant future.

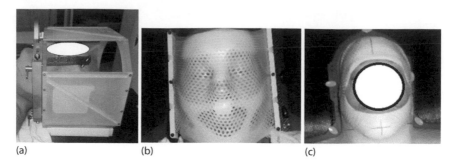

(a)   (b)   (c)

FIGURE 26.2   (a) Head frame. (b) Closed mask that covers the face. (c) Open mask that leaves the face uncovered. The ability to continuously monitor with surface imaging and to quickly stop treatment if the patient moves has allowed less restrictive immobilization to be tested and used with minimal risk.

26.2 NEAR FUTURE

26.2.1 The Importance of Treatment Time

Time is not typically a concern when evaluating radiotherapy plan quality. However, as will be described below, several recent studies have shown that time may be a key driver of plan quality.

The first of these studies considered 30 breast patients that were treated with similar setups.[7] The patients were continuously monitored for 831 treatment sessions with surface imaging. Intra-fraction motion was analyzed for the first 15 minutes of each session. Figure 26.3 shows the

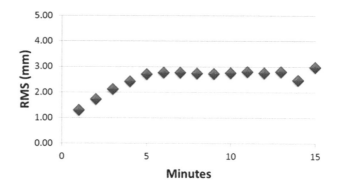

FIGURE 26.3   Displacement averaged over each minute for all patients. (From Wiant, D.B. et al.: Surface imaging-based analysis of intrafraction motion for breast radiotherapy patients. *J. Appl. Clin. Med. Phys.* 2014, 15, 147–159. Copyright Wiley-VCH Verlag GmbH & Co. KGaA. Reprinted with permission.)

mean magnitude of the linear displacement over all patients and fractions for each minute of treatment. The mean displacement increases sharply for the first 5 to 6 minutes of monitoring then levels off through 15 minutes.

The second study examined 29 pelvis patients treated supine with leg immobilization and similar bladder and rectal preparation.[8] The patients were initially set up by laser alignment to tattoos. A static surface image was then acquired and used to make fine linear adjustments to the patient position. Additional surface images were acquired at the beginning, middle, and end of each treatment to evaluate intra-fraction motion. No corrections were made for patient motion after the initial setup. This process was performed for a total of 792 fractions. Intra-fraction setup variations between the time points are shown in Figure 26.4. The mean time for the middle images was about 8 minutes into treatment. The setup variations between the initial and middle points were significantly larger than the variations between the middle and end points.

Two different sites, breast and pelvis, both showed very similar behavior in that displacement increased rapidly for the first 6 to 8 minutes of treatment then leveled off. This suggests that a shorter treatment time will

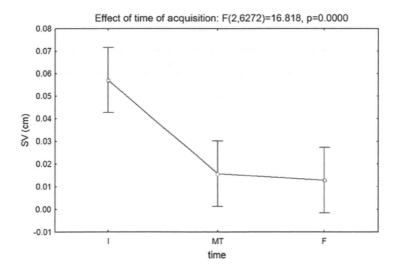

FIGURE 26.4 Mean setup variation (SV) as a function of time for pelvis patients, where I = initial time, MT = mid-treatment time, and F = final time. The MT occurred about 8 minutes into treatment. (With kind permission from Springer Science+Business Media: *Radiol Med.*, Three-dimensional surface imaging for detection of intra-fraction setup variations during radiotherapy of pelvic tumors, 121, 2016, 805–810, Apicella, G. et al.)

directly lead to more accurate localization and treatment across multiple sites. Similar behavior has also been seen for prostate patients monitored by implanted radiofrequency beacons.[9] The idea that time affects treatment delivery accuracy has implications in many areas that beg to be considered. The number of fields per plan and any dosimetric improvements in plan quality must be weighed against loss of accuracy due to the increased treatment time. Similarly, the time taken to register and evaluate treatment setup images must be balanced against the patient motion that occurs during the evaluation, i.e., will 0.1 mm changes in image position be "washed-out" by patient motion during that time. In addition to its impact on localization, time affects patient experience and clinical throughput. With all these considerations, time should become an important consideration in radiation therapy value.

In addition to providing data on the importance of treatment duration and patient motion, SGRT may also help to mitigate some of the negative effects that treatment duration can have on treatment accuracy by providing faster patient setup and continuous position monitoring. Setups guided by SGRT do not require detailed analysis of image registration to determine position corrections as with radiographic images, which will reduce the amount of time the patient is in the treatment position. Continuous monitoring with SGRT may allow for rapid intra-fraction position correction that improves treatment accuracy and quality.

26.2.2 Patient Surface Similarity

The Joint Commission continues to make correct patient identification its number one hospital national patient safety goal on an annual basis.[10] This seemingly mundane task has caused issues for hospitals across the world, and it will only become more difficult in the future as we see increased patient loads, more complex treatment techniques, and shorter treatment timelines.

A review of patient hazard mitigation strategies by Hendee and Herman[11] found that automation and computerization were much more effective tools than policies, procedures, or checklists to limit risk. In this vein, surface imaging may provide a robust, automated method to verify radiotherapy patient identity immediately prior to treatment.

A study of 16 left-sided breast cases, treated with similar setups, tested the ability of surface imaging to uniquely identify patients.[12] Each of the patients in this work had a reference surface created from the body contour on their treatment planning computed tomography (CT) scan, and a daily surface was acquired at each fraction. The similarity of the daily acquired setup surface images to the reference surface images was evaluated by scoring the percentage of points on the daily setup surface images that fell within either 3 or 5 mm of the reference surface images after optimal alignment. Each patient had their reference surface images compared to: (1) 10 different daily acquired setup surface images randomly selected from their treatment and (2) 10 different daily acquired setup surface images from each of the other patients, for a total of 160 comparisons.

The results of the comparisons are shown in Figure 26.5. On a per patient basis, there is no overlap between the same patient scores and the different patient scores. If all patients are considered together, the mean same patient comparison scores were 83% and 94% for the 3 and 5 mm criteria, respectively. The mean different patient scores were 26% and 40% for the 3 and 5 mm criteria, respectively. With all patients included there is a slight overlap between the same patient and different patient scores. However, if a 55% threshold is used to separate same and different patients for the 3 mm group, it yields about 1% false-positives and 1% false-negatives. These findings suggest that, with no post-processing or additional information, surface imaging can identify a breast patient with a relatively high degree of accuracy.

A closer examination of the overlap regions in Figure 26.5 found that the much of the overlap resulted from low same patient scores, and that the low scores were a result of patient posture change or shape deformation.

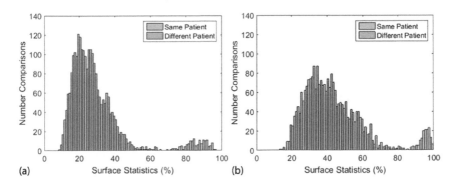

FIGURE 26.5 The number of comparisons for a given surface statistic (or similarity score) for distances to agreement of (a) 3 mm and (b) 5 mm. (From Wiant, D.B. et al.: A novel method for radiotherapy patient identification using surface imaging. *J. Appl. Clin. Med. Phys.* 2016, 17, 271–278. Copyright Wiley-VCH Verlag GmbH & Co. KGaA. Reprinted with permission.)

Patient posture changes between the reference surface and the acquired surface likely indicates a poor setup that could be adjusted to improve treatment accuracy, while patient shape change could result from inflammation or weight changes that should be clinically examined. Thus, even if the low score does not indicate that the wrong patient is being treated, it may still provide valuable clinical information.

Patient surface imaging similarity scores have been able to correctly identify a patient over a cohort of breast patients, and they may be indicative of several clinically important treatment issues. These scores are powerful pieces of information that could easily be incorporated into the daily treatment process with no additional time or effort by the therapy team.

The chest, and in particular the female breast, typically offers smoothly varying features and similar anatomy across patients. Since patient identification was successful in this challenging anatomic region it strongly suggests that areas such as the face, with more distinct features will be amenable to surface imaging-based identification. Currently, available surface imaging systems have adequate resolution to accurately model the face and should be able to easily perform face-based patient identification, which would open up this technology to additional classes of patients not currently treated with SGRT.

26.2.3 Less Restrictive Immobilization

The main function of SGRT is to help setup and monitor patients during treatment. However, surface imaging may end up making its biggest impact on areas of radiotherapy that are not directly related to patient setup and treatment. The idea that SGRT can help form new thoughts around the role of time in radiotherapy was one example of the broad reach of surface imaging technology. Another example is the effect that SGRT has had and will have on treatment immobilization devices.

Prior to SGRT, there were no reliable methods to directly monitor patient posture and position in real-time. Since patients' positions could not be accurately monitored in real-time it was difficult to reliably stop the treatment when motion occurred. Because of this, efforts were made to limit motion as much as possible, which often meant rigid, uncomfortable immobilization devices. When SGRT is used, the treatment team has the ability to both monitor and correct for motion during treatment, which offers the opportunity to use less restrictive immobilization. The best example of this so far is the use of open-face mask immobilization for intracranial treatments.

Noninvasive intracranial immobilization has typically been achieved with a mask that covers the entire face in order to limit motion to the highest degree possible. Recently, masks and support devices that leave the face uncovered or open have been developed (Figure 26.2). The idea behind these open masks is to make the patient more comfortable, which should lead to less motion and an improved patient experience. These thoughts were well-described in Cervino et al.'s work with frame-less and mask-less treatment where their stated aim was to "minimize patient discomfort while maintaining high precision treatment."[13] This sentiment was supported by patient surveys from clinical experiences that showed that patients preferred open masks to closed masks, and that open masks may limit anxiety and claustrophobia.[14,15]

Patient comfort and experience as primary design goals in immobilization is a drastic shift in focus from rigid motion suppression. Intracranial immobilization is one of the first areas to see changes, but with a new focus on comfort and experience, many disease sites should soon see less restrictive immobilization options, such as head and neck (H&N), as was discussed in Chapter 17.

Also, surface imaging offers a powerful tool to help develop new devices and techniques. Surface imaging can be used to rapidly iterate through new designs in a test environment with minimal risk to the test subjects. The ability to test on healthy volunteers greatly reduces the risk new devices pose to patients and greatly speeds up the development cycle. With SGRT there is no reason patient comfort and experience cannot be improved without a loss in treatment accuracy.

26.2.4 Collisions

Collisions between the patient, gantry, couch, accessories, or immobilization devices have always been a worry in radiotherapy. Fear of collisions has regularly limited the use (or necessitated testing) of certain gantry and couch combinations in the planning process, which has a direct impact on plan quality. Collisions that are not discovered until treatment time will likely cause a delay while a new plan can be created, again negatively impacting the quality of care and the patient experience.

A number of different approaches have been employed to help limit collisions. Most institutions have restrictions on the types of patient immobilization, setups, and treatment geometries permissible, in order to limit the risk of collisions. An example of this is the arm position of breast patients, which is often closely monitored in the simulation step.

In other cases, certain gantry, couch, and accessory combinations are not allowed, for example, inferior gantry positions are not allowed at certain couch heights with radiosurgery cones. The inclusion of the couch in treatment planning systems is another way to help prevent collisions. While the primary aim of the couch model is to improve dosimetry, its inclusion can also help to evaluate collision probabilities. These methods, and others, have helped to reduce collisions, but they have not completely eliminated them.

The best way to prevent collisions in the pretreatment phase is to create a model of the situation, including the patient surface, treatment machine, and accessories. The treatment system and accessories do not vary between cases, and can be easily modeled from a library. An accurate map of the patient surface is the most difficult component to acquire. A patient surface can be generated from the treatment planning CT, but these images are often truncated by the CT field of view and only give a partial model of the patient surface (Figure 26.6). Furthermore, as arc therapy becomes more popular and feasible treatment geometries move into the 4π treatment space,[16–18] a partial surface from a CT scan will be inadequate to fully predict and prevent collisions.

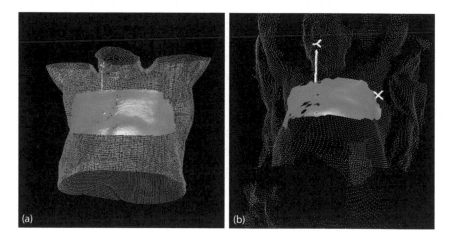

FIGURE 26.6 (a) A surface created from a body contour in a treatment planning scan. (b) A surface captured by a dedicated surface imaging system. The dedicated surface imaging system is easily able to acquire a patient range that is sufficient to accurately predict collisions, while the treatment plan scan does not provide enough information for collision predication.

Surface imaging is a ready-made solution to this problem. It allows for a full body surface to be acquired prior to treatment planning with no additional radiation dose to the patient. A great deal of work has been done in this area.[19–24] The current standard for surface imaging-based collision prediction is to use multiple cameras to capture a surface prior to treatment planning that can be used to check for collisions at each isocenter, gantry, and couch combination. The next steps in this process are to make the collision model available during treatment and to give the model access to the positions of the treatment machine in real-time, so that unexpected collisions can be prevented.

26.2.5 Quality Assurance with Surface Imaging

Due to its accuracy, relative portability, and passive sampling, surface imaging (or more generally photogrammetry) has been used for quality assurance (QA)/quality control in many industries, such as aerospace, shipbuilding, automotive, satellite, and power generation. However, surface imaging has not yet found many QA applications in radiotherapy.

Commercial SGRT systems allow for easy acquisition of reference surfaces and have reported accuracy with phantoms on the order of 0.1–0.2 mm at static couch positions[25] and much less than 1 mm and 1° with couch motion.[26] Many linac mechanical QA procedures are time-consuming manual processes that involve measurements of small displacements. These measurements have high levels of inter- and intra-operator variability and are subject to gross human errors. This type of measurement could clearly benefit from the use of surface imaging.

One simple example of linac mechanical QA is couch motion tests, which are done to verify that couch movements are accurately interpreted by the linac. Couch movement tests typically use lasers and rulers or pre- and post-movement radiographic images to validate linac couch movements. Surface images could easily be inserted into the process to evaluate motion with a high degree of accuracy and consistency (and likely also enable a reduction in the time required to complete the test).

As the use of high-precision, hypofractionated treatments increases, the demands on the treatment delivery system will also continue to increase. These delivery systems will need to function with increasingly higher levels of accuracy and consistency. It follows that machine QA will also need to improve so that the QA tests can be performed quickly, consistently, and accurately. Surface imaging could become a key part of these QA strategies.

26.3 DISTANT FUTURE

26.3.1 Real-Time Position Correction

In the current state, SGRT patient shifts must be manually applied by the user when treatment is stopped. While this solution may help to reduce inter- and intra-fraction localization errors, it does not take full advantage of the SGRT system. As described above, commercial SGRT systems are very adept at capturing movement relative to a recently acquired reference surface. This ability could be leveraged to ensure that the patient position is maintained very near to the reference surface during each fraction by automatically applying couch motions to minimize SGRT-suggested shifts.

In principle, the idea is quite simple: the patient is constantly monitored during treatment with surface imaging and the couch is constantly moved to keep the SGRT-suggested shifts as close to zero as possible. In practice there are several questions that need to be addressed before real-time position correction becomes reality. The first question is whether surface imaging is a reasonable surrogate for target motion. For breast patients, surface imaging directly monitors the target volume, so it is likely a reliable indicator of intra-fraction target motion. For H&N and intracranial treatments, there is a very limited range of motion available to the targets relative to the skin surface that is being monitored, so real-time surface imaging-based corrections for these areas also seems feasible. Treatment sites with internal targets (e.g., lung, prostate, abdomen, and pelvis) are able to move relative to the skin surface, so the value of surface imaging as a motion surrogate is questionable. Limited single institution studies using radiographic imaging to monitor fiducial markers in lung tumors have shown intra-fraction correlation between surface imaging and target motion. Additional studies using intra-fraction radiographic imaging and/or implanted radiofrequency beacons are needed to confirm that surface imaging is a suitable surrogate for targets that can easily move relative to the skin surface.

The second question is whether the latency of surface image-guided couch motion can be made small enough to allow for clinically meaningful position corrections. Real-time corrections involve: (1) surface image acquisition, (2) surface image registration to the reference image, (3) calculation of the couch motion needed to minimize the displacement, and (4) execution of the couch movement. (There will also need to be verification that the couch movement will not produce collisions, but this may be done pretreatment by setting a global shift limit or a control point shift limit, for example.)

In order to provide clinical value, all of these steps will need to occur before the patient moves to a new position. The time scale for these steps will vary depending on whether the goal is to track target motion, correct for positional drift, or apply plan directed shifts (like couch rotation). In most cases, compensation for patient drift or planned shifts can likely occur over several seconds and result in a net improvement in patient position accuracy. Tracking applications will likely require the entire process to take a second or less. In order to accomplish this, predictive motion models, similar to those used in commercial systems,[27] will need to be put in place.

Disease sites, such as breast, that experience relatively slow drift over the course of a fraction and where the surface image is a suitable surrogate for target motion are candidates for real-time surface image-guided motion correction with current technology and capabilities. Sites with internal targets and/or more complex motion need additional work to verify the efficacy of the surface image as a motion surrogate and to reduce couch motion latency.

26.3.2 Big Data

As the medical field matures, vast stores of discrete data are being accumulated. The analysis and application of this "big data" is essential to streamline the health care process, improve quality, and reduce costs. Specific examples in radiotherapy where big data could be beneficial are contours and treatment plans. If patients with similar anatomy, disease, and treatment goals have already been treated, why must new plans and contours be created each time? One way to answer this question is with knowledge-based planning, which leverages past plans to form a dosimetry model for a given anatomy and apply it to new patient plans. Similarly, atlas-based automated contouring searches an atlas of contours from past patients with similar characteristics and then deforms the atlas contours to the anatomy of the new patients.

The atlas subject selection process can be based on a wide range of criteria. One of the most frequently used and successful criteria is anatomy. Anatomic similarity is often evaluated by performing rigid or deformable registration between volumetric images. Volumetric image sets can take up huge amounts of disk space, and the registration process can be time consuming. These factors limit the practical size of an atlas, and subsequently the accuracy of the atlas.

Another option is to evaluate anatomic similarity with surface imaging. Since the surface image takes up less space than a volumetric image, the registration process can be 10–1000 times faster. In theory, the use of surface imaging could allow for atlases that are thousands of time larger than what is in current use. These huge atlases offer the possibility to generate highly accurate contours and plans with minimal human interaction in just minutes.

The implementation of this idea is relatively simple and straightforward. However, work needs to be done to determine if surfaces are a reasonable identifier of anatomic and plan similarity, and to find out if extremely large atlases will translate to accurate contours and plans.

26.3.3 Treatment Based on Diagnostic Images

The majority of patients that present to a radiation oncology clinic will have prior CT or magnetic resonance (MR) imaging. These patients will typically consult with a radiation oncologist, undergo a simulation CT scan, have a treatment plan created, and then begin treatment. There are a number of limitations to this process: (1) Each of these steps take time. A recent study showed that delays in diagnosis to treatment directly impacts outcome,[28] so efforts should be made to reduce time to treatment; (2) the need to show up for multiple appointments and to repeat volumetric imaging negatively affects the patient experience and may give unnecessary imaging dose; (3) the fixed time from simulation may limit the treatment planning options and may cause stress to the staff.

If the patient already has diagnostic volumetric images, it may be reasonable to ask why they are not used to create a treatment plan. The main factors currently preventing this are concerns over the geometric accuracy/fidelity of outside imaging devices, dose calculation on image sets acquired on images devices not characterized for electron density, and reproducing the diagnostic patient position at treatment. Each of these factors can be reasonably addressed.

The majority of imaging centers and devices are required to maintain certain levels of quality by either national regulations or insurance payers. For example, sites in the United States of America are required to maintain accreditation by the American College of Radiology (or equivalent accreditation body) to receive certain reimbursements. A part of the accreditation process is to ensure that proper QA is performed and that the minimum required levels of geometric accuracy are achieved among other criteria. This adds one level of assurance that the geometry of diagnostic images are accurate. Another level of protection is that radiographic imaging at the initial treatment can be used to validate the geometric accuracy of outside image sets.

CT scanners in radiation oncology departments are very accurately characterized, so that electron densities can be determined to calculate electron transport in the treatment planning process. Electron density calculations based on outside image sets will likely not be as accurate as those made from characterized radiation oncology-specific images. However, the use of synthetic or pseudo-CT data can help to overcome this issue. There has been a great deal of work in this area, particularly around the generation of a pseudo-CT from MR.[29-36] The pseudo-CT images have been shown to yield sufficiently accurate dose calculations for radiotherapy planning. Creation of a pseudo CT from an outside CT scan is less challenging on a technical level than pseudo CT generation from an MR and would be an easy tool to develop.

Another option to verify geometry and dose is to recalculate and evaluate dose on the cone-beam CT (CBCT) from the first treatment.

Recreation of the diagnostic image position is perhaps the most difficult part of this process. One reason is that most diagnostic scanners use concave, rather than flat, couch tops. From a technical standpoint there is no reason a rounded couch top could not be used for radiotherapy. A rounded couch top actually has many advantages, including increased comfort and stability for the patient and helping to keep the patient inside the limited CBCT field of view. If rounded couch tops are not adopted, many different devices that overlay on a flat couch to create a rounded base could be imagined. For example, the use of milled or machined foam to recreate the patient posture on the diagnostic table from images has been proposed.[37] Once the couch top issue is addressed, surface imaging could be used to help move the patient to the diagnostic position and then to create immobilization devices to support this posture.

There are certain disease sites that might not require prior diagnostic imaging or where the diagnostic images are of low quality. Other sites might benefit from a position or posture that is not typically used for diagnostic scans. Some of these cases could be remedied through partnership with the diagnostic radiology departments to acquire more or higher quality images in different postures, while other cases would still require a treatment planning CT.

For patients that typically undergo diagnostic studies in a position suitable for radiation therapy delivery (e.g., brain, head and neck, lung, spine, prostate), the union of surface imaging and diagnostic images could allow for them to go immediately from consult to treatment with no additional simulation or marks. This would be a truly disruptive change to current radiation therapy practice, benefiting these patients and the field of radiation oncology as a whole.

26.3.4 Cherenkov Radiation

Cherenkov radiation occurs when a charged particle (e.g., an electron) moves with a phase velocity faster than the speed of light in a dielectric medium (e.g., the human body). Cherenkov radiation is emitted at wavelengths ranging from ultraviolet to the near infrared. The spectral intensity of the Cherenkov radiation is described by the Frank-Tamm's formula (Equation 26.1):

$$\frac{dE}{dx} = \frac{q^2}{4\pi} \int_v \mu(\omega)\omega \left(1 - \frac{c^2}{v^2 n^2(\omega)}\right) d\omega \qquad (26.1)$$

where $\frac{dE}{dx}$ is the energy emitted per unit length the particle travels, q is the particle charge, μ is the permeability of the medium, n is the index of refraction of the medium, v is the speed of the particle in medium, c is the speed of light in vacuum, and ω is the frequency of the emitted radiation. The Frank-Tamm's formula shows that the intensity of the Cherenkov radiation is proportional to frequency (or inversely proportional to wavelength), so the observed radiation is weighted toward the blue end of the spectrum.[38] The Cherenkov effect has been observed in biological materials with incident electron energies above a threshold of about 220 keV.

Electrons generated from therapeutic X-rays have sufficient energy to produce Cherenkov radiation. It has been shown that if the condition of charged particle equilibrium is assumed, the dose deposited by therapeutic X-ray beams is directly proportional to the locally released Cherenkov radiation.[38] This makes Cherenkov radiation an attractive option to monitor dose to patient skin, verify treatment delivery, or to perform QA procedures at the surface of a water tank. The ability to monitor skin dose would be attractive for patients that have targets close to the skin or patients that might expect to develop non-negligible skin toxicity from treatment. Real-time skin dose could also be used as a QA tool to help evaluate patient shape, posture, and treatment delivery accuracy on a daily basis. Cherenkov radiation could also be used as a general beam quality QA tool. For instance one could imagine quick checks of beam flatness and symmetry with this tool.

The main hurdle to the development of Cherenkov tools is the relatively low intensity of the signal produced during radiotherapy, which makes detection and localization difficult. As described by Andreozzi et al., there are very specific camera requirements to acquire Cherenkov radiation in a radiotherapy setting: (1) the camera must have sufficient gain to image at low light levels, (2) the camera must be able to capture the signal in ambient light conditions, and (3) the video frame rate must be greater than 5 fps.[39]

The setup of commercial SGRT systems within the treatment room leaves them almost perfectly positioned to acquire and process Cherenkov radiation. The current generation of SGRT cameras meets the above recommendations to varying extents depending on the exact model. The opportunity exists to modify current SGRT cameras to improve Cherenkov acquisition or to add additional cameras to the platforms to focus specifically on Cherenkov radiation. The addition of Cherenkov dose maps to patient surfaces could create a powerful tool to calculate and/or measure skin dose, which is currently very difficult. An accurate daily map of skin dose could provide much needed insights on how to predict and manage skin toxicity. Also, skin dose could serve as a daily QA tool that quickly verifies that the patient is in the correct position, the correct accessories have been used, and the beam delivery is correct.

26.4 SUMMARY

SGRT has made a significant impact on the field of radiation oncology. For the first time, the radiation therapy team is able to see patients for every second of every treatment, and to be sure that the patients are in the right positions for those treatments. If surface imaging technology stopped evolving today, it would continue to be a key piece of the radiation oncology workflow for years to come. However, the growth and development of SGRT is nowhere near finished. The future is bright, with the items discussed in this chapter just giving a glimpse of what may be to come.

KEY POINTS

- SGRT has become an important part of radiation therapy setup and delivery, and the scope of SGRT applications will increase in the future.

- SGRT will improve patient safety by providing automated patient identification prior to treatment.

- SGRT will help the development of more comfortable, less restrictive immobilization devices.

- SGRT will provide accurate collision models.

- SGRT will unlock of greater understanding of the impact of treatment delivery time on quality.

- There are many possible disruptive future applications of surface imaging and SGRT, such as real-time patient position correction, the use of Cherenkov radiation to assure accurate treatment delivery, and using diagnostic images to treat without a radiation oncology simulation scan.

REFERENCES

1. Shah AP, Dvorak T, Curry MS, Buchholz DJ, Meeks SL. Clinical evaluation of interfractional variations for whole breast radiotherapy using 3-dimensional surface imaging. *Pract Radiat Oncol.* 2013;3(1):16–25.
2. Chang AJ, Zhao H, Wahab SH, et al. Video surface image guidance for external beam partial breast irradiation. *Pract Radiat Oncol.* 2012;2(2):97–105.
3. Padilla L, Kang H, Washington M, Hasan Y, Chmura SJ, Al-Hallaq H. Assessment of interfractional variation of the breast surface following conventional patient positioning for whole-breast radiotherapy. *J Appl Clin Med Phys.* 2014;15(5):4921.
4. Gierga DP, Turcotte JC, Sharp GC, Sedlacek DE, Cotter CR, Taghian AG. A voluntary breath-hold treatment technique for the left breast with unfavorable cardiac anatomy using surface imaging. *Int J Radiat Oncol Biol Phys.* 2012;84(5):e663–e668.
5. Rong Y, Walston S, Welliver MX, Chakravarti A, Quick AM. Improving intra-fractional target position accuracy using a 3D surface surrogate for left breast irradiation using the respiratory-gated deep-inspiration breath-hold technique. *PLoS One.* 2014;9(5):e97933.

6. Tang X, Zagar TM, Bair E, et al. Clinical experience with 3-dimensional surface matching-based deep inspiration breath hold for left-sided breast cancer radiation therapy. *Pract Radiat Oncol.* 2014;4(3):e151–e158.

7. Wiant DB, Wentworth S, Maurer JM, Vanderstraeten CL, Terrell JA, Sintay BJ. Surface imaging-based analysis of intrafraction motion for breast radiotherapy patients. *J Appl Clin Med Phys.* 2014;15(6):147–159.

8. Apicella G, Loi G, Torrente S, Crespi S, Beldì D, Brambilla M, Krengli M. Three-dimensional surface imaging for detection of intra-fraction setup variations during radiotherapy of pelvic tumors. *Radiol Med.* 2016;121(10):805–810.

9. Tong X, Chen X, Li J, Xu Q, Lin MH, Chen L, Price RA, Ma CM. Intrafractional prostate motion during external beam radiotherapy monitored by a real-time target localization system. *J Appl Clin Med Phys.* 2015;16(2):51–61.

10. The Joint Commission, National Patient Safety Goals Effective January 2018. Ambulatory Health Care Accreditation Program. https://www.joint-commission.org/assets/1/6/NPSG_Chapter_AHC_Jan2018.pdf. Accessed August 24, 2018.

11. Hendee WR, Herman MG. Improving patient safety in radiation oncology. *Med Phys.* 2011;38(1):78–82.

12. Wiant DB, Verchick Q, Gates P, Vanderstraeten CL, Maurer JM, Hayes TL, Liu H, Sintay BJ. A novel method for radiotherapy patient identification using surface imaging. *J Appl Clin Med Phys.* 2016;17(2):271–278.

13. Cervino LI, Pawlicki T, Lawson JD, Jiang SB. Frame-less and mask-less cranial stereotactic radiosurgery: a feasibility study. *Phys Med Biol.* 2010;55(7):1863–1873.

14. Wiant D, Squire S, Liu H, Maurer J, Hayes TL, Sintay B. A prospective evaluation of open face masks for head and neck radiation therapy. *Pract Radiat Oncol.* 2016;6(6):e259–e267.

15. Li G, Lovelock DM, Mechalakos J, Rao S, Della-Biancia C, Amols H, Lee N. Migration from full-head mask to "open-face" mask for immobilization of patients with head and neck cancer. *J Appl Clin Med Phys.* 2013;14(5):243–254.

16. Dong P, Lee P, Ruan D, Long T, Romeijn E, Yang Y, Low D, Kupelian P, Sheng K. 4π non-coplanar liver SBRT: a novel delivery technique. *Int J Radiat Oncol Biol Phys.* 2013:85(5):1360–1366.

17. Victoria YY, Landers A, Woods K, Nguyen D, Cao M, Du D, Chin RK, Sheng K, Kaprealian TB. A prospective 4π radiation therapy clinical study in recurrent high-grade glioma patients. *Int J Radiat Oncol Biol Phys.* 2018;101(1):144–151.

18. Woods K, Kaprealian TB, Lee P, Sheng K. Cochlea-sparing acoustic neuroma treatment with 4π radiation therapy. *Adv Radiat Oncol.* 2018;3(2):100–107.

19. Becker SJ. Collision indicator charts for gantry-couch position combinations for Varian linacs. *J Appl Clin Med Phys.* 2011;12(3):16–22.

20. Cardan RA, Popple RA, Fiveash J. A priori patient-specific collision avoidance in radiotherapy using consumer grade depth cameras. *Med Phys.* 2017;44(7):3430–3436.

21. Humm JL, Pizzuto D, Fleischman E, Mohan R. Collision detection and avoidance during treatment planning. *Int J Radiat Oncol Biol Phys.* 1995;33(5):1101–1108.

22. Padilla L, Pearson EA, Pelizzari CA. Collision prediction software for radiotherapy treatments: collision prediction in radiotherapy. *Med Phys.* 2015;42(11):6448–6456.

23. Tsiakalos MF, Schrebmann E, Theodorou K, Kappas C. Graphical treatment simulation and automated collision detection for conformal and stereotactic radiotherapy treatment planning. *Med Phys.* 2001;28(7):1359–1363.

24. Yu VY, Tran A, Nguyen D, Cao M, Ruan D, Low DA, Sheng K. The development and verification of a highly accurate collision prediction model for automated noncoplanar plan delivery: collision prediction model for noncoplanar radiotherapy delivery automation. *Med Phys.* 2015;42(11):6457–6467.

25. Li G, Ballangrud Å, Kuo LC, Kang H, Kirov A, Lovelock M, Yamada Y, Mechalakos J, Amols H. Motion monitoring for cranial frameless stereotactic radiosurgery using video-based three-dimensional optical surface imaging. *Med Phys.* 2011;38(7):3981.

26. Wiant D, Liu H, Hayes TL, Shang Q, Mutic S, Sintay B. Direct comparison between surface imaging and orthogonal radiographic imaging for SRS localization in phantom. *J Appl Clin Med Phys.* 2019;20(1):137–144.

27. Sothmann T, Blanck O, Poels K, Werner R, Gauer T. Real time tracking in liver SBRT: comparison of CyberKnife and Vero by planning structure-based γ-evaluation and dose-area-histograms. *Phys Med Biol.* 2016;61(4):1677–1691.

28. Khorana AA, Tullio K, Elson P, et al. Increase in time to initiating cancer therapy and association with worsened survival in curative settings: a U.S. analysis of common solid tumors. *J Clin Oncol.* 2017;35(suppl; abstr 6557).

29. Andreasen D, Van Leemput K, Hansen RH, Andersen JA, Edmund JM. Patch-based generation of a pseudo CT from conventional MRI sequences for MRI-only radiotherapy of the brain. *Med Phys.* 2015;42(4):1596–1605.

30. Johansson A, Karlsson M, Nyholm T. CT substitute derived from MRI sequences with ultrashort echo time. *Med Phys.* 2011;38(5):2708.

31. Paradis E, Cao Y, Lawrence TS, Tsien C, Feng M, Vineberg K, Balter JM. Assessing the dosimetric accuracy of magnetic resonance-generated synthetic CT images for focal brain VMAT radiation therapy. *Int J Radiat Oncol Biol Phys.* 2015;93(5):1154–1161.

32. Price RG, Kim JP, Zheng W, Chetty IJ, Glide-Hurst C. Image guided radiation therapy using synthetic computed tomography images in brain cancer. *Int J Radiat Oncol Biol Phys.* 2016;95(4):1281–1289.

33. Sjölund J, Forsberg D, Andersson M, Knutsson H. Generating patient specific pseudo-CT of the head from MR using atlas-based regression. *Phys Med Biol.* 2015;60(2):825–839.

34. Uh J, Merchant TE, Li Y, Li X, Hua C. MRI-based treatment planning with pseudo CT generated through atlas registration. *Med Phys.* 2014;41(5):051711.
35. Yang Y, Cao M, Kaprealian T. Accuracy of UTE-MRI-based patient setup for brain cancer radiation therapy. *Med Phys.* 2016;43(1):262–267.
36. Zheng W, Kim JP, Kadbi M, Movsas B, Chetty IJ, Glide-Hurst CK. Magnetic resonance–based automatic air segmentation for generation of synthetic computed tomography scans in the head region. *Int J Radiat Oncol Biol Phys.* 2015;93(3):497–506.
37. Gensheimer MF, Bush K, Juang T, Herzberg B, Villegas M, Maxim PG, Diehn M, Loo BW. Practical workflow for rapid prototyping of radiation therapy positioning devices. *Practl Radiat Oncol.* 2017; 7(6):442–445.
38. Zhang R, Glaser AK, Gladstone DJ, Fox CJ, Pogue BW. Superficial dosimetry imaging based on Čerenkov emission for external beam radiotherapy with megavoltage x-ray beam: superficial dosimetry imaging based on Čerenkov emission. *Med Phys.* 2013;40(10):101914.
39. Andreozzi JM, Zhang R, Glaser AK, Jarvis LA, Pogue BW, Gladstone DJ. Camera selection for real-time *in vivo* radiation treatment verification systems using Cherenkov imaging: camera selection for real-time *in vivo* radiation treatment. *Med Phys.* 2015;42(2):994–1004.

Index

Note: Page numbers in italic and bold refer to figures and tables, respectively.

T - #0803 - 101024 - C516 - 234/156/23 - PB - 9781032173757 - Gloss Lamination